The History
of
Project Management

The History of Project Management

By Mark Kozak-Holland

First Edition

Oshawa, Ontario

The History of Project Management
by Mark Kozak-Holland

Managing Editor:	Kevin Aguanno
Typesetting:	Peggy LeTrent and Charles Sin
Cover Design:	Mark Kozak-Holland
eBook Conversion:	Charles Sin and Agustina Baid

Published by:
Multi-Media Publications Inc.
Box 58043, Rosslynn RPO
Oshawa, ON, Canada, L1J 8L6

http://www.mmpubs.com/

Hardcover	ISBN-13: 9781554890965
Adobe PDF ebook	ISBN-13: 9781554891061

Published in Canada. Printed simultaneously in the United States of America and the United Kingdom.

CIP data available from the publisher.

Table of Contents

The Late Middle Ages (1300 - 1450) 237

The 15th Century and Renaissance (1450 - 1500) 269

Acknowledgments

This book is part of a series that has taken many years to complete. I would like to thank Kevin Aguanno for helping pull this book together and for driving the initiative. I would also like to thank Glenn Le Clair and John Byrne, from the Lessons from History team, for their early reviews.

Overall, I am indebted to my wife and family who have been so gracious in allowing me to continue with this writing project at the expense of our valuable time together. I would like to mention that the design and layout of front cover was completed with the help of Jamie and Evie Kozak-Holland.

Preface

Why this Book?

Project management is one of the fastest growing professions worldwide in the 21st Century. The perception held by many people is that project management started in the 20th Century - after all, the major project management organizations were founded in the sixties.[1] Yet how were all the great projects of the past delivered, like the Giza Pyramid, the Parthenon, the Colosseum, the Gothic Cathedrals of Europe, the Taj Mahal, and the Transcontinental Railroad? Was project management used? Were the concepts of project management understood? Answering these questions is the fundamental reason for this book. If project management has been around for thousands of years, why has no one attempted to quantify historical examples, and link them back to today's concepts of project management?

This type of book is somewhat unique probably because of the broad scope of the subject (the different types of projects) and the lack of readily-available material. The writing of this book has not been easy, taking over four years. My publisher suggested it as a natural fit to the Lessons from History series. But the turning point and the reason why it was finally written came through two things: first, the Lessons-from-History.com website, where the most popular pages were those related to the History of Project Management; and second, in writing the book *Project Lessons from the Great Escape*, I purposely looked at this Second World War case study from the perspective of a modern project to determine whether it followed all the modern project management concepts. Surprisingly, it did which made me ask what about other case studies that went further back in time. As a student of history, I have viewed history as a natural progression where things just don't suddenly appear but evolve. Similarly, project management has evolved with time. So, were the historical mega-projects as complex as today's? Yes and possibly more, because of the larger size of the project work forces, inferior technologies, and less scientific knowledge.

Why the history of Project Management is Important?

The Project Management Institute (PMI) has seen double digit growth and much evolution throughout the last decade. People entering the field are looking at the past to better understand the background to project management today. Using examples from the past helps make sense of today's projects, and puts into context the steady evolution in projects.

What Will This Book Help you Do?

The general perception in today's project management communities is that project management started in the 20th Century, specifically after the Second World War. This book calls this modern project management. The book will demonstrate that project management started at least 4,500 years ago, with the building of the Giza Pyramid. It will dispel myths like ancient projects lacked the concept of scope management, or had few time dependencies, or employed slave labor, or had no formal project manager. It will show categorically that ancient projects had to manage a budget, materials (equipment), and a project workforce to create deliverables within a specific time frame, to a predefined scope, and level of quality. It will demonstrate that little has changed with time in that the same factors that affect today's projects impacted the past, and the overall approach is similar. It will review the nine Knowledge Areas[1] from the *Guide to the Project Management Body Of Knowledge (PMBoK® Guide)* and show how relevant these were in every selected project from the past millennia and centuries.

What We Can Learn from the Past?

One example taken from the Giza Pyramid project relates to Hemienu's overall approach. Heminu was the overseer of construction for the Giza Pyramid. Scope, cost, and time management are recognized as modern project management concepts yet Hemienu understood all three and managed these in the project. Scope was not just the process of erecting two million blocks over a period of 20 years. Hemienu knew that certain complex features would have a significant impact on the project scope. For example, the king's burial chamber (61 meters or 200 feet above the foundation), or the system of internal and external ramps, or the raising of the cap stone. These all had to be carefully thought through, well in advance, or the project would not be completed in the life time of the pharaoh. By understanding the scope, Hemienu could build and manage the schedules accordingly. For example, one of the first critical path activities was dispatching a quarrying team of 500 men 800 kilometers (500 miles) to Aswan to quarry the massive granite blocks that would take ten years to deliver.[2]

In Summary

If project management is as old as the beginning of civilization then we might consider changing our definition of it so it better reflects this. I believe that project management has been a continuous evolution. It has absorbed the best practices from every field that it has come in contact with. This book aims to be the first in linking the project management of the past with the present.

The challenge of managing projects today is to combine the technology of the near-future with the lessons from the past. This is a principal objective of the Lessons from History series. Hence the research in this book establishes a baseline by which historical projects can be used for comparative purposes to today's projects, and with that a wealth of lessons learned. But if nothing else I hope the journey through the book entertains as much as it educates.

Finally, on finishing this book you will see that the historical projects covered, and others beyond these, should be reinterpreted again. This should be done from a project management perspective with today's knowledge base, tools, and concepts. This will help dispel many of the myths surrounding these projects and establish the continuum of project management from the past to today, and into the future.

Introduction

Studies in the history of project management are rarely seen which is somewhat surprising considering the mega projects of the last 4,500 years. These were not anomalies in history but projects delivered in a systematic way with similar characteristics to today's projects. Typically, they had what we would call today a project charter, and a business justification. They followed a life-cycle of phases, and they incorporated the equivalent of the Project Management Process Groups (initiating, planning, executing, and closing), and the nine *PMBoK® Guide* knowledge areas (or PRINCE2® as identified in the mapping to *PMBoK® Guide* in Appendix E).

For most people, the starting point of reference in the history of Project Management is the project constructing the Great Pyramid at Giza. This is a monumental structure for its time, 2550 BCE (Before the Common Era), and conjures up images of thousands of slaves serving a merciless pharaoh and toiling in inhospitable conditions. In reality, labor was not an inexhaustible supply but came at a higher price. There is little evidence to suggest the use of slave labor in any of the projects presented, with the exception of peasant labor. For the next 2,700 years, most significant projects, highlighted by the architectural masterpieces of the Greek and Roman eras, were in the construction of edifices and structures. Over time subtle changes were being made in the use of ever-improving materials like brick, concrete, and iron which provided the project architects with more design options and greater flexibility in these structures.

Other types of projects flourished with the development of new technologies. For example, the European Renaissance led to the great European voyages of exploration during the 15th Century, with advances brought about by the invention of grids/maps, the astrolabe, the compass, the lateen sail, and the improvements in ship building (the Caravel).

In the last 400 years, there has been a further evolution in project types. This was first influenced by the birth of the scientific age that provided important discoveries and inventions which impacted Western society. The advances in science during the 17th Century created a better understanding of materials and spurred advances in the field of engineering and the industrial revolution. For example, advances in technology changed the cost of materials so that by 1840, iron became so cheap it started to replace clay and wooden pipes.

The Industrial Revolution of the 18th Century was a catalyst for change and had an impact on projects not only with new materials, that could be assembled more quickly, but also where steam replaced muscle. This improved the productivity of labor and advanced all sorts of machines and equipment to handle materials and heavy loads. This in turn fueled a Transportation Revolution in the 19th Century which saw the extensive building of canals, bridges, roads, and rails.

Over time, the drivers for projects have included religious, political and commercial factors, and often a mix of several of these. Many ancient projects were driven by religion, for example, construction projects like the Giza Pyramid, Stonehenge, and the Gothic cathedrals. However, with the Gothic cathedral projects there was a commercial side to as they did improve the economy in a town. Florence Cathedral was all about prestige and commerce. The Colosseum was political in nature and was used to promote the Roman Government and provide employment. It also stabilized the Roman Empire for a time. The great voyages of exploration were projects driven by commerce, but in the name of religion and the monarchy. In the 18th and 19th centuries, practically all projects during the Industrial Revolution were commercial in nature like the Iron Bridge, Transatlantic Cable, Crystal Palace, etc. Some projects were very strategic, and political, like the Suez and Panama Canal projects, or the U.S., Canadian, or Russian Transcontinental Railroad projects, but they were also very important commercial successes as well. In the 20th Century public works projects emerged to address the growth in unemployment such as the Hoover Dam and Golden Gate Bridge in the U.S., Autobahns in Germany, ocean liners (Queen Mary and Queen Elizabeth) in the UK, and the Maginot Line in France. These projects were not much different from the Parthenon and Colosseum, 2,000 years earlier, which were also major public works projects.

In writing this book, the challenge has been to try and identify the sequence of world developments and to determine what to include as a project. For example, whilst Europe struggled through the Dark Ages, other cultures and civilizations flourished and significant projects continued to evolve outside of Europe like Hagia Sophia, the Cathedral in Constantinople (Istanbul).

More difficult was the debate on what not to include. For example, this book is not a history of inventions. Yet some inventions developed as projects and are very significant. As an example, the development of Hollerith's calculating engine and mechanical computer both had a massive impact on history. In this area, there are many omissions such as the phone, wireless radio, and television.

Even more difficult was the debate on who to include as a project manager. For example, Magellan's voyage around the world was delivered before its time, with levels of technology and knowledge that were really inadequate. Was he a great project manager? The pro argument is that he completed the project objective. The con argument is that he used brute force, and caused the loss of life among many of his crew, and contributed to the loss of his own life. Another candidate is the architect Hemienu at Giza who overcame major logistical and technical challenges to deliver a very unique edifice, which was not surpassed in height for 4,000 years. The last section of this book covers selected project leaders/managers for the significant projects included through the book.

This book covers projects up to the start of the Second World War. After this, there were so many significant developments from 1940 onward that it would substantially increase the scope of this book. This latter period is seen as the birth of modern project management.

CHAPTER 1

Project Management

Project management has existed in some form for thousands of years. After all, anything that requires an approach where humans organize effectively to a plan and achieve specific objectives can be loosely defined as a project. How else would have humans achieved its stunning wonders and achievements? Here are some examples: the Great Pyramid of Giza (2550 BCE) and the Great Wall of China (221-206 BCE). But, how were things done before today's "essential tools" existed such as laptops, spreadsheets, and cell phones? People had much different tools then, but the approach was the same namely planning and executing. Today, we have a deep scientific knowledge of the world around us that is essential for today's projects; however, the ancients could manage without this understanding.

Abraham Maslow[1] eloquently pointed out that for man to complete certain functions and achievements requires a hierarchy of needs to be in place, represented through a pyramid consisting of five levels. In examining significant projects across history there are a number of assumptions that can be made. For example, for a society to carry out major projects it must have a number of factors, similar to Maslow's model; a specific purpose for the project with an objective, the right skills available, a level of societal stability, maturity of process and organization, equipment (tools) and techniques, and the availability of finances and resources.

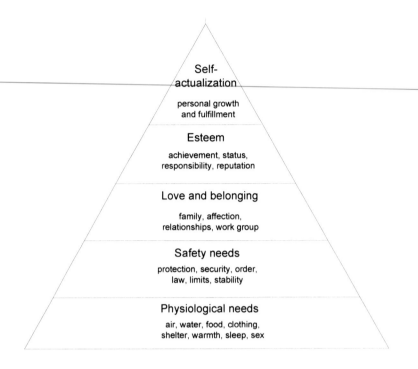

The Evolution of Project Management

The evolution of project management has been very much driven by external factors like war and economic forces where, for example, there have been business pressures to organize resources and meet end goals. The Industrial Revolutions of the last 300 years created new technologies which solved problems in transportation, manufacturing, and communication. Often, these solutions required increasingly-complex projects to manage their completion and delivery. Similarly, by the mid to late 20th Century, software development projects became so complex they required sophisticated management techniques.

An important question is when did modern project management appear, or what aspects of it appeared when? For example, the *PMBoK® Guide* to knowledge areas like Risk Management was developed in 1983. Yet, risk management has been practiced since projects began and is exemplified in the insurance industry that began with insurance for ships and ship cargos as early as 1800 BCE to help finance ship voyages.

The Term Project Management

The Latin word projectum means "to throw something forward."
The word "project" originally meant "something that comes
before anything else is done." When the word was initially
adopted, it referred to a plan of something, not to the act
of actually carrying this plan out. Something performed in
accordance with a project was called an object. This use of
"project" changed in the 1950s when several techniques for
project management were introduced: with this advent the word
slightly changed meaning to cover both projects and objects.
However, in certain projects, there may still exist so-called
objects and object leaders, reflecting the older use of the words.

What is Project Management?

One modern definition of project management:

> "The application of knowledge, skills, tools and techniques,
> to a broad range of activities to meet the requirements of the
> particular project. For example, the use of methodologies, project
> life cycles and plans, and tools like Gantt and Pert charts."

> —Unknown

Today's Definition (post 1983)

A project is a temporary and a one-time endeavor undertaken to
create a unique product or service, which brings about beneficial
change or added value. This property of being a temporary and
one-time undertaking contrasts with processes, or operations,
which are permanent or semi-permanent, that drive ongoing
functional work to create the same product or service over and
over again. The management of these two systems is often very
different and requires varying technical skills and philosophy,
requiring the development of project management.

What Makes up a Modern Project?

Project management knowledge and practices are best described
in terms of their component processes. According to the *PMBoK®*
Guide these processes can be placed into five process groups
(initiating, planning, executing, controlling and closing) and

nine knowledge areas (project integration management, project scope management, project time management, project cost management, project quality management, project human resource management, project communications management, project risk management and project procurement management). These are all modern terms; in fact, they were only fully defined in 1983 as part of the *PMBoK® Guide*. So, the question is what happened before that? That is what the book will examine.

Different Types of Projects

Where do you start with a book of such daunting scope? How do you approach this subject? Projects broadly fall into different types that emerged during different periods of history:

1. Structural construction projects - edifices, structures, and buildings.

2. Transportation related projects - of people, vehicles, water and this includes the building of roads, bridges, aqueducts, canals, ships and railroads.

3. Project expeditions - voyages, journeys, or explorations.

Examples of these Types of Projects

1. Structural construction projects – like Giza, Colosseum, Gothic cathedrals, Taj Mahal, Crystal Palace, Hoover Dam, and Empire State Building.

2. Transportation (and other) related projects – Hadrian's Wall, Iron Bridge, Transcontinental Railroad, Panama Canal, Golden Gate Bridge, and London's sewer projects.

3. Project expeditions – Columbus, Magellan.

Era/ Period	Type 1	Type 2	Type 3
Pre	Great Pyramid of Giza, Stonehenge, Ziggurat of Urnammu,	Great Wall of China	Necno circumnavigating Africa
5th - 1st BCE	Parthenon, 6 Wonders of World, Petra		Alexander's Campaign in the East
100 BCE – 500	Coliseum, Pantheon	Caesar's Rhine Bridges, Hadrian's wall, Aqueduct - Pont du Gard	
500 – 1050	Cathedral at Hagia Sophia, Toltec Pyramid	Grand Canal of China	Caesar's Campaigns
1050 – 1300	Gothic cathedrals, Gothic Castles, Angkor Wat	Grand Canal of China	Norman Invasion, First Crusade - Holy Land, Viking voyages to Vineland
1300 – 1450	Florence Cathedral Duomo		Marco Polo, Admiral Zheng's Voyages
15th			Columbus Voyage of Exploration
16th	St. Basil's Cathedral, St Peter's Basilica		Magellan Circumnavigation
17th	Taj Mahal, Palace of Versailles, Blue Mosque, Hardwick Hall,	Canal Du Midi	
1ST Industrial Revolution 1750-1840 Phase 1		The Iron Bridge, Menai Suspension Bridge, First Railway (Stockton and Darlington)	James Cook's Circumnavigation
1ST Industrial Revolution 1840-1890 Phase 2	Crystal Palace	Transcontinental Railroad, SS Great Britain, Suez Canal London Sewers, Transatlantic Cable, Hollerith's Mechanical computer	
Second Industrial Revolution 1890-1940	Empire State Building,	Panama Canal, Titanic, Hoover Dam, Golden Gate,	Race to North Pole Race to South Pole

Table 1.1: The succession of historical projects selected.

Projects driven by Institutions versus Commerce

For the most part, significant projects up to the 17th Century were driven by institutions like the state (government) or religious organizations (temple or church). For example, the Roman Government collected taxes that were then applied to projects. Likewise, the Medieval Church sponsored the building of church-

es and cathedrals although the latter became critical to the economy of towns as they attracted pilgrims. The economic success
spurred a competition between towns to build these structures.

Commercial projects became more prominent with the
Renaissance and the move to a money economy from one
based on barter. This transformation further evolved with the
formation of corporations (commercial organizations) in the 17th
Century and the development of national banks.

Different Views of Projects

In examining historical projects, it is essential to view the
project from the lens of that period. Undoubtedly the projects
were executed with very different mind sets, by different
cultures, and with different belief systems. The following were
significant projects from different periods:

1. The Giza Pyramid workforce of Egypt (2550 BCE) lived
    in a state-controlled hierarchical society. In this world
    the pharaohs had the stature of gods, priests held a
    status that was greatly higher than the populace beneath
    them. The elite lived off the masses who lived a basic
    lifestyle.

2. In the Gothic cathedral workforce of medieval times
    (1050-1300), most people lived within a seven-mile radius
    of where they were born for most of their lives. The
    church was the centre piece for a population that relied
    on the spoken word (99% illiteracy). The church was a
    conduit of information for villages and the priest was had
    great authority.

3. The Titanic workforce at Harland Wolff in Belfast
    (1908-1912) was working for the premier ship builder
    in Europe. In the middle of the second Industrial
    Revolution, they exuded great confidence as new
    technology was pushed to its limits in creating the
    Olympic class of ships. The workforce saw themselves as
    craftsmen, with skills unparalleled anywhere, and with a
    job for life.

4. The Hoover Dam was built during the Great Depression.
    The large workforce was transient, willing to travel long

distances for work, and lived in substandard conditions at risk to themselves and their families.

Program vs. Project

A "Program" is a term for a broad set of projects initiated to fulfill common objectives and strategy. Sometimes, a program can contain many nearly-identical tactics which are the same. A good example of this is when an institution takes out a program that may last centuries, like the Roman Road Building Program where the same set of activities is repeated continuously. A program is not a project.

The Evolution of Project Management in Civilizations

Project management started in the field of construction and quickly incorporated developments in new materials, technologies (equipment), and building techniques (through guilds). For example, in the Roman era, both the evolution of concrete and iron had a significant impact on project planning and execution (see Colosseum page 134).

Beyond construction, different types of projects originated with the great voyages of discovery and exploration across the world. Through pioneers like Christopher Columbus, Vasco de Gama, Cabral, and Magellan, voyages opened up new trading routes and shortened traditional overland journeys.

In Europe, up to the 15th Century, books were copied by hand and were laborious and expensive to produce. There was a reliance on memorization of facts without the mass printing of the written word. The development of printing was a massive catalyst for change as knowledge became widely disseminated. It had a major impact on projects as the project team could now consult a book rather than find an expert on any technical subject.

The birth of printing and the scientific age (16th/17th Century) had an impact on project management as it moved from a purely non-empirical (derived from observation or experiment) approach to one based on scientific and mathematical calculation. This period saw important discoveries and inventions, and subsequent developments in the field of modern engineering.

The Industrial Revolutions of the 18th and 19th centuries saw the evolution of the machine age, and the mechanization of society. The industrial age and the whole mode of life adapted to the pace and capabilities of the machine,[2] and the mechanization of life we have today. This era saw project management deliver significant projects in transportation, and infrastructure. It also saw a new phenomenon with projects that were initiated by an evidence-based approach and scientific results were used for justifying the project. For example, the London Sewer project was based on empirical evidence and was then widely copied by other major municipalities.

The Role of Knowledge in the Evolution of Projects

Through history, as the overall volume of knowledge expanded, more complex problems were solved-especially those that required different disciplines of engineering. For example, could the South Pole have been conquered 200 years earlier? Could the Panama Canal also have been built during the same time frame? Probably not, as the ability to solve problems is based on having the right tools and technology, and a repository of knowledge which is also a major factor in managing risk to an acceptable level.

The Uneven Adoption of Technology

Since the Victorian era, the development of technology in the West has been linked to progress. However, the spread of technology has been uneven and its rate of adoption is affected by the culture. Sometimes a new technology exists in a particular society but is inconsequential to it. For example:

- The Greeks developed a primitive steam engine, but its further development and application were inhibited by the low cost of labor. In today's project terminology, there was no return on investment (ROI) for it.

- The plough and the horse collar had little impact on farming in the Mediterranean, but was a major factor in the medieval agricultural revolution in Northern Europe. In the Mediterranean, an ox pulling a primitive stick was good enough to cut through the loose, dry top soil. In Northern Europe only a steel blade could cut through the heavy wet soils.

- In the 16th Century, the Japanese acquired firearms and then abandoned them for 300 years. Their preference was for precisely crafted steel Samurai swords.

- The Chinese had gunpowder, the compass, and lateen sails centuries before the Europeans. Yet they did not put these together to aggressively explore new lands and conquer native populations.

- The steam shovel was developed in the UK, but only adopted in the U.S., as cheap labor was available for digging canals in the UK (through the Navigators). The U.S. Panama Canal project could not have been built without this technology.

Basic Essentials to Projects

Certain essentials need to be in place for a project to be feasible, but there are always exceptions:

- Level of stability in the culture.

- Governance.

- Ability to plan and execute.

- Ability to measure.

- Financial support.

- Availability of knowledge and technology along with skilled people who can use these.

- Verbal and written Communication.

Ability to Measure a Catalyst for Projects

The ability to measure is fundamental to almost any project, and the ancient projects were able to do this to a high degree of accuracy. For example, many early projects provided large scale irrigation, such as the Persian tunnels, and the Roman aqueducts one of which had a gradient of 17 meters (56 feet) over 35 kilometers (21 miles) of tunnels, requiring the ability to measure very accurate gradients over long distances. The Egyptian Cubit evolved as a standard during the pyramid building era.

The early voyages of Stone Age Columbus (20,000 BCE)[3] or the Vikings (1000) relied on ice or land hopping across the Atlantic Ocean to North America. This was a slow and difficult way of exploring, but it was within the limitations of knowledge and technology (carracks and long boats) available.

The Athenians that built the Parthenon understood the optical illusion of line sag. They adjusted it by beveling up horizontal lines very slightly, and adding a curve to the columns.

As the Romans established an empire a road building program was essential to ensure the rapid movement of troops and supplies. Land surveying was fundamental to this vast initiative. With this basic skill the Romans were able to set up a massive construction program consisting of thousands of building projects.

With the Renaissance and the development of perspective drawing, where a grid was used to frame a scene, Renaissance explorers developed the ability to measure and travel long distances. The grid was then applied to maps to give distances and this helped spur the age of exploration where a ship would venture into a large body of water rather than island hop hugging the shoreline for security. However, sailing south of the equator was a barrier for European ships as astrolabes could no longer sight the North Star as it disappeared off the horizon. Eventually, it was overcome by Portuguese sailors who in 1434 finally braved the passage, only after fifteen voyages had turned back.

In 1492, Columbus had some measurements related to earth's circumference and with these he calculated on reaching Japan some 2,000 miles after leaving Europe. His theory was almost correct. He travelled the distance but it was the West Indies he reached instead. He was very fortunate.

The declination of the sun could be used to calculate latitude, but on rolling ships this proved difficult, and this was not adequately solved until Harrison's Chronometer in the 18th Century. The Renaissance explorers were taking massive risks in sailing large oceans without accurately knowing the latitude.

Modern Terms in a Historical Context

The terms "scientific," "engineer" and "technology" are relatively modern and problematic in the context of projects of classical antiquity.[4] The term "engineer" (derives from the Romance languages) was incorporated into the English language in the 14th Century. "Science" is derived from the Latin scientia (knowledge) and has also been in use in English since the 14th Century. However, the term "scientific" did not appear in general usage until the late 16th Century. The English word "technology" is derived from the Greek technología.[5] In the 17th Century, the term generally referred to a "discourse or treatise on an art or arts; the scientific study of the practical or industrial arts." By the early 18th Century it also meant the mechanical or the practical arts. The word "scientist" was not coined until the first half of the 19th Century.[6]

Engineers versus Builders versus Project Managers

It is important to note that the term "Project Manager" is recent whereas "Engineer" is a term used for close to a thousand years.

The term project manager may not have existed even in the 19th Century but the role on major historical projects has always been fulfilled by an individual but under a different title. Engineers, architects, master-builders, and sponsors have all played a project leadership role within projects.

By definition, Roman architects were significantly different from their modern counterparts, acting as engineers, architects, artists, and craftsmen combined. Vitruvius was very much a combination of these, a fact reflected in De Architectura where he defined the characteristics of an architect (see page 136).

Right up to the medieval period, the master mason had two responsibilities, that of architect, as in designer; and builder, as in the contractor. In the Middle Ages, the builders of catapults, battering rams, and engines of war were referred to as *ingeniators* by Latin writers.

"....the name "engineer" itself originated in the 11th Century from the Latin ingeniator, meaning one with ingenium, the ingenious one. The name, used for builders of ingenious

*fortifications or makers of ingenious devices, was closely related
to the notion of ingenuity, which was captured in the old meaning
of "engine" until the word was taken over by steam engines and
its like. Leonardo da Vinci bore the official title of Ingegnere
General. His notebooks reveal that some Renaissance engineers
began to ask systematically what works and why."* [7]

*"The first engineers were irrigators, architects, and military
engineers. The same man was usually expected to be an expert at
all three kinds of work. This was still the case thousands of years
later. During the Renaissance, Leonardo, Michelangelo, and
Dürer were not only all-around engineers but outstanding artists
as well. Specialization within the engineering profession has
developed only in the last two or three centuries."* [8]

Scribes played an important role in ancient societies and
projects; for example, the ancient Egyptian scribe was educated
in the arts of writing and arithmetic, and was part of a middle
class elite, employed in the bureaucratic administration of the
state, army, and temples. They drew up everyday legal documents for commerce and money lending, and they maintained
archives. Sons of scribes were brought up in the same scribal
tradition, were sent to school and, by entering the civil service,
inherited their fathers' positions. Monumental buildings were
erected under their supervision, and administrative and economic activities were documented by them.

Villard de Honnecourt, who lived in 13th Century Picardy
in northern France, provided writings through his model-book [9]
insights on a wide range of topics. These were part of the tool kit
of a cathedral master-builder and included architectural plans,
elevations and details, ecclesiastical objects and mechanical
devices like a mill-driven saw, a number of automata, and lifting
devices. Villard's detailed sketch book of construction work was
based on his travels through many cathedral building sites.

Figure 1.1: Villard de Honnecourt[10] through his model-book provided a tool kit for the cathedral master-builder.

Evolution of Materials

One factor that has affected the history of projects has been the transition from natural to man-made materials, and then the subsequent evolution of those man-made materials. For example, building with stone and timber materials reached a plateau, that is a maximum in height and stability of structures, until other man-made materials were developed like brick and concrete. Significant Roman projects benefited from these new materials, like the large viaducts and Colosseum. The development of these materials was driven by several factors; primarily, a better scientific understanding and the ability to measure things accurately. To make concrete and bricks strong and flexible required a level of scientific knowledge. How to put these materials together into large and possibly complex structures required an increasing sophistication in mathematics such as practical geometry. The Roman's extensive use of arches allowed for vast structures to be built with less material, like the Colosseum. In the 17th Century, cheap glass became available, was manufactured in large sheets, and was used prominently in construction projects. Similarly, in the 18th Century, cheap iron became available and by 1840 it was cheap enough to replace clay and wood pipes. As a result, the low cost of iron opened all sorts of engineering possibilities, and construction projects like the Crystal Palace (1851). The advent of Bessemer steel in 1856 had an important impact on building projects with its use as a skeletal structure within buildings with concrete (1756). However, the modern skyscraper was only viable with the development of elevators (1860).

The Main Branches of Engineering

Engineering has evolved into a professional discipline over several hundreds of years to include:

- Civil Engineering - The design and construction of public and private works, such as infrastructure, bridges and buildings. The first engineering school was opened in France in 1747.

- Mechanical engineering - The design of physical or mechanical systems, such as engines, power-trains, kinematic chains, and vibration isolation equipment The Institution of Mechanical Engineers in England was founded in 1847.

- Chemical Engineering - The conversion of raw materials into usable commodities. In 1885, a course in "chemical engineering" was offered at Imperial College, London.

- Electrical Engineering - The design of electrical systems, such as transformers, as well as electronic goods. Darmstadt University of Technology is the first faculty of electrical engineering worldwide in 1882.

- Aerospace Engineering - The design of aircraft, spacecraft and related topics (early 20th Century).

The timing coincides with significant changes in project management. First, projects became empirically based, as precise mathematics and physics was available in the 17th Century. Second, materials (glass, iron, concrete) become available in quantity and at a lower price point.

Commonality across Projects

The world history of projects reveals that most projects, whether construction, transportation, or voyages of exploration, have strong similarities in terms of the phases or stages in the sequence of steps. They follow a method and have a set of requirements, designs, and constraints. They depend on communication, problem solving, decision making, and creative processes. They involve a schedule, a budget, and a recipient, customer, or audience. Importantly, they combine the effort of a team into a singular coherent output useful to someone. Undeniably, there is a core set of concepts that most projects share.

Structure of the Book

The chapters of this book are organized for *a given historical period* along a time-line that runs through eras and centuries with some overlap:

- From Village to City (2550-510 BCE).

- Roman Republic (510 – 100 BCE).

- The Roman Empire (100 BCE - 456).

- The Middle Ages.

    ○ The Early Middle Ages, lasting from about 350 to about 1050.

    ○ The Central Middle Ages, lasting from about 1050 to about 1300.

    ○ The Late Middle Ages, lasting from about 1300 to about 1450.

- The 15th Century and Renaissance and age of discovery (European).

- The 16th Century and the Modern Age of Engineering.

- The 17th Century and the First Scientific Revolution.
- The 18th Century and the Industrial Revolutions.
  - First Industrial Revolution 1700-1840 (Phase 1).
- The 19th Century.
  - First Industrial Revolution 1840-1890 (Phase 2).
  - The Second Scientific Revolution.
- The 20th Century.
  - Second Industrial Revolution 1890-1940.

Structure of Chapters

The sections include:

- Brief synopsis of trends/changes (societal, economic, ecological, technological).
- Impact of changes on society.
- Major events - natural disasters, wars, epidemics.
- New tools, techniques and breakthroughs.
- Regions significant to the historical period.
- Significant projects that arose (1 or 2) and notable mentions.
- Key players of the historical period.
- Chapter Wrap-up.
  a. Conclusions.
  b. Key Lessons.
  c. Tips for educators.

Trends and Changes

A trend is a general tendency or direction, in a civilization or culture, or a market or industry, in which something tends to change. For example, paper making originated in China in 100, and then spread to Tibet 650, India 670, Central Asia 750, Baghdad 794, Cairo 850, Tripoli 1000, Valencia 1151,

Italy 1276, France 1326, and to England in 1490.[11] In another example, during the 11th Century the Chinese Government introduced an early ripening rice from Indochina which allowed two or even three harvests per growing season. The upsurge of rice cultivation sweeping across South China, coupled with terracing and irrigation, produced large surpluses which doubled the population from 50 million (800) to 123 million in (1200). Urbanization rocketed and a middle-class grew which increased trade and manufacturing.[12]

Over time, the speed of adoption of new technologies started to increase dramatically. In 14th Century Europe, the adoption of print had many spin off effects and it can be seen as a rolling wave sweeping across Europe from Mainz to Cologne in 1464, Basel in 1466, Rome in 1467, Paris, Nuremburg and Utrecht 1470, Milan, Naples, and Florence 1471, Augsburg 1472, Lyons, Valencia and Budapest 1473, Cracow and Bruges 1474, and to London in 1480. By the end of the 15th Century there were 73 presses in Italy, 51 in Germany, 39 in France, and 25 in Spain.[13] Through history, change comes along in spurts driven by:

- Introduction of new practices or technology.
- Warfare.
  - Catalyst that spurs changes in technology or evolution of it.
  - Results in conquest and collapse of empires.
- The environment, climatic changes, extraterrestrial (solar flares, comets).
- Natural disasters like earthquakes, or volcanoes.
- Diseases, plagues and epidemics.
- Gains or increase in knowledge.

Impact of Changes

Typically, there is a reaction to any trend/change which has an impact on the civilization or culture. For example, the Medieval Ice Age (1400 to 1750) had a long term impact to only a small shift in temperature. Likewise, a dramatic fall in the population during the 14th Century was a reaction to the bubonic plague.

Major Events

During the outlined historical period, what major events happened? For example, the collapse of Empires like the Roman or Mayan, and the birth of something new, or a discovery, or the development of a new technology.

The fall of Constantinople to the Ottoman Empire and the closure of the trade route to Asia were catalysts for the 15th Century voyages of discovery and exploration; Most notably, Columbus' voyage west to establish a trade route to the Far East.

The 20th Century Space Race became a reality when President Kennedy made a commitment to complete the U.S. project and adequate funding was made available. NASA got a substantial 4% of the national budget. Looking back, Kennedy's speeches were a major catalyst in securing the funding and changing the history of modern space projects.

New Tools Techniques and Breakthroughs

The development of significant tools, templates, and techniques was a catalyst for many significant projects that could only be attempted when this breakthrough was reached. For example:

- A building of the scale of the Roman Colosseum was wholly dependent on a design driven by the arch and barrel vault, and vast quantities of concrete and brick.

- The 15th Century voyages of exploration could only be undertaken with a better understanding of what was beyond the visible horizon. This required tools like the astrolabe, and maps that were scaled, which were becoming available with grid-lines and so provided a guide to possible distances.

- Practically all ancient projects were completed on a non-empirical[14] basis, as precise mathematics and physics did not appear until the 17th Century.

- The 19th Century London city sewer project (1860) was only undertaken when there was overwhelming statistical evidence that cholera was directly linked to sewage polluting the Thames River, which was used for drinking water.

Regions Significant to the Historical Period

In the outlined historical periods, certain regions may have had a more significant impact on the evolution of projects, even though the book takes a global perspective of projects. This may be because there was an evolution of a certain technology driven by a specific need in a region.

Significant Projects

So, how were the significant projects identified for the book? A project had to be discernable as a project with a clear objective, was predefined by a degree of planning, and was led by a recognized leader (project manager). It had to be completed in a specific time frame, or was just faced with many challenges along the way like the lack of key resources, or physical obstacles.

A significant project is one that is successful beyond all expectations, a ground breaker, a catalyst for change, and will have other projects following in its footsteps. These significant projects are generally recognized as great achievements.

The output of a project has to be significant. With the Giza Pyramid or the Gothic cathedrals, this is very obvious. But successful project management brings together a combination of scope, resources and time, often shown in a triangle. For example, the Great Escape from Stalag Luft III[15] was significant not just in its output but for the hostile environment and the perceived lack of resources that surrounded the project. The Colosseum had a significant output, aggressive time frame, and large project workforce.

Determining Significant Projects

The criteria used for determining a significant project are as follows:

1. Reasonable supporting evidence of authenticity.

2. Significant output or project objective.

3. Remarkable use of resources, efficiently and within budget.

4. Completed within a specified time frame.

5. Inspiring to other projects in how it was delivered.

6. Completed in a hostile environment.

7. Unique as possible, a first that sets a direction.

A project that measures up well to most of the above criteria can be classified significant. Some examples include the Transcontinental Railroad project (page 410). This project had a significant objective, and depended on limited resources as it was a government-instituted venture and depended on share issues. It had to be completed in a time frame to see a commercial return for investors. It also was inspiring to the watching world in its rate of progress, was completed in an inhospitable environment (across the Nevada mountains), and was unique in its massive scale. The government had to underwrite it to the point that contractors demanded to be paid in advance.

In another example, the Colosseum project (see page 134) had a significant objective, and had to be completed in a time frame to prop up a shaky political regime. It also was inspiring to the watching world, and was completed in a challenging environment. However, it was not built with limited resources; in fact, huge amounts of resources were poured into the project to make it a success. Was it always on course? The answer is no. The Emperor Vespasian had to underwrite it to make sure it was successful.

The Palace of Versailles (see page 341) has been listed as a significant project as parts of it were widely copied, and the output, the hall of mirrors, is viewed as a masterpiece. However, the project was an out of control disaster, and the overall debt load from this project was a contributing factor to starting the French Revolution.

Key players

This section discusses some of the key players of the historical period. Some were not necessarily related to specific projects, but who influenced the era; for example, Vitruvius or Villard de Honnecourt.

Chapter Wrap-up

Conclusions

This section will list the conclusions for each chapter with a focus on the significant projects of the era.

Key Lessons

This is a close examination of the nine *PMBoK® Guide* Knowledge Areas against the significant projects to determine their usage and importance to the project.

Educators

- Discussions in the context of today's projects, important aspects of project management, designed to help educators.

Other

Projects not Considered for the Book

There has been a deliberate omission of certain projects because they do not meet the criteria. A project should be unique to qualify for consideration. Reasonable supporting evidence and documentation about the project itself needs to exist. For example, the Toltec Pyramid at Cholula is remarkable in scale as the total volume is almost one third larger than the Great Pyramid of Giza. Yet little is known about the construction project from archaeological records or ethno history. It has been classified as a "Notable Project," (see page 185). Similarly, other projects follow this classification where the documentation is sparse and sporadic or nonexistent.

The First Project Managers

These were irrigators, architects, engineers and technicians (typically military), and generally multi-skilled generalists adept at dealing with many situations. For example, the Parthenon project was under the general supervision of the sculptor Phidias, who also had charge of the sculptural decoration. The same man was usually expected to be an expert at organizing, planning and directing work and the project work force. The first project managers needed:

- ~~Skills in communicating a vision.~~

- Theoretical knowledge in the subject area, for example, construction, or mechanics.

- Experience in transforming a vision into reality.

- Skill in negotiating and implementing the work, knowing the activities and the Work Breakdown Structure.

- Organizational skills; specifically, that of a project workforce.

As an example the following had many of the skills and traits:

- The Egyptian architect as exemplified by Hemienu.

- The Roman architect as laid out by Vitruvius.

- The medieval master-builder as exemplified by Villard de Honnecourt.

- The Renaissance architect as exemplified by Brunelleschi.

- The Renaissance explorer as exemplified by Columbus or Magellan.

- The Victorian engineer as exemplified by Darby or Stephenson.

For more information on these project managers see page 527.

Modern Project Management

This book ends at the dawn of modern project management which started in the 1940s; in many ways, the Second World War was a catalyst for the transition. Modern project management borrows from elsewhere and is rooted in Empires and later in Western civilization, as well as the output of the three Industrial Revolutions and the development of scientific techniques. In 1983 these century year old practices were distilled into the *PMBoK® Guide*.

CHAPTER 2

Evolution of PMBoK® Guide Knowledge Areas

What are the *PMBoK® Guide* Knowledge Areas

In today's projects, each project phase is completed by the delivery of one or more deliverables. The *PMBoK® Guide*™¹ advocates that not only phases and deliverables but also knowledge areas need to be used through the course of a project (see Table 2.1 below).

Knowledge Areas	Description
Integration Management	Integrates all eight knowledge areas and includes project plan development, integrated change control, and project execution
Scope Management	Plans and defines the scope, identifies major deliverables and the work breakdown structure (WBS)
Time Management	Defines the activities in the project, completes the activity sequencing, and develops the schedule
Cost Management	Defines estimates, develops a budget, and controls cost
Quality Management	Plans the approach to quality in a project, identifies the required quality characteristics, and builds the quality assignments into the schedule

Human Resource Management	Manages the stakeholders and the team throughout the life cycle, identifies skill requirements for assignments, and sets team development and rewards
Communications Management	Determines stakeholders, plans communications, sets expectations, distributes information, reports performance, and manages stakeholders
Risk Management	Makes initial assumptions that affect the project, develops the risk management plan, identifies and analyses risk, and plans response and implementation of the risk and contingency plans
Procurement Management	Plans and solicits bids, assesses make-or-buy decisions, and negotiates contract, administration, and closeout

Table 2.1: The nine knowledge areas that are essential to a project as Advocated by PMBoK® Guide.

Significance of PMBoK® Guide Areas

The *PMBoK® Guide* plays a significant role in projects today and brings a framework to the profession, both discipline and structure. Its use is generally accepted as best practice within the project management discipline and it is recognized as an international standard (IEEE Std 1490-2003). It describes the fundamentals of project management, irrespective of the type of project, be it construction, software development, engineering, automotive, etc.

PMBoK® Guide recognizes five basic process groups and nine knowledge areas typical of almost all projects. The basic concepts are applicable to projects, programs and operations.

The five process groups are:

1. Initiating
2. Planning
3. Executing
4. Controlling
5. Monitoring
6. Closing

The following section provides an initial definition for each knowledge area (listed in Table 2.1).

Integration Management

PMBoK® Guide Definition - Integrates all eight knowledge areas and includes project charter and plan development, project execution, monitors and controls the project, integrated change control, and closing the project. The integration management processes are useful for project managers to start, control and shutdown the project.

Origins

Integration management is important as it acts as the glue for the other knowledge areas. The written project plan was probably not developed until the 20th Century although through history the chief architect or master-builder would have been very familiar with the sequence of activities. Most projects were targeting a project end date, often for political reasons, so the concept of the schedule and plan has always been in place. Like today's project most projects had a significant return on investment whether through indirect revenue, like the Gothic cathedral projects which drew in pilgrims into the towns and boosted the economy; or like the Parthenon project which stimulated the local economy, small businesses, and reduced unemployment.

Scope Management

PMBoK® Guide Definition – collects requirements, plans and defines the scope, identifies major deliverables and the work breakdown structure, and verifies and controls scope. The project scope management processes are useful for project managers to determine what is, and is not, included in projects, and to control project/product scope during the project life cycle.

Origins

Scope management is fundamental to project management in that it is hard to imagine how any project can move forward without this element. There is strong evidence that the master-builders on ancient projects had a comprehensive grasp of the totality of their projects, the entire building operation, including all the major activities and the concept of a work breakdown structure. From the earliest ancient projects scope had to have been understood and used then as it is used today with very little difference. Otherwise projects like the Giza Pyramid and the Colosseum would have been very difficult to manage and deliver in the required time frames. The main exceptions to this were projects that were clearly out of control.

Time Management

PMBoK® Guide Definition - Defines the activities in the project, completes the activity sequencing, estimates the activity resources and durations, and develops and controls the schedule. The project time management processes are useful for project managers to apply and explain the steps needed to create the project schedule, and the ability to plan and finish the project in a timely manner.

Origins

The ancient building projects of the past, as do projects today, required effective administration and management of many activities in a complex schedule. The master-builders had to have a comprehensive grasp of the totality of building operations and all their ramifying complexities so they could sequence the activities most efficiently. The projects did not only consist of assembly and erection activities but also the preparation of the site and the logistics for the project. Logistics scheduling evolved in the military. In 2000 BCE, Sun Tzu wrote about strategy and *scheduling* from a military perspective. This is one of the earliest descriptions of a project management function.

Cost Management

PMBoK® Guide Definition - Defines estimates, develops a budget, and controls costs. The project cost management processes are useful for project managers to apply and explain cost management and cost control, including earned value.

Origins

Cost management has played a significant role in all projects since the development of money, and its origins are in the development of banking. Ancient projects relied on investments and loans as do projects today. The workforce required some sort of a regular payment. The Gothic cathedral projects are an excellent example of very effective cost management. These projects ran for decades if not centuries and they had to successfully manage and control their budgets. This was done through very effective governance structures, that consisted of chapters (boards) and overseers of works, and their priority was to keep the project going.

Quality Management

PMBoK® Guide Definition - Plans the approach to quality in a project, identifies the required quality characteristics, and builds the quality assignments into the schedule. It performs quality assurance, and control. The project quality management processes are useful for project managers to apply required quality to projects, and to apply metrics to improve project performance.

Origins

A major feature of quality control is the establishment of self control where a party producing a product has the means and knowledge to determine compliance with specifications and, furthermore, is empowered to initiate corrective actions when needed. An individual performing the work conducts an inspection of his product. Implementing self control implies a comparison of actual results to specifications.

Within ancient projects a non-empirical approach to the project was the only option available as science and mathematics had not evolved enough to predict empirically potential outcomes. Experience with materials and construction was paramount and continuous inspections of materials, assemblies and structures was part and parcel of the project. The inspections would look for defects, anomalies, decayed areas, and overall structural quality. Appraisals employed scale models of proposed buildings and visual observations, and assessments of failures in existing buildings.

Human Resource Management

PMBoK® Guide Definition – Develops the human resource plan, acquires and develops and manages the project team throughout the lifecycle. It identifies skill requirements for assignments, and sets team development and rewards. The Project Human Resource Management processes are useful for project managers to get and keep productive staff for projects, and improve project performance.

Origins

Human resource management addresses both the theoretical and practical techniques of managing a workforce who individually and collectively contribute to the achievement of the objectives of an organization. Within the *PMBoK® Guide*, it is not just acquiring but developing and maintaining a qualified workforce.

Human resource management has always been critical to projects. A common perception today is that projects of ancient times relied on an almost inexhaustible, low cost labor pool. This is not the case. The reality is the labor pool was limited and composed of skilled and unskilled labor, the former being considerably smaller and much more valued. As far back as 3000 BCE in the early Empires of the Nile, Euphrates, Tigris, and Indus valleys there were separate classes of craftsmen concentrated in the cities, skillful in metallurgy and working with wood and stone. A strong central government was required to coordinate the labor pool for projects.

Communications Management

PMBoK® Guide Definition - Identifies stakeholders, plans communications, distributes information, manages stakeholder's expectations, and reports performance. The Project Communications Management processes are useful for project managers in continuously managing all project communications with stakeholders.

Origins

Ancient projects with an extremely large workforce (in excess of 10,000 and vast by today's standards) needed very effective communications management. On an ancient project the communications had to exist between all the places of work including the offices, construction site, workshops, quarries and workers villages. One aspect of communication was rapidly transferring design ideas to all areas of the project team. This was done through simple templates and miniature models, reducing the need for lots of documentation.

Risk Management

PMBoK® Guide Definition - Makes initial assumptions that affect the project, develops the risk management plan, identifies and analyses risk (qualitative and quantitative), plans response and implementation of the risk and contingency plans, and monitors and controls risk. The Project Risk Management processes are useful for project managers to manage and track all the project risks holistically, and to be better prepared for the unexpected.

Origins

Risk management is probably the oldest of the *PMBoK® Guide* areas. A simple examination of ancient projects shows that over time civilizations took on projects knowingly with increasing amounts of risk. The project architects were getting more comfortable with the increased level of risk and more effective in managing it. For example, Romans pushed concrete technology in their buildings to the limit, with extensive arches, barrel vaults, and domes as seen in the Colosseum and Pantheon.

The Romans like the Greeks had a trial and error approach to construction based on experimental construction knowledge that was not theoretical. For example, they did not know the exact structural strength of stone. This non-empirical approach would qualify for a qualitative risk analysis based on experience, rather than a quantitative risk analysis based on metrics.

Procurement Management

PMBoK® Guide Definition - Plans and conducts procurements, soliciting bids for products and services, assesses make-or-buy decisions, and negotiates a contract. It also administers, and closes-out the procurements. The Project Procurement Management processes are useful for project managers to take a more comprehensive approach to procurement in tracking the various activities, and ensuring contracts are fully delivered.

Origins

Procurement management has its roots in trading and can be traced back to very early civilizations. It has been constantly critical from ancient to more recent projects, unsurprising with the volume of materials required for some of the projects like the Giza Pyramid and the Colosseum. Both had extensive supply chains moving many millions of tons of materials. The Colosseum project was delivered by four contractors through contracts with requirements for guarantees. More recent projects like the Hoover Dam were completely tendered out and the contract was driven by bonuses and penalties, attached to the delivery.

Conclusion to *PMBoK® Guide* Knowledge Areas

The book will reveal that all nine *PMBoK® Guide* areas were practiced throughout ancient projects through to the 20th Century. The main difference between then and now is that these have not been formulated and documented in terms of the conceptual way that these areas exist today. The main changes through the historical eras were the evolution of new materials, and their usage in projects. This was coupled with an increased knowledge, scientific and engineering which in turn introduced empirical methods into projects. The transition was often a challenge as with the Iron Bridge project (page 371) that had no precedent and where existing practices were used. The pieces of the iron that were assembled into the Iron Bridge were modeled on wooden pieces that were based on the principles of carpentry. Because of the property difference in materials they were somewhat inefficient.

CHAPTER 3

From Village to City (2550 - 510 BCE)

T his historical period was dominated by the very early civilizations that were forming.

"The city is almost the most defining characteristic of civilization."[1]

High civilizations emerged in lands between rivers in both China and Mesopotamia (Tigris-Euphrates Rivers). With these civilizations came the early projects and these were driven by a "temple economy" (grain silos held in religious structures) where the ruler initiated communal work projects as the head of administrative and religious activities. These were typically construction projects using stone, and newly emerging materials like bricks, using metal tools like bronze, copper, and later iron. The edifices quickly evolve in scale and complexity.

In this Historical Period

- Brief synopsis of trends/changes (natural disasters, wars, technology, epidemics).

  a. Agricultural revolution - 10,000 years ago man went from hunting and food-gathering to raising crops and taming animals.

  b. Communal living and the creation of early communities.

- Impact of changes

  a. Agriculture led to several major changes.

    i. It created food surpluses.

    ii. It allowed far larger population densities.

    iii. This led to the birth of the first towns and cities.

  b. The creation of large hierarchical class based societies ran by large-scale government.

  c. Large hierarchical societies could afford to explore and develop sciences to create new materials (like metals), tools, and writing through script.[2]

- Major events

  a. Rapid evolution of autocratic societies in Egypt and China.

- New tools, techniques and breakthroughs

  a. The development of simple tools like wheels, levers, and wedges, around 3000 BCE.

    i. The wheel evolved from the potter's wheel but there is no single common geographical origin, 3200-3500 BCE.

  b. The harnessing of animal labor in carrying materials.

  c. The forging of materials specifically bronze 3000 BCE and iron 1200 BCE.

  d. The discovery of Geometry, (ability to measure), in ancient Babylonia 3000 BCE which covered lengths, angles, areas, and volumes for surveying, construction.

  e. The earliest recorded postal system in Egypt 2000 BCE.

  f. The oldest record of writing in China on bones and tortoise shells 1400 BCE.

  g. The first encyclopedia is written in Syria 1270 BCE.

    h. The foundation of Rome in 753 BCE.

    i. The Greeks start the very first library 530 BCE.

- Regions

    a. Fertile Crescent (Levant (Eastern Mediterranean) and Mesopotamia, and Egypt).

- Significant projects (primarily structural construction projects).

    a. The Great Pyramid of Giza (2,550 BCE) was in a succession of evolving monumental large scale projects in Egypt.

- Other Notable Projects

    a. The Great Wall of China (600 BCE - 206 BCE).

    b. Stonehenge (2,550 BCE).

    c. Necno voyage around Africa (610 BCE).

    d. Ziggurat of Urnammu, Ur.

- Key players

    a. Imhotep.

- Chapter Wrap-up

Trends and Changes

The world population in this period grew from 25 to 100 million.

Agricultural Revolution

10,000 years ago man went from hunting and food-gathering to raising crops and taming animals based on discoveries that enabled a square mile of fertile land to support 20 to 200 times as many people, through the control of water through irrigation channels. This revolution first took place in the hills around the north of Iraq and Syria and spread to the valleys of the Nile and the Indus to become centers of cultural radiation.

Early Communities

A community can be defined as a group of interacting people, united by ties of descent from a common ancestor, living in a common location with common customs and traditions.

Impact of Changes

The agriculture revolution led to several major changes simultaneously. First, it created, and allowed for, the storage of food surpluses that could support people not directly involved in food production. This allowed for far larger population densities, up by a hundred fold. This in turn allowed for the creation and development of the first cities which led to civilization.

The earliest surviving business records can be traced back more than 7,000 years ago to the Chaldean-Babylonians, Assyrians, and Sumerians.[3]

Food surpluses freed people for other specialized occupations and features of civilizations. This led to the creation of large hierarchical societies ran by large-scale government, ruled from the top. Autocratic societies evolved in Egypt and China where priests held and controlled knowledge and held a high position in society. Skilled and unskilled labor was used for monumental large scale projects. These large hierarchical societies had surplus resources and materials. They could therefore afford to explore and develop sciences to create new materials (like metals), tools, and writing.

Major Events

From Village to City

The first communities developed around the Fertile Crescent driven by the agricultural revolution. New technology spread through the Middle East as metal replaced stone tools, and oxen replaced human muscle. The period saw the rapid evolution of the autocratic society of Egypt (then later China) along the banks of the Nile river around 3150 BCE Ancient Mesopotamia emerged 3000 BCE hosted by the Sumerians, succeeded by the Akkadians, then Babylonians, Assyrians, and Persians.

New Tools Techniques and Breakthroughs

Simple Machines

The development of simple machines[1] like the lever, inclined plane, and wedge around 3000 BCE was a catalyst for early simple construction projects. For example, the diversion of water using dams and channels. The wheel evolved from the potter's wheel but the two shared no common geographical origin, 3200-3500 BCE. These helped man move from a hunter/gatherer, to a farmer/city dweller.

Metal Tools

The forging of materials, specifically bronze around 3000 BCE, and then iron around 1200 BCE, impacted the farmer/city dweller by providing metal tools. Bronze immediately improved the durability and efficiency of hand tools. The first metal tools were knives, important for hunting and survival, and weapons (swords) for warfare.

Improvements in Agriculture

Farming revolved around planting seeds which requires breaking up the ground. In the early days this was achieved by scraping with a pointed implement, like a deer antler, or a sharp stick. But to grow a surplus required a useful furrowed field. The first ploughs consisted of sharp pointed timber, hardened in a flame or tipped with flint, projecting downwards at the end of a long handle. This type of plough was sufficient to break up the earth and form a shallow trench in the light soil of Egypt and Mesopotamia. The plough could be dragged by a couple of men but the process was greatly sped up by the harnessing of plough and draught animals, which first appeared from 3000 BCE. In Egypt the centralization of harvests in state warehouses also led to the development of a system of banking.

The harnessing of animal labor in carrying materials had a significant effect on construction projects as it reduced the size of the workforce significantly.

Ability to Measure

It was not until the discovery of Geometry, and the ability to measure, in ancient Babylonia in 3000 BCE that heralded construction projects of edifices, temples, and civilian buildings. Early geometry was a collection of empirically discovered principles concerning lengths, angles, areas, and volumes. These were used to meet practical needs in surveying, and construction.

Ability to Write

The Egyptian papyrus, an aquatic plant, appeared in 3000 BCE and has remained in regular use longer than any other material in the history of written documents. Writing transferred from clay tablets to papyrus. Lightweight papyrus could be easily carried, in volume, made records less cumbersome and permitted a wider use of supporting documents.[5] This catalyzed the development of the earliest recorded postal system in Egypt 2000 BCE. Egyptian authorities required that records of transactions be kept by two different scribes, whose records had to agree. This was an example of early internal control procedures.

The oldest record of writing in China was on bones around 1400 BCE. The first encyclopedia was written in Syria 1270 BCE, and the Greeks started the very first library in 530 BCE.

Keeping Time

The invention of the sundial provided a crude mechanism for breaking the day up into morning and afternoon. As early as 1500 BCE, the Egyptians had developed an advanced sundial. A T-shaped bar placed in the ground and was calibrated to divide the interval between sunrise and sunset into 12 parts. Although the hours within a given day were approximately equal, their lengths varied during the year, with summer hours being much longer than winter hours. This didn't get around the problem of time variation. In projects, scheduling was done using the day as the basic unit of measure, although the day length varied through the calendar.

The next major breakthrough arose around 1400 BCE, when Egyptians used a simple water clock to break down the day into hours. A simple stone vessel with sloping sides allowed water to drip at a constant rate from a small hole. There were twelve separate but differently marked columns on the inside that measured hours. Each monthly column had differently spaced markings that allowed for the seasonal variations of the hours.

Regions

The pace of development started and continued in and around the Fertile Crescent (Levant (Eastern Mediterranean) and Mesopotamia, and Egypt).

Significant Projects

All were civil construction projects based on large labor pools, that required vast resources and materials, and spanned across decades. Simple machines were used predominantly.

The Great Pyramid of Giza Project (2550-2530 BCE)

Although Egypt was subject to outside influences the principal defining characteristics of its culture remained homogeneous throughout the course of its long history.

Background

The Nile, the heart of Egypt, with its yearly floods provided an abundance of food that helped establish a wealthy society. Egyptian society was dominated by a rigid class system with little social mobility - although it was possible to move based on merit, luck or patronage. Women worked and were valued individually, but held subservient. Overall labor was the property of the community who at different times contributed to the construction of the pyramids, the building of temples, cities, roads, the banks of the Nile, and other public works. Some people were exempt as they were required in areas of support for social and state interests. Egypt's foundation was the poorest class. Although slaves were common in Egypt, for the most part, they were the property of rich and not the state.

The excess food of Egypt could readily feed the population of one million and it allowed Egypt to have a standing professional army. War was part of the Egyptian economy, and with this army Egypt was successful in its unification to become the first nation in history (3200 BCE) and the greatest nation at the time. With unification the first king or pharaoh became ruler but, who was also considered a god (sun). When a new pharaoh was crowned the calendar was reset to day one for that dynasty.[6]

The ancient Egyptians believed that death was the start of a journey to the next life. The king was embalmed and entombed within the pyramid to protect him and allow for this transformation and ascension.

Initiating

Prior to the Giza Pyramid (2550 BCE) there was a succession of pyramids that progressed over a few centuries. The first pyramid ever built, was Djoser's Step Pyramid (2668 BCE) in Saqqara, by Mahnud Hotep (Imhotep). Imhotep was a royal architect and physician, but more importantly he was the top government administrator (equivalent to a prime minister) of Egypt. He was in a unique position as he had all the money he needed, and he didn't have to worry about approval from committees. If the Pharaoh Djoser (2668–2649 BCE) wanted something it would be built.

Imhotep had a very clear project charter. He was going to build a very unique tomb out of stone as he wanted a tomb that would stand for all of time. It was the largest stone building ever constructed and impressive in that there were no smaller structures first. It was a 70 meter (240 foot) pyramid that used blocks of stone less than 100 pounds (45 Kilograms). It was surrounded by a stone city of houses, chapels and statues.

Figure 3.1: The Step Pyramid of Djoser (Zoser) was an early attempt of Pyramid building. Painting by Johann Frey (19ᵗʰ Century).[7]

The Step Pyramid (2630–2611 BCE) kicked off a frenzy of projects. The Meidum Pyramid (2613 BCE) was also a stepped pyramid but some time near the end of the project the steps were filled. The casing blocks were not securely tied and the pyramid was abandoned. The Pharaoh Sekhemkhet (2649–2643 BCE) was never buried in it. It is now in a ruinous state, and was quarried for stone by the locals.

The next two Pharaohs were Khaba (2643–2637 BCE) and Huni (2637–2613 BCE), and there is much uncertainty about them. Khaba is commonly associated with the Layer Pyramid, an unfinished pyramid that rose to about 42-45 meters (140 feet) in height. There is no pyramid associated with Huni.

The next Pharaoh in line was Sneferu (2613–2589 BCE) who wished to be buried in a pyramid. At this time the Egyptian's approach to pyramid building was experimental. They didn't have the scientific calculations to determine load bearing capacity of various materials. It had to be worked out by trial and error, or prototyping. The next pyramid, whose modern name is the Bent Pyramid, was the first pyramid disaster. It started as a true pyramid with a 54 degree slope. About two-thirds up the face 45 meters (147 feet) the angle of incline changes to a flatter 43 degree slope. The reason for the disaster was one of the corners was built on a layer of gravel rather than bedrock. As the pyramid grew the mass increased and there was a problem with stability and subsidence as the angle proved too steep. The inner burial chamber cracked and the pyramid was stabilized by adjusting the angle. The reduced angle brought increased stability.

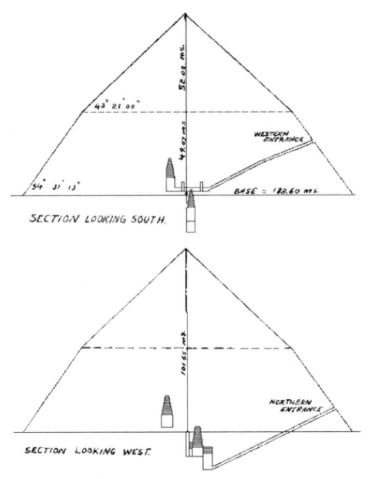

Diagram showing the dimensions and angles of the Bent Pyramid at Dahshur. (Hassan Mustapha.)

Figure 3.2: the Bent Pyramid of Sneferu showing the two build cycles in the project.[8]

Sneferu's next attempt was the Red Pyramid (at Dahshur) built with a very gentle 42 degree slope. It proved to be very successful and he was laid to rest in it. This was an intense period of pyramid projects, with the largest structures ever built. Through the process the Egyptians gained skills in cutting and transporting large blocks of stone.

The Egyptians now had the experience of several centuries of pyramid projects and a somewhat successful track record. They were ready to scale up their efforts and deliver a truly monumental pyramid.

Egyptian construction knowledge was not theoretical but experimental based on trial and error, for example, the structural strength of stone. They could make schematic drawings, plans and cross-sections for their colossal structures with very simple tools like a builder's thread to delineate vertical lines, an angle, a measuring arm 0.52 meters (1.7 feet) long and a straight edge.

By Ptolemaic times, the Egyptians were known to track building projects and create a very formulated approach to new projects. For example, when it came to building an Egyptian temple, they consulted the "Book of Foundations of Temples," which listed complicated and ever-present ritualistic formulas. These carried the proper procedures and operations for each step in minute detail. Every temple had its archives preserved with instructions and best practices. The formulas were rigidly observed as the priests demanded that their buildings conform to the regulations laid down[9] by earlier generations.

The next Pharaoh in line was Khufu (2589–2566 BCE). The Giza Pyramid (2550 BCE) was initiated under the architect Hemienu, Khufu's brother and Sneferu's son, following a tradition of young princes taking the architects role. The funding for the project was never an issue through the project time line as the Pharaoh's stature was that of a god (sun).

Planning

Planning for the Pharaoh Khufu's pyramid started as Sneferu's body was being embalmed. Khufu was 40 years old so there was not a lot of time to complete the project.[10] Hemienu's first task was to select the building site, so together with the overseers of quarries and transport, they sailed along the Nile. They could fairly readily determine the suitability of a site as they knew the volume and the external dimensions of the pyramid, based on the location and size of the pre-determined internal chambers.

The technical challenges to the project were going to be enormous. First, a perfectly level base was required, inaccuracies of centimeters at the bottom translated into meters at the top. Second, the burial chambers would have to be supported by large granite blocks to prevent collapse. But at 60 tons each and 42 meters (140 feet) above the base the maneuvering and positioning of these blocks to a high degree of accuracy was going to be difficult. Third, ramps were required to move the material to the required height. The completion of the last third of the pyramid was going to be particularly challenging with the placing of the cap stone.

The work breakdown structure laid out the project with the following activities: site preparation, construction, removal and ramp demolition.

When Hemienu's party saw the Giza plateau they recognized it had several advantages in that it was a quarry itself, and a harbor could be readily built for handing the huge volume of in-coming materials. The Giza Pyramid was to become the centre of an extensive supply chain as millions of tons of materials poured onto the site. Copper and bronze were needed for the chisels, in all about 250 tons, which had to be shipped in from the red sea, on a nine day journey. Around half a million tons of gypsum was mined and transformed into a mortar, used to bind the core rubble. There were several other materials that were needed in quantity like the smooth white facing blocks, backing blocks, and rough blocks for core rubble. Large quantities of timber were needed for equipment (sled, scaffolding, rollers, etc). As part of the planning the team also worked out the logistics for:

- Site preparation - determination of true north, creation of a perfect square of limestone blocks, within which was a natural limestone core cut into shape.

  o Perfect leveling of the base as being off one inch would mean yards at the top. Each side is 230 meters (767 feet) in length and almost perfectly level (to within 15 millimeters).

- About 10% of the pyramid is part of a limestone
  outcrop that was shaped to look like the rest of the
  blocks.
- Quarry operations
  - Limestone, 80% of it came from Giza quarries (on-
    site), with two off site quarries in Tura, eight miles
    upstream. The Tura stone was milk-pure with fewer
    fossils and required a workforce of 500 in the first
    years.
  - Robust copper/bronze chisels were used by the gangs,
    these blunted after 100 blows so sharpening teams
    were needed. Stone pounders and wooden wedges
    were also used.
  - Sleds were used for the limestone blocks as wheels or
    rollers would sink in the sand.
  - Miles of hand woven thick rope was required.
  - The granite blocks, including the massive
    sarcophagus required dolerite pounders to hack it
    out. It would take a workforce of 500 a decade to
    extract the required quantity.
- The transportation of the finished stone
  - Limestone (2 ton blocks) from Tura (8 miles or 17
    kilometers away).
  - Granite (60 ton blocks, 4 x 4 x 24 feet) from Aswan
    (500 miles (800 kilometers) away).
- The creation of a harbor and canal
  - Right up to site to float granite blocks up the canal
    for half a mile. Wooden sledges would not have been
    effective and would have sunk into the sand.
- The creation of a workers' village
  - For permanent skilled workforce, of up to 25,000,
    including houses, bakeries and breweries. Ranches
    were set up around Egypt to provide meat in vast
    quantities. It was a luxury item of the middle classes.
- The construction of the ramps

- To deliver the blocks to a precise position in the structure.
- The performance of the finished work.
- The removal of the ramps at the end of construction.

All the experience in pyramid projects over the previous centuries helped work out the problems of social organization which the Egyptians would run into at Giza. According to one report by the 5th Century BCE Greek historian Herodotus,[11] at Giza there were two gangs, each of 100,000 men (at peak periods). Each gang was then divided into five zaa's of 20,000 men. The zaa's were further sub-divided into crews of approximately 2,000, and then sub-divided into named gangs of 1,000. Recently-found graffiti shows that the workers named themselves the 'Friends of Menkaure' and the 'Drunkards of Menkaure'. These gangs were divided into phyles of roughly 200. Finally, the phyles were split into divisions of maybe 25 workers, who were allocated their own specific task, like transporting stone from quarries to site, which were being delivered at a rate of one every two minutes. Each gang had their own project leader, a soldier for a foreman.[12] Thus a 20,000 workforce could be separated into efficient, easily monitored, units and a seemingly impossible project, the raising of a huge pyramid, became an achievable ambition.

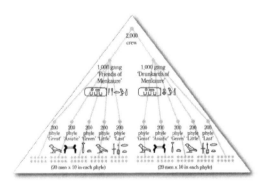

Figure 3.3: The breakdown of the hierarchical organization.[13]

The skilled workforce was made up of many trades like scribes (to track the stone blocks), stonecutters, mason, surveyors, draughts men, mortar makers, and carpenters (for the equipment and sleds), potters, and cooks. They (4,000 to 5,000) worked all year around either on site or in the quarries. Recent excavation of tombs of dressed stone, belonging to the wealthier classes, found titles such as Director of the Draughts men, Inspector of the Craftsmen, and Overseer of the Masonry. These men led the skilled workforce.

The unskilled workforce, mostly farmers, were conscripts who worked between July and November when their fields were flooded. Around 20% of the adult male population could have been available for conscription.[14] The logistics were substantial as the workforce had to be housed, fed and supplied with equipment. Food and beer were produced on an industrial scale. This project workforce was rotated and worked in a modular, team-based kind of organization.

> "...as barracks housing for a rotating labor force, perhaps as large as 1,600 to 2,000 workers. There were slaves in Egypt but the discovery that pyramid workers were fed like royalty buttresses other evidence that they were not slaves at all, at least in the modern sense of the word."
>
> —Wrote Mark Lehner at the end of the 2002 season[15]

The workers village, uncovered recently, highlights the workers were well looked after. The excavated bones from workers' graves were mineralized indicating the workers ate meat which was the food of the Egyptian middle classes. There are scores of bakeries flanking the galleries, as well as an abundance of animal bones from sheep, and cows.

The 100,000 work force figure seems high in the light of what we know today. DMJM (Daniel, Mann, Johnson, & Mendenhall)[16] surmised that there was a full-time workforce of about 4,000 to 5,000, not including the workers responsible for cutting limestone at the distant quarries, transporting it to Giza, and bringing it to the site. DMJM concluded that the total project required an average workforce of 13,200 persons and a peak workforce of 40,000, over the 10-year duration. The study has been criticized for only using 2 million, rather than the 2.5

million blocks, that are believed to have gone into the Great Pyramid.

Executing

The project faced a series of technical challenges that dictated the scope and schedule. The first technical challenge, to create a perfectly level base, was addressed by digging a thin trench around the perimeter and filling it with water. A base line was then traced on the water mark and the perimeter leveled with this line.

The quarry gangs formed a production line as a steady stream of 2 ton blocks was hacked out of the on-site quarry, and transported to site and placed in position. The process demanded backbreaking efforts and considerable endurance and patience on the part of the gangs working in harmony and unison. It is likely the gangs competed between themselves as to how many blocks were pulled up in a day. The gangs from the off-site quarries rowed stone-carrying boats across the Nile River, and then pulled the blocks overland to their destination.

After ten years the first of the 43 granite blocks (30 to 60 tons) arrived from Aswan, for setting up the king's burial chamber. The team had to address the second technical challenge. Nine of these large blocks would make up the roof. They had to be raised 42 meters (140 feet) onto the pyramid, and then maneuvered and positioned with a very high degree of accuracy.

The third technical challenge, required ramps so as the material was moved to the correct height. An external ramp up to the 61 meters (200 feet) level was sufficient and relatively easy to build. Beyond 61 meters (200 feet) and up to a height of 146 meters (480 feet) the ramp would reach an 8% grade, too steep to pull blocks. It would also stretch out at least a 1.6 kilometers (1 mile) beyond the pyramid. This would not have been feasible as the volume of material for the ramp would be equal to the pyramid. It would have added years to the project in terms of the ramp construction time, and added considerable time in the transportation. An external corkscrew ramp was more feasible but the one flaw with this theory is it would have been difficult to make repeated measurements along the edges to make sure the pyramid angles were constant. The ramp would have obscured this. An alternative is the internal ramp theory which is getting acceptance today since first introduced in 2005.[17] The internal ramp at a height of 4.6 meters (15 feet) would have started at the bottom and snaked up the sides at a 7% slope, making fourteen turns and stretching out to 1.6 kilometers (1 mile). The internal ramp would be coupled with a 61 meters (200 feet) external ramp, made up of smaller blocks.

Figure 3.4: Workers pulling up granite blocks along the external ramp of the Giza Pyramid.[18]

The number of blocks by various estimates is around 2.5 million blocks. The four quarries supplied a total of 100,000 blocks per year.

*Figure 3.5: Giza in the 19th Century. The height used was **201** steps high, or **146** meters (**480** feet).[19]*

Controlling and Monitoring

As the project neared completion the short external ramp would have been dismantled and then pulled through the internal ramp to finish the top half of the pyramid. Hemienu was concerned with the structure of the pyramid`. He was worried about the king's burial chamber holding up as limestone blocks were piled up above it. Hemienu had stonemasons chisel a narrow tunnel through the limestone at the top of the Grand Gallery so he could assess the chamber. Small cracks were found in the granite roof beams, it wasn't serious but could get worse. The cracks were plastered, to monitor them so further expansion would be apparent. Over time the roof beams held. Recent investigations speculate[20] that one side of the burial chamber settled too quickly when the external ramp was dismantled, causing a drop in the wall and the ceiling to crack.

Closing

The output of the project was significant, the only one of the seven wonders of world to have survived to today. The scale of a 40 story building has inspired generations, and remained the tallest building for 4,000 years until the Eiffel Tower was erected in 1870. It was completed within a specified time frame of twenty years, remarkable when considering that it required 81 two ton blocks to be laid per day.

There was a remarkable and efficient use of resources within the budget to support the project. It was a limited project workforce which required a high degree of organization. It had to work in a very constrained and complex environment for example, the maneuvering of mammoth blocks up small ramps. The project was remarkable in the simplicity of the equipment used like wooden mallets and wedges, copper saws and chisels, ropes, and wooden sledges. The masons would have used very simple tools to finish the shaping of the blocks namely, set squares for right angles, plum-bobs and a-frames for leveling.

The Giza Pyramid was constructed over a twenty year period in a hostile desert environment. The project highlighted how careful planning could create a highly effective social organization. The work was so broad it touched almost everyone in Egyptian society such that it helped the nation consolidate into a true state, the first in history.

With time, our understanding will continue to increase about this project, as we have not discovered any documents on the construction details.

Other Notable Projects

The Great Wall of China Project (600-206 BCE).

Background

With the threat of "barbarian" invasions or raids, the Great Wall of China was built to serve as a site for lookout posts, and to provide the armies a swift warning system, and to create an elevated military roadway through rugged terrain. The Wall safeguarded the trade routes like the Silk Road, and facilitated safe transmission of information and transportation. The Great Wall also provided a protection to the overall economic development and cultural progress.

The wall was part of a rigorous defense project composed of countless passes, watchtowers, garrison towns, beacon towers and blockhouses. It would assume the shape of a horseshoe and be built on an unprecedented scale.[21]

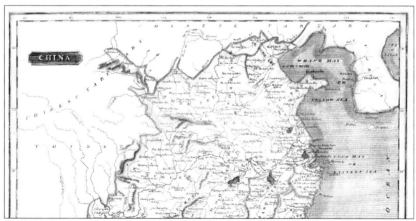

Figure 3.6: Great Wall (19ᵗʰ Century print).[22]

Initiating

Qin Shi Huang conquered all opposing states and unified China in 221 BCE, when he established the Qin Dynasty. He was the first Emperor of a united China and connected a number of existing defensive walls, from the disparate border fortifications and castles of individual Chinese kingdoms, into a single system. He ordered the building of a new wall to connect the remaining fortifications along the empire's new northern frontier, that terminated at the shores of the Yellow Sea.

Planning

During the Qin Dynasty (221BCE - 206BCE), 300,000 soldiers were redirected to build the wall, and fight if necessary, after General Meng Tian conquered the Huns. It took nine years to finish the work.

Figure 3.7: Great Wall in the 19ᵗʰ Century.[23]

Executing

Transporting the large quantity of materials required for construction was difficult, so the projects always tried to use local resources. Stones from the mountains were used over mountain ranges, while rammed earth was used for construction in the plains. In the deserts, sanded reeds and juniper tamarisks were put to use. The mortar in the wall used a rice adhesive to better hold it together.

The project was completed in a hostile environment, under attack by the Xiongnu people from the north. The regime was brutal and hundreds of thousands were killed in its creation, mainly by the cold. It had a very high death rate of 25%, mainly in the winter. These bodies were just added to the wall. After Shi Huang's death and the fall of the Qin dynasty, the wall was left largely ungarrisoned and fell into disrepair.

Controlling and Monitoring

The next major construction projects, which were nearly as extensive in scope as the building projects of the Qin dynasty, were launched in 552 by the Northern Zhou, the Bei (Northern) Qi kingdom (550–577) to strengthen the northern frontier and prevent invasion from the west.

Three years later the emperor ordered the recruitment of a project workforce of 1.8 million to repair and extend other sections. The workforce was made up of three groups: soldiers, common people, and criminals, of which soldiers made up the main construction group.

A section of Sui's (581 - 618) wall in Inner Mongolia required more than 1 million men for the project. Criminals were forced into labor as punishment. During the day, they were ordered to take turns patrolling the wall, while at night, building and extending the wall. They were required to serve four years of hard labor according to laws at that time.

Closing

The great wall produced something that was very significant. It was built, rebuilt, and maintained between the 6th Century BCE and the 16th Century. Initially it was not planned as one wall, but different sections built over time. The total length reached 6,700 kilometers (4,160 miles). Today remaining sections of the wall rise to a height of 10 meters (33 feet) and are 4.5 meters (15 feet) wide with a paved road running along the top.

It could be argued that the Great Wall had a negative influence on technological development as the enormous project consumed vast numbers of people, resources, materials, and equipment, but without stimulating innovation.[24] It is the only man made structure seen that can be seen from space.

Stonehenge Project

Background

Stonehenge, in England, means hanging stone and was a formidable technological achievement. A high level of precision was involved, not just in the construction but the positioning itself. The output was significant, a huge calendar, and the stones had to line up precisely with celestial events. The architects laid out the monument in a true circle using practical geometry and a standard measure the megalithic yard.

Initiating

The workforce was likely seasonal and required surplus stores of food to feed it. A centralized authority would have distributed food and supervised the project. The labor requirements are estimated to range to 30 million man-hours,[25] equivalent to an annual productive workforce of 10,000 people (8 hours x 300 days). The 100 meter (330 foot) circular ditch produced 3,200 cubic meters (3,500 cubic yards) of dirt.

Planning

A good deal of planning had to go into the project particularly with scheduling the delivery of the stones, specifically the larger 25 ton blue stones that had to be quarried and transported a great distance. The site itself had to be very carefully laid out with a high degree of precision, to achieve the objective of creating a celestial calendar where the stones had to line up precisely with celestial events.

Executing

Stonehenge consists of an outer circle of 30 upright sacred "blue" stones, 4.1 meters (13.5 feet), each weighing 25 tons were transported 438 kilometers (300 miles). The 30 lintels weighed 7 tons each. An inner circle of Standing Stones up to 7 meters (22 feet) high, weighing up to 45 tons each, were moved about 40 kilometers (25 miles) using sledges and ropes as transportation by water would have been very difficult. Altogether it needed 1000 tons of stones. Today's calculations show that it would have taken 500 men, using leather ropes, to quarry and pull one large stone, with an extra 100 men needed to lay the huge rollers in front of the sledge.

One of the most challenging aspects was aligning the massive stone structure so that on various calendar days, the sun would shine precisely from the outer ring openings to the central point on main days, such as equinoxes and solstices. The principal alignments are on the midsummer solstice sunrise and the midwinter solstice sunset.

Closing

Stonehenge is an example of a project that was planned and executed, over a long time frame. The output of the Stonehenge project was significant as it was used as a calendar to help measure the seasonal cycles and plan agricultural and religious events.

The overall project was completed over 1,600 years in three phases with specific outputs for each phase. What we can learn from this project is not that different from Giza, there are many parallels, with the quarrying, transportation, and placement of stones. **A society that puts resources to an end goal, surplus food, and takes care of its workforce by rotating seasonally can achieve astounding results.**

Figure 3.8: Stonehenge in the 19th Century.[26]

Necno (610 BCE)

According to Herodotus, the 5th Century BCE Greek historian, the Pharaoh Necno II, who reigned c. 615-595 BCE, was determined to see if Africa could be circumnavigated (east to west). He commissioned a number of ships manned by Phoenicians for the task to sail down the Red Sea and along the East Coast of Africa. Every year they settled for a while on the coast, cleared a strip of land, planted a crop and, when they had harvested it, continued on their journey. In the third year they sailed through the Pillars of Hercules and back to Egypt again. They reported that as they sailed around Africa they had the sun on their right. Herodotus refused to believe this possible 'but perhaps others may.' This is good evidence that such a voyage was made.[27]

Ziggurat of Urnammu Project (between 2113-2096 BCE)

The vast scale of the Ziggurat at Ur, built by the Sumerians, qualifies it as the output of a major project, although there is not a lot of documentation about the project.

"The third dynasty of Ur strictly supervised the economy. Huge numbers of laborers and craftsmen were employed in the service of the state in the 'grand households,' which included the great temples and palaces. The chancelleries produced documentation which bears witness to complex administrative processes. A standardized form was established for the high temples-multi-storied structures with a central flight of steps-called Ziggurats, made of mud bricks. This form was used for religious edifices erected by and for the kings."[28]

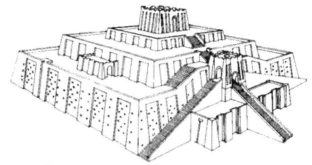

Reconstruction of the Ziggurat at Ur

Figure 3.9: The Ziggurat Temple and Shrine of Ur in Iraq, up to 91 meters (300 feet). Today only the bottom layer survives.[29]

Key Players

Imhotep (2650–2600 BCE)

Egyptian meaning "the one who comes in peace," was an Egyptian Polymath (broad, or varied knowledge), who served under the Third Dynasty king, Djoser, as chancellor to the pharaoh and high priest of the sun god Ra at Heliopolis. He is considered to be the first documented architect and physician in history known by name. As one of the officials of the Pharaoh Djoser, he designed the Pyramid of Djoser (the Step Pyramid, Figure 3.1) at Saqqara in Egypt in 2630–2611 BCE. As an instigator of Egyptian culture, Imhotep's idealized image lasted well into the Ptolemaic period (332–30 BCE).

Chapter Wrap-up

Conclusions

The technical challenges of the Giza Pyramid project were enormous, from creating a perfectly level base, to building the burial chambers, to completing the last third and mounting the cap stone. The history of pyramid building shows a clear evolution in learning, and this was applied to the Giza Pyramid project. The Egyptians had such a comprehensive grasp of the whole project, work breakdown structure and schedule, that at the project start they sent teams 800 kilometers (500 miles) to Aswan to quarry granite. They understood the difference in properties between granite and limestone, and the use for each. They knew it would take 10 years to hack out the 62 tons granite beams needed for the roof of the king's burial chamber. This was scheduled accordingly to stay within time-lines as they had calculated that within this 10 year span the pyramid would be 48 meters (150 feet) high and ready for the granite beams.[30]

In contrast the Great Wall of China was a completely different project, punitive in nature. It really was a protracted building program that improved and extended an existing infrastructure. This was done in spurts as the pressure of external invasion mounted. As a result, the project pressed groups into forced labor notably soldiers, and criminals. A point of note is this project had the greatest number of laborers ever used (1,800,000) in a single period.

Key Lessons

PMBoK® Guide Knowledge Areas:

- Integration Management

  - The Giza Pyramid project had a very clear charter and deadline to deliver the project before the pharaoh died which was approximately 20 years. The project was pushing the limits of technologies, materials, and resources in terms of the heights and stability of the evolving structures, and the ability to organize a vast workforce in a confined environment.

  - The project had a well defined governance structure. It also had a mechanism for course adjustment and controlling change. Lessons from the Bent Pyramid demonstrated how these were used to rectify design problems and deliver the project.

  - There was less consideration for a business justification as the project was built for the pharaoh the all powerful ruler. However, there was not unlimited state funding available for the project. There was a significant payback in the way the project helped to unify the nation and provided a boost to the economy, as its wide logistics and supply chain touched everyone through Egypt.

- Scope Management
  - ○ With the Giza Pyramid project a great deal of planning had to take place and Hemienu would have had ground plans and three-dimensional models. Planning included preparation and logistics. Preparation required the physical planning of the site including surveys, site selection, layout, and initial steps in the process. For this surveying skills were required, and the Egyptians had established these because of the inundations and flooding of the Nile.
  - ○ The functional requirements were principally centered around the king's granite burial chamber.
  - ○ The non-functional requirements included creating a secure site that would deter grave robbers. Much attention was paid to concealing entrances and creating obstacles and traps.
  - ○ The scope in terms of materials, and equipment was dictated by:
    - The height of the pyramid following established angles.
    - The availability of materials, quarrying, transporting of stone.
    - The hacking and transporting of the granite blocks was particularly challenging.
  - ○ The work breakdown structure laid out the project with the following:
    - Site preparation with determination of true North, and creation of a perfect square foundation of limestone blocks.
    - Construction of ramps, and delivery of blocks to a precise position in the structure, and creation of a harbor and canal, workers' village for 25,000 including houses, bakeries and breweries.
    - Construction of the structure.

- Ramp removal and demolition, incorporating the rubble into the pyramid core.

  ○ The scope was decomposed down to a level where a small team of 10 could undertake a task, like the quarrying and transporting of single blocks.

  ○ The scope was decreased by building in a quarry. Similarly, the Great Wall used local materials were possible varying from stone, to tamped earth, to sanded reeds.

- Time Management

  ○ The Giza Pyramid project activities were well understood, and the estimating was reliable and based on experience. The project required careful sequencing and scheduling of activities so that the skilled on-site workforce was kept busy year around, either on site or in the quarries. For example, enough stone had to be quarried out and brought to site for work to continue without interruption. As the project team shrank in the winter there had to be enough work to keep the skeleton skilled workforce busy.

  ○ The Egyptians worked with granite and limestone, and understood the difference in properties and how each should be used. It would take 10 years to hack out the granite beams needed for the roof of the king's burial chamber that were up to 62 tons. Copper tools were ineffective and dolerite pounders were used. Because of this, quarrying had to be scheduled to stay within timelines. The project manager had calculated that at the end of this 10-year span, the pyramid would be 46 meters (150 feet) high and ready for the creation of the king's burial chamber.[31]

  ○ The extensive supply chain had to work in lockstep with the project schedule. For example, the late delivery of copper and timber materials was a showstopper to the schedule, as well as food and provisions for the project workforce.

  ○ The Giza Pyramid had to be built within the lifetime of the pharaoh. The project followed a schedule with a target end-date. Hemienu, as the pharaoh's, nephew was close enough to his uncle to be aware of his health and well-being. He therefore could crash the project schedule if required to meet the end-date.

- ○ At Stonehenge the enormity of the project required independent teams to be coordinated to work in parallel and independently to quarry, transport, and deliver the stones.

- Cost Management

  - ○ Based on 200 years of Pyramid projects, both successful and not (Red, Bent, and Step Pyramids), analogous estimating was used. Giza was a continuation in this line of projects but a scale up in size.

  - ○ Determining the budget was relatively easy based on the accuracy of the cost estimates.

  - ○ Giza was built by a project workforce of tens of thousands of skilled and unskilled workers (farm laborers). They camped near the pyramids and worked for a salary or as a form of paying taxes, and were paid in food and clothing[32] until the construction was completed. The large unskilled workforce worked between July and November when their fields were flooded by the Nile. Project accounting was thorough and completed by a system of scribes working for the state.

  - ○ The agrarian economy of Egypt, with a population of 1.5 million, could provide the resources to support a project of such an undertaking over a period of time.

  - ○ Cost control would have been relatively straightforward as the principal activities were consistent year to year.

  - ○ The project employed tens of thousands of skilled and unskilled workers (farm laborers) who camped near the pyramids. They worked in lieu of paying taxes as well as pay through food and clothing. The large unskilled workforce worked between July and November when their fields were flooded by the Nile. Project accounting was thorough and completed by a system of scribes working for the state.

- In the Great Wall of China project the government accounting reached a peak of sophistication with the introduction of timely-reports. A lunar calendar was used dividing the year into twelve months, and each month into three ten day periods called hsun. Accordingly, ten-day, monthly, and annual reports were prepared throughout the budget cycle."

- With the Great Wall of China project, some sections were built in a hostile environment were the workforce was under the threat of attack. This would have increased the cost.

- Quality Management

  - With the Giza Pyramid project quality was paramount in site preparation, and ensuring firmness of the base. An imperfect leveling of the base out by one inch would mean yards at the top. The Egyptians were able to build with incredible accuracy and precision even by today's standards.

    - The site was leveled within a fraction of an inch over the entire 13.1-acre base.

    - The four sides of the base of the Great Pyramid are perpendicular to within 58 millimeters (2.28 inches) in length, over 230 meters (767 feet) in length on each side, and 1 minute in angle of a perfect square.

  - The cap stone and limestone casing fitted walls, joined at a consistent angle, on all four sides, to create a singular smooth surface.

  - Another area of quality control was positioning the nine granite blocks, 60 tons, on the roof of the king's burial chamber. Great care had to be taken not to collapse the structure.

  - The project adopted an approach to quality based on continual inspections. At the top of the Grand Gallery stonemasons chiseled a narrow tunnel through the limestone so Hemienu could assess any damage. Small cracks were found in the granite roof beams, it wasn't serious but could get worse. The cracks were plastered, to monitor' them so further expansion would be apparent. The roof beams held and further tunnels didn't have to be cut above.

○ Likewise the gradient of the ramp could not be steeper than 8% otherwise it would have been impossible to pull the blocks up the slope.

○ The Egyptians had readily transferred the concept of inspection to their construction projects. For centuries they had used "Inspectors" to survey work done for the water-houses or irrigation units that cut the dykes and canals, and kept these under constant repair.

○ Stonehenge required great accuracy in aligning the massive stone structure to equinoxes and solstices on various calendar days. The architects had accurately laid out the monument in a true circle using practical geometry and a standard measure - the megalithic yard.

○ The ancient Egyptians tracked building projects with a very formulated approach to building an Egyptian temple. They consulted the "Book of Foundations of Temples," which listed complicated and ever-present ritualistic formulas.

  • These carried the proper procedures and operations for each step in minute detail. Every temple had its archives preserved with instructions and best practices. The formulas were rigidly observed as the priests demanded that their buildings conform to the regulations laid down.[33]

○ Similarly, the Sumerians adopted a quality approach in their projects. A standardized form was established for the high temples, multi-storied structures made of mud bricks with a central flight of steps-called Ziggurats. The chancelleries produced documentation which outlined the complex administrative processes. This form was used for religious edifices erected by and for the kings.[34]

- Human Resource Management

  ○ The Giza Pyramid project required an average workforce of 13,200 workers and a peak workforce of
  40,000, over its duration. The 200 years of experience
  with pyramid projects helped in identifying the size of
  the project workforce and the mix of trades and skills.

  ○ The logistics were substantial as the workforce had to
  be housed, fed and supplied with equipment.

  ○ The skilled project workforce was made up of of
  between 4,000 to 5,000 skilled craftsmen,[35] typically
  stonecutters, masons, surveyors, mortar makers,
  and carpenters. They worked all year around either
  on site or in the quarries. They benefited from a
  system of privileges. The unskilled workforce, mostly
  farmers, worked for five months between July and
  November when their fields were flooded by the
  Nile. This was known as the annual "inundation"
  one of the three seasons. The other two seasons were
  winter and summer, each four months long[36]. Each
  month consisted of three weeks of 10 days each. The
  workforce of up to 20,000 was made up of excavators,
  carriers, and haulers that provided the labor for
  moving the vast number of blocks from quarry to site.

  ○ The project workforce was well looked after, recently
  excavated bones were mineralized indicating the
  workers ate meat which was the food of the Egyptian
  middle classes.

  ○ There is evidence that the workers received good
  medical care as one skeletal remain shows a worker
  who suffered several leg fractures healed well and
  straight, another shows evidence of brain surgery,
  and others had broken hands treated by binding.
  These were not expendable slaves.

  ○ Egyptian workers stopped working when they were
  not being paid enough. Workers in Egypt were also
  known to have approached authorities with grievanc-
  es, which resulted in the punishment of their supervisors who had treated the workers unfairly.[37] The Giza

Pyramid project highlighted how careful planning could create a highly effective social organization.

- ○ The workforce extended well beyond the site as an extensive supply chain poured in millions of tons of materials (copper, gypsum, limestone, marble).

- ○ The organization of the workforce into gangs encouraged a healthy competition to develop between these, and a team spirit to evolve.

- ○ The workforce was sourced from communities across Egypt that were required to provide a steady worker supply on a rotational basis.

- ○ They were organized into gangs of 25 men to transport the stone from quarries to site with a soldier for a foreman. Typically, 10 men would pull one block.

- ○ In ancient construction projects where operations were carried out over a period of time and in a remote area (quarry, logging camp) the master-builder was responsible for housing the project workforce, its security, as well as the worker's tools, equipment and food.[38]

- ○ The Great Wall project required the organization of a huge project workforce force (1,800,000) made up of soldiers (the main construction group), common people, and criminals.

- ○ Criminals were forced into four years of hard labor as a punishment. During the day, they patrolled the wall. While at night, they worked on it.

- Communication Management

  - ○ The Giza Pyramid project was a national project, designed to unite a nation, and therefore it had a very high visibility with the Egyptian population. It was popularized by having a large number of stakeholders amongst the villages and towns, who supplied the laborers and craftsmen.

  - ○ The workforce had to be well organized for it to function effectively. It was divided into crews of 2,000, subdivided into gangs of 1,000, and then into phyles of roughly 200, and then divisions of 25 workers. A hierarchical organization that was quasi-military could be led through top down oral/audio communications, the oldest form.

- ○ Other forms of written communication included an organized courier service for written documents, where the pharaohs diffused their decrees (2400 BCE was the first documented use of this).

- ○ Another common form of written communication, where important information had to be distributed, was through the mason's marks. This was common in stone construction projects that required accurately cut, closely fitting and usually unmortared blocks, like at Giza. These ciphers communicated a record of the work done, the identity of the stonecutter so he could be credited for the work, and held accountable for its craftsmanship.

- Risk Management

  - ○ With the Great Wall project there were huge risks to the project workforce (25% death rate) from external attacks and raids, to the cold conditions in winter.

  - ○ In contrast, the Giza Pyramid project workforce was housed in well organized villages and well looked after.

  - ○ Risk management was absolutely critical with the project, as the extensive supply chain that spanned quarries, workshops, and construction site was fraught with dangers. The risk to the project workforce was high because of the quantities of materials being moved with relatively simple equipment (sleds, ropes, levers). Huge granite blocks had to be pulled out of quarries and up ramps into position. There was little room for maneuver on the site itself. All these were identified and managed. Losing workers through injury or death was very bad for the morale of the project.

  - ○ In an example of risk monitoring the king's burial chamber had to be reinforced with massive 30 to

60 tons granite blocks to prevent collapse. When cracks appeared in the granite roof beams, they were monitored by plastering over them to see if any further expansion would occur.

○ This was a public works project which improved its popularity with the Egyptian population. It also increased the range of stakeholders and this broader support helped mitigate some of the project risks.

- Procurement Management

  ○ Procurement management has its roots in trading and the earliest surviving business records can be traced back more than 7,000 years ago to the Chaldean-Babylonians, Assyrians, and Sumerians.[39] In the 5th Century BCE the rich farming area between the Tigris and Euphrates Rivers saw a civilization grow where small businesses were established and an extensive trade grew. The city of Babylon was labeled the "queen of commerce."

  ○ Scribes were the predecessor of today's accountants. Their duty was to record the business transactions on commercial tablets, for example, the types and quantities of goods as they arrived at storehouses. They also prepared inventories of assets on hand and charge-and-discharge[40] type summaries of commodities received and paid out.

  ○ After a contract or an agreement had been negotiated, written and signed, the scribe took a new piece of clay and wrapped it completely around the original tablet providing a carbon copy where any tampering could immediately be detected.

  ○ Logistics encompasses procurement and is a military term for procurement and acquisition, shipping and distribution, maintenance (upkeep), and replacement of materials and personnel. The military has always been concerned with logistics going back to ancient times, and this is where it has its roots.

  ○ Procurement Management was core to the Giza Pyramid project as the extensive supply chain moved millions of tons of procured materials, food and provisions to the project site. The materials that poured into the site included copper (250 tons), gypsum (half a million tons), smooth white Tura facing blocks, backing blocks, rough blocks, and granite. The logistics in keeping the project going was enormous as this workforce had to be fed, watered, and housed.

- ○ Families in the Delta and Upper Egypt sent 21 buffalo and 23 sheep to the plateau every day just to feed the workers. The worker's village housed bakeries and breweries that could produce on an industrial scale an expansive production line of food.

- ○ The Egyptian agrarian economy, with a population of 1.5 million, provided the resources to support the project over the period.

Educators

- · Discuss in the context of today's projects the importance of what the Giza Pyramid project achieved?

- · Did the project have any benefits beyond its main objective, a grandiose tomb for the pharaohs?

- · Discuss which of the nine *PMBoK® Guide* Knowledge Areas were most significant to the project?

CHAPTER 4

Roman Republic (510 - 100 BCE)

T his historical period was dominated by the Greek and Roman empires in Europe and those of Mauryas in India, and Han in China. From a project management perspective construction projects continued to be driven by a temple economy (grain silos held in religious structures) where the ruler initiated communal work projects. These were typically construction projects using stone, brick, and new materials like concrete. There was an increased scientific knowledge specifically with geometry, and a more sophisticated use of equipment such as lifts and cranes.

In this Historical Period

- Brief synopsis of trends/changes (natural disasters, wars, technology, epidemics).

    a. Urban revolution, development of cities and evolution of city-states into republics and empires.

    b. Ionian (early Greeks) explorers created a different society away from autocratism that mapped and explored the known world.

- Impact of changes

    a. Development of rational and logical thought.

97

    b. Birth of democracy in the Greek city-state.

    c. Development of cities leads to a golden period for Greece with the construction of magnificent formal buildings.

    d. Growth of standing professional armies.

- Major events

    a. Greek wars.

    b. Roman Republic created in 509 BC.

    c. Roman wars.

- New tools, techniques and breakthroughs

    a. The Greek machines invented by 500 BCE included the screw, the ratchet, the pulley, the water wheel and the aeolipile (Hero's turbine).

    b. Coins and money.

    c. Cement.

    d. Maps.

    e. Philosophy (Plato, Aristotle).

- Regions

    a. Mediterranean.

    b. China.

- Significant projects

    a. Parthenon in Athens, Greece, 477 to 438 BCE.

- Other Notable Projects

    a. Terra Cotta Warriors (246-210 BCE).

    b. This category covers ancient wonders and structures, buildings, and towers. It includes five of the seven wonders of the ancient world.

- Key players

    a. Pericles.

- Chapter Wrap-up

Trends and Changes

The world population in this period grew from 100 to 150 million.

Cities and City-States

The urban revolution continued and saw the development of cities and evolution of city-states. The city-states of ancient Greece were quarrelsome and fought one another repeatedly with citizens fulfilling military requirements.

Similarly, the city-state of Rome became more powerful and was constantly at war with neighboring peoples in Italy. Eventually Rome transitioned to a republic but continued its life and death struggle with rival powers like Carthage and the Hellenistic kingdoms (in 146 BC).

Ancient Greek Explorers

As the population of ancient Greece grew beyond its limited food production the ancient Greeks started to explore and settle colonies in all directions from about 750 BCE. By 300 BCE they had mapped and explored most of the known world.

Impact of Changes

Change in Thinking

The Greeks through the Greek city-states were instrumental in the development of reason, inquiry, and rational thought. They were the first Europeans capable of analyzing and explaining political institutions, and considering alternatives.

Path to Democracy

To prevent Athens from becoming a Spartan puppet state, an Athenian noble Cleisthenes proposed to the citizens of Athens to share power regardless of status and create a democracy. The democracy solved many problems for Athens, and helped repel a Spartan-led invasion. Magistrates were elected by all freemen, and jurors in trials were paid fees, so democracy wasn't just for the rich. The democracy did not include women, slaves, or foreigners.

Growth in Buildings

This was a golden period for Greece and then Rome, a classical era with the construction of magnificent formal buildings. Today's sense of an ideal city is derived to a great extent from Athens around 450 BCE.[1]

Professional Armies

The city-state armies were made of citizen soldiers and each time war broke out raising an army was inefficient, and for Rome this was disruptive to the economy. It was also harder to keep the levels of discipline in a citizen army. By the 1st Century BCE Rome had a standing professional army of career soldiers, recruited from the poor and equipped by the state. This allowed for the development of military skills, guaranteeing further Roman conquests as the republic became an empire.[2] These were based on the Assyrians who were the first to organize a professional army (745-727 BCE).

The foremost features of this Assyrian army were its logistics arrangements, supply depots, columns, and bridging trains. Central supply depots held war materials of all sorts, with chariots and horses. This was the first long range army that could campaign 300 miles (500 kilometers) from base and move at advanced speeds of up to 30 miles a day. It was paid regularly and supplied with weapons. The army was organized into units led by professional generals. The recruits were multi-ethnic with foreign mercenaries and prisoners of war making up the major elements of the ranks. The Syrians were experts in siege: attacking foundation walls, digging mine shafts, building siege ramps, and working on siege engines. Many successor armies of other empires were modeled on this army including the Roman legions.[3]

The Persians took the concept of specialization further as they assembled specific ethnic groups like Greek mercenaries for infantry, Medes and Scythians as horsemen, and Phoenicians as sailors. Strict training welded these forces into disciplined armies.

Major Events

Greek wars

The city-states of ancient Greece fought one another. This was done to prevent any one city-state from dominating, like Athens or Sparta. They also contested external threats like the Persians which they united into a democratic military alliance to push back these invaders. Athens was determined to maintain its democratic society

Roman Republic (509 BCE)

The Romans were determined to prevent dictatorial kings from taking complete power. They set up a complex system of controls were power was shared by two consuls, elected yearly. One was in Rome and the other was with the army. Other officials were elected as well.

Roman Wars

Similar to the Greeks, the Romans fought to preserve a fledgling democracy - a republican form of government. Rome survived a series of civil wars and wars against its neighbors, notably the Etruscans in the 5th Century BCE.

New Tools Techniques and Breakthroughs

Simple Machines

The Greeks were renowned inventors and over a thousand years of empire they contributed the screw, the ratchet, the water wheel and the aeolipile (Hero's turbine).

Figure 4.1: The Water Screw, a simple machine used in projects.[4]

In the 4th Century BCE the pulley was developed. This was a very useful device in construction projects for lifting heavier objects. A major mechanical advantage was achieved when two or more wheels were used together. This explains their use in machines like yardarms (pulleys with ropes) for sailing vessels and for moving moderate weights.

Cement

In 200 BCE the Greeks evolved cement as a structural material, to replace weaker mortars such as gypsum plaster or bitumen. The use of lime, a new material, could bind sand, water and clay.

Astrolabe

In 140 BCE Hipparchus, a leading Greek astronomer, created the astrolabe (meaning star taker) to measure the angle of the sun or a star above the horizon and provide a chart showing the heavens at differing latitudes and times. The altitude of the Pole Star revealed the observer's latitude, in relation to the position of sun and stars at the time of day or night. It is the world's oldest scientific instrument and was to become critical for navigation.

Accounting

Athens in the 5th Century BCE was a city run by its citizens as they "possessed real authority over government finance and the official bureaucracy."[5] There were 10 publicly chosen state accountants, who recorded all revenues as they came into the city and compiled lists of government debtors. The fact that Athens was run by popular sovereignty introduced the idea of financial disclosure for the first time in history. "To ensure maximum publicity, certain accounts were even engraved in stone and placed on public view."[6] Fraud was common and this created a requirement for the records of government officials to be examined by auditors at the end of the official's term.

Origination of Risk

The term risk can be traced back to classical Greek where the word risk, meant root, stone, cut from firm land (later used in Latin for cliff). The original term was used in Homer's Rhapsody M of Odyssey "Sirens, Scylla, Charybdee and the bulls of Helios (Sun)." Odysseus tried to save himself from Charybdee at the cliffs of Scylla, where his ship was destroyed by heavy seas generated by Zeus.

Standardizing the Quality of Gold

The touchstone[7] (Schist, flint-like black stone) was first used in ancient Greece and Anatolia circa 500 BC. It standardized the quality of gold, and made it into a tradable commodity that could be used as a currency. This was the first time something could be trusted and it had a massive impact on trade.

Coined Money

The Greeks began issuing coined money in about 630 BCE. Records were kept in money form and this medium of exchange also guaranteed consistency. Although they could record business transactions in money terms often they did not. At times they represented property in physical quantities and monetary assets in money terms with inventory and the cash equivalent of assets on the books of any company.

Banking System

Athens in the 5th Century BCE was the first state to make financial disclosures to create transparency with public projects. The Greeks had a highly developed banking system[8] that changed and loaned money, accepted deposits, acted as intermediaries and trustees, and arranged cash transfers for clients through correspondents in distant cities. Bankers kept account books that had to be produced as evidence in court.

Regions

There was significant prosperity in Greece and Rome that saw growth in thriving cultures around the Mediterranean. In the East, China started to emerge as a significant power when the first Emperor of China harnessed the resources of the nation and started to pull it together.

Significant Projects

All the significant projects in this historical period were civil construction projects that had an increasingly more sophisticated architecture, delivered by a smaller and more skilled labor pools. These projects spanned years and sometimes across decades.

Parthenon Project (447-438 BCE)

Background

Greece was devastated by the Persian invasion of 481 BCE, where the Persian forces sacked Athens and demolished the buildings on the Acropolis. Under the leadership of Athens, one hundred Greek city-states joined forces to form the Delian League, a democratic military alliance to push back the invaders. Following victory Pericles, the military leader victorious at the Battle of Mycale, took the initiative to restore the destroyed parts of the city.

Figure 4.2: The Beginnings of Historic Greece 700-600 BCE outlines the city states.[9]

Initiating

The primary objective for Pericles, the principal stakeholder, was to rebuild the ancient shrines destroyed by the Persians and to architecturally adorn the Acropolis with a magnificent temple dedicated to the goddess Athena, known as the Parthenon. The Acropolis is a limestone hill that rises 150 meters (about 500 feet) above sea level. The agora beneath the Acropolis was the commercial and administrative center of the city.

The Athenians had notions of perfection that they called *Symmetria*, the harmonious relationship of part to part and of the part to the whole. As part of the project charter the Parthenon was to be an expression of these ideas, a perfectly symmetrical building, where the harmony depended on a certain mathematical system of proportions.[10]

Pericles initiated the building program. He proposed to the Assembly (primary stakeholders) that the project financing came from a war chest (the treasure of the Delian League) from any part that was not used for the common defense. He argued it should be used to beautify what he deemed the legitimate capital of a magnificent empire. For the Athenians this was quite acceptable although they were loathed to spend great amounts on adorning the city. "Very well," responded Pericles, "let nothing be charged to the public treasure, but all to my own estate, and I will dedicate the public buildings in my name." Whether it was surprise in his show of spirit or a desire to get in on the glory, the Athenians shouted their approval, "Spend on and spare no cost until all is finished," (from Plutarch). The Athenians were seasoned record keepers and believed in the freedom of information, which they published in the form of inscribed marble slabs atop the Acropolis. The Parthenon accounts are now fragmentary but tell us about the funding.[11]

Planning

The Acropolis and other public buildings were part of this public works program. Pericles devised a plan to rebuild ancient shrines, and his human resource plan would utilize the flourishing artistic talent along with the unemployed[12] Athenians, so that every Athenian had food on his table. It would keep the unemployed workers off the streets, and stimulate the Athenian economy by creating a number of important feeder industries[13] by putting the whole city into state-pay. This was an important part of the project charter and it helped get buy-in to the project on a broad scale. The population of Athens was around 250,000–300,000 people in this period.

The workforce consisted of artisans and tradesmen including smiths and carpenters, molders, founders and braziers, stone-cutters and stonemasons,[14] dyers, goldsmiths, ivory-workers, painters, embroiderers, turners. The workforce was organized on military lines. Where a captain had a company of soldiers, every trade had its own hired company of journeymen and laborers belonging to it, and banded together as in an array.[15]

A driving and critical requirement was the esthetic qualities of the building. It was to be an object of beauty and inspiration. But this had to be incorporated into the project without increasing the scope or the cost.

The plans laid out a perfectly symmetrical building where symmetry was a 9:4 ratio present in various dimensions like the length of the stylobate (the base of the building) to its width, and to the height of the column (see figure 4.3).

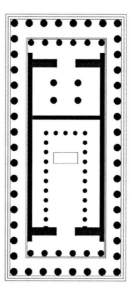

Figure 4.3: Floor plan of the Parthenon showing the Doric columns.[16] The temple measures over 7,000 meters² (23,000 feet²) at the level of its top step.

In fact, the Parthenon incorporated so-called optical refinements where deviations from the perfectly horizontal or vertical, and the straight and perpendicular were in fact intentional. The slight beveling in the blocks was intended to create a curvature of the steps. A perfectly straight line carried over a long distance would appear to sag. The upward curvature of the long steps would counteract that optical illusion, making the line look perfectly straight horizontally and vertically.[17] Quality was planned for from the start.

Executing

Work began on the Parthenon in 447 BCE. The Athenian Empire was at the height of its power. The project team is known today. Phidias had the oversight of all the works, and was the surveyor-general, though other great masters were employed upon the various portions. The main architect was Iktinos, and the master builder was Callicrates. Kalamis was in charge of the design of the sculptures and decorations.

The whole temple, from the base to the roof tiles, was built of marble although other materials included brass, ivory, gold, ebony and cypress-wood. The quarries were located about 16 kilometers (10 miles) from the site. Quarry workers used iron tools like wedges to split natural fissures in the marble, with iron headed mallets, and iron levers. These were then fashioned in the quarry into partially finished blocks, so as to withstand any damage on-route to site. The blocks weighed 11,800 kilograms (26,000 pounds or 13 tons) and were pulled out of the quarry on sleds using a combination of ropes, pulleys, winches, levers and rollers. The sleds were maneuvered with ramps onto heavy wagons (Tetrakykyle) which were pulled by mules.

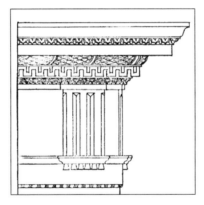

Figure 4.4: Doric Order of the Parthenon.[18]

Some of the financial accounts that survive today show the largest single expense was transporting the marble from Mount Pentelicus, about 16 kilometers (10 miles) from Athens, to the Acropolis. The day long journey required the wagon to be pulled up a steep slope. The approach used pulleys and ropes attached to an empty wagon going downhill pulled by mules.

The project workforce cut 100,000 tons of marble, about 70,000 pieces. Quarrying and transporting the marble from the quarry to site were likely the most significant cost items, but this was quite common to most ancient projects.

Figure 4.5: Art Hum Section: Parthenon (447-438 BCE) Athens, Acropolis, 19th Century photograph, the west end of the Parthenon Item ID: 1651.[19]

On site the partially finished blocks went through the final stages of carving, and polishing into column drums. The Doric columns were assembled from thousands of these.

The workforce used highly sophisticated tools like the antique mason's device called the Pantograph with which they were able to record the three-dimensional shape of an object (piece of stone), and then precisely replicate it. This allowed them to achieve an incredible level of precision when fitting similar pieces together.

On site the master stonemasons did all the carving by hand. They used red clay to ensure that their pieces would fit together precisely that is, mating the two pieces. They applied red clay to the inside surface of one of the blocks to ensure there was a perfect match between pieces. Often the difference between a perfect fit was down to just a tenth of a millimeter, the thickness of a human hair. The quality standards were very high and unique for their time. The masons were then able to sand the marble pieces to within 1/20th millimeter accuracy by using a metal plate and sand to grind the surfaces.

The workforce used ingenious axial dowels to align massive the column drums (2 meters (6 feet) in diameter and 0.75 meter

(2½ feet) thick) into the columns. Near perfect alignment was not just there for esthetic purposes. Accuracy of structural judgment was integral were thrust had to be balanced by opposing thrust. The quality requirements were continually audited at a local level to assure the quality.

The most complex part of the project was the completion of the Doric columns which measured 1.9 meters (6.2 feet) in diameter and were 10.4 meters (34.1 feet) high. The corner columns were slightly larger in diameter. The Parthenon had 46 outer pillars and 19 inner pillars in total.

Controlling and Monitoring

The propylaea, or entrances, were finished in five years' time. The Parthenon took nine years to complete (447-432 BCE), yet would survive some 24 centuries. The building was substantially completed by 432 BCE, but work on the decorations continued until at least 431.

The building's original function was as a temple dedicated to the goddess Athena, but upon its completion the structure housed the Delian League's financial reserves. Some experts maintain that the early Parthenon was used as a treasury.

Figure 4.6: Art Hum Section: Parthenon (447-438 BCE) Athens, Acropolis, 19th Century photograph, view from the southwest Item ID: 1659.[20]

Closing

The Parthenon became the glory of the city and the legacy of this building is its contribution to Western architecture and the many buildings that have followed in its influence. In particular, the harmonious and integrated way it is architected, the steps curve upward, the columns tilt inward, the metopes (marble panels) tilt outward, the columns swell, the corner columns of the building are slightly thicker than the other columns of the building. All of these refinements are combined masterfully.

"Greek Architecture (and by this I mean the architecture of the sixth and fifth centuries BCE) remains one of the great outstanding facts in the history of the Architecture of the Western world."

—By Sir Reginald Blomfield, F.S.A., R.A.

The project was remarkable in that such a refined and sophisticated building could be built to such incredible specifications in the time frame and with the level of technology available.

Other Notable Projects

Worthy of an important mention are the following projects.

Terra Cotta Warriors (210 BCE)

The first Emperor of China, Qin Shi Huang, conquered and united China from a collection of warring states. He built the Great Wall and amassed great works of art. He obsessed with a quest for the secret of immortality. As result, a project workforce of 70,000 built his tomb with over 8,000 life-size clay warriors to protect it, with 130 chariots, and 520 horses and 150 cavalry horses. The warriors are unique life-size and individually modeled figures in clay with a remarkable level of detail. This includes body armor with heads of rivets. The hands and the heads were made separately, and each is different and individual.

The Grand Canal of China Project

This is the world's oldest and longest canal, surpassing the Suez and Panama Canals and was begun in 486 BCE. It was extended in 605 (see page 220).

The City of Petra

This was carved out of rock in the foothills of modern Jordan, in the 6th Century BCE by the Nabataeans. The project workforce created a sophisticated system of clay pipes that brought water into the city for a population of 30,000. They created tunnels and channels to prevent floods from destroying the buildings.

Figure 4.7: El Deir Petra (6th BCE). Picture by David Roberts created 8th of March 1839.

Seven Wonders of the Ancient World

The list is very subjective as few of these wonders (the project outputs) remain today, and details about the projects are scant and only now coming to light through archaeology. The most astonishing fact about these projects is the scale of the deliverable (the construct itself) considering the main resource available was human capital. Social organization played a big part as did ingenuity. These wonders include:

1. Giza Pyramid, 2580-2560 BCE (see previous chapter).

2. Hanging Gardens of Babylon, 604-562 BCE, near modern
    day Baghdad. No confirmed finds exist, only accounts by
    a Babylonian priest from the 2nd Century. The gardens
    were constructed with massive walls 25 meters (80
    feet) thick that were stepped to form the terraces of a
    structure that was 122 meters wide x 122 meters long
    (400 x 400 feet), and 30 meters (100 feet) high. The
    garden was sloped like a hillside and thickly planted with
    trees of every kind, and hidden water machines supplied
    water from the river.

3. Mausoleum at Halicarnassus, 353 BCE, in Turkey, a
    huge marble tomb for Mausolus, a Persian official. It
    initiated the trend of calling all large tombs mausoleums.
    The Mausoleum was almost square, 125 meters (411 feet)
    and bounded by 36 columns. The top formed a 24-step
    pyramid surmounted by a four-horse marble chariot with
    base dimensions of about 40 x 30 meters (120 x 100 feet).
    The total height was 45 meters (148 feet). Fragments of
    the sculptural decoration survive today.

4. Colossus of Rhodes, 304 BCE, near the harbor of Rhodes
    on the Aegean Sea stood a huge bronze statue which
    stood 45 meters (148 feet) high and upon a 12 meters (50
    feet) pedestal near the harbor mole. It was constructed
    in 12 years and was destroyed by an earthquake in 224
    BCE. The Statue of Liberty is roughly the same size.

5. Lighthouse of Alexandria, 270 BCE, in Alexandria har-
    bor, Egypt, stood on the island of Pharos, and the word
    "pharos" came to mean lighthouse. The total height of the
    Lighthouse, including the foundation, was 117 meters
    (384 feet), equivalent to a 40-story modern building. It
    guided sailors into the city harbor for 1,500 years and
    was the last of the six lost wonders to disappear. A shaft
    in the internal core lifted fuel for the fire used during the
    night. A mirror reflected the sunlight during the day, and
    could be seen more than 50 kilometers (35 miles) away.
    The building was constructed of marble blocks with lead
    mortar. A statue of Poseidon adorned the summit of the
    building. It is likely that it was lost to earthquakes.

6. Statue of Zeus at Olympia, circa 2nd Century at Olympia, Greece, likely perished with the temple when burned in 425. The base of the statue was about 6.5 meters (20 feet) wide and 1.0 meter (3 feet) high. The height was 13 meters (40 feet), equivalent to a modern 4-story building. The Olympic Games, in honor of the King of the gods, were first started in 776 BCE and were held at the shrine to Zeus.

7. Temple of Artemis at Ephesus, 262, was destroyed by a flood in the 7th Century. The Temple was 80 x 130 meters (260 x 430 feet) in plan. The 127 columns were 20 meters (66 feet) high with Ionic capitals and carved circular sides.

Key Players

Pericles (495 – 429 BCE)

Pericles was an able general but after the war he devoted his energies to the beautification of Athens, and creating a stable peace time economy and environment. Sculpture, architecture, drama, and philosophy all flourished wondrously during the thirty years of his leadership.

Chapter Wrap-up

Conclusions

The Parthenon set the pattern for future large-scale projects, or public works, where the objective was to stimulate the local economy and help the unemployed. What Pericles achieved though went much further as he created one of the most stunning buildings ever built, a massive contribution to Western architecture which was copied for millennia to come. **This set the notion it was possible to deliver in a short time frame (9 years) esthetically pleasing buildings within the scope of a modest project budget.**

Key Lessons

PMBoK® Guide Knowledge Areas:

- Integration Management

  - ○ Although the Parthenon project costs were high, the business justification was based on stimulating the local economy and helping the local population. The project charter stated that the project would incorporate local small businesses, and unemployed workers in its vast workforce.

  - ○ The project charter was voted upon by the Ekklesia (the assembly of the citizens of Athens), so that the final deliverable would be a monument of the democracy and not of one man.

  - ○ The building was larger and more opulent than any temple that had been constructed on the Greek mainland before. The building was to become a showcase to the world and a symbol of the culture and sophistication of Athenian society. Sophisticated public buildings were important as they underpinned the Greek public institutions and commitment to democracy. Today it is one of the most representative symbols of the ancient Greeks. This was an intangible benefit but very important to the business case. Many of Seven Ancient Wonders of the World followed a similar kind of logic.

  - ○ The project was run on the principals of freedom of information and financial disclosures. These were traits of Athens, and the project had a well defined governance structure.

- Scope Management

  - ○ A driving requirement for the project was the esthetics of the building. It required a lasting beauty pleasing to the eye and an enduring legacy. The project had to follow this requirement which typically would add tremendous scope to any project. This was built into the project and stemmed from a design of perfect proportions, but importantly it was kept within a budget and time frame.

- The scope in terms of materials and equipment was dictated by:
  - The height, breadth, and width of the structure.
  - The quarrying and transporting of marble to site.
  - The ornate pieces (sculptures) on the exterior.
- The work breakdown structure laid out the project with the following:
  - Site preparation creation of a solid foundation with resilience to earthquakes.
  - Construction of ramps, and delivery of marble blocks to site.
  - Construction of the structure.
  - Assembly and precise positioning of the blocks within the structure.
  - Removal and ramp demolition, and site clean up.
- The project's scope was increased by the distance of the quarry, and the location of the site on a hill.

- Time Management
  - The high profile and prestigious project had to be completed in a single lifetime (that of Pericles), as it was a showcase building built for political purposes and to put Athens on the map as the center of an empire.
  - Scheduling of activities was critical as a long supply chain of marble had to continuously feed the on-site assembly teams. The rate of marble quarrying had to arrive in lockstep with the project construction schedule. The site itself did not have a lot of room for temporary storage of blocks which would be at risk of being chipped or damaged.

- Cost Management
  - The project was initiated only after there was consensus that the finances would be available for the project from what was left over in building the

city defenses. Some of the financial accounts for the Parthenon survive today and show that the largest expense was transporting the stone from the quarry 16 kilometers (10 miles) from site.

- ○ The cost of the esthetic requirements was built into the project within the project budget.

- ○ Analogous estimating was used based on a tradition of Temple projects such as the Temple of Zeus which was finished 10 years earlier.

- ○ The Greeks evolved sophisticated accounting techniques in an effort to create transparency with public projects, through the disclosure of accounts. These techniques helped manage the project budget.

- ○ The cost of extracting and transporting the marble was the most expensive major activity in the project and cost over 400 silver talents, the equivalent of 400 fully-equipped Greek warships.[21] This was found inscribed in part on a stone placard from 434 BCE. The gross annual income of the City of Athens was 1000 talents. Athens had a democratic system of government so that the Greeks required the expenditure of public monies be made public.

- Quality Management

  - ○ The quality management plan was critical and incorporated quality control at a local level with the craftsmen. The quality requirements were continually audited at a local level to assure the quality of work.

  - ○ The temple was not geometrically straight and not built from interchangeable parts. Deviations from the perfectly horizontal or vertical lines were in fact intentional. The upward curvature of the long steps counteracted an optical illusion that made the line look perfectly straight. Likewise the columns tilted inward and the corner columns were slightly thicker than the other columns.

  - ○ The Greeks used a number of tools that helped achieve quality. For example, the use of red clay in mating two stone pieces, the pantograph in recording the shape of an object, and ingenious axial dowels to align the massive column drums 2 meters (6 feet) in diameter and 0.7 meters (2.5 feet) thick in the columns of the Parthenon. Near perfect alignment was not just there for esthetic purposes but safety as well.

- ○ The quality standards of the project were taken to a new level, not seen before in projects, with a perfection in the craftsmanship that provided a standard of excellence for Western civilization.

  - For example, marble pieces could be sanded to 1/20th of millimeter accuracy, so the fit was precise. The surface was perfectly connected and airtight.

- ○ Pheidias was a sculptor who was made the overseer of the project. This ensured an emphasis and higher standards on the esthetics parts of the project.

- ○ Stonemasons were recruited from throughout the Greek islands which had many different systems of measurement. Without a common standard, coordinating this workforce would have been a logistical nightmare. The project used the Salamis Stone which represented all the competing ancient Greek measurements: the Doric foot, the Ionic foot, and, for the first time, the Common foot—virtually the same measurement used today.[22]

- • Human Resource Management

  - ○ There was a clear assignment of responsibility with individual roles dividing the work between architects, designers, over-seers and sponsors.

  - ○ The project was part of a public works program to keep unemployed workers off the streets, and stimulate the Athenian economy by creating a number of important feeder industries.

  - ○ The project workforce was organized on military lines where every trade had its own hired company of journeymen and laborers.

- *"And every trade in the same nature, as a captain in an army has his particular company of soldiers under him, had its own hired company of journeymen and laborers belonging to it banded together as in array, to be as it were the instrument and body for the performance of the service. Thus, to say all in a word, the occasions and services of these public works distributed plenty through every age and condition."*

- This is clear evidence of the existence of guilds. *"As then grew the works up, no less stately in size than exquisite in form, the workmen striving to outvie the material and the design with the beauty of their workmanship, yet the most wonderful thing of all was the rapidity of their execution."*[28]

- The trades included smiths and carpenters, molders, founders and braziers, stone-cutters, dyers, goldsmiths, ivory-workers, painters, embroiderers, and turners. They wrought and fashioned materials such as stone, brass, ivory, gold, and ebony cypress-wood.

  - From minute differences in the chisel marks, the distinctive workmanship of about 200 different stonemasons has been recently identified. The stonemasons were so experienced that they may have only required minimum direction.

  - The project had to take a disparate pool of skilled and unskilled workers and hone them into a team that could follow very high standards of quality.

  - The increased use of simple machines, like yardarms with pulleys, reduced the workforce size in certain areas of the project.

- Communication Management

  - Pericles was recognized as the most distinguished orator. Indeed, it was with him that oratory became a political force of the first magnitude.

- ○ Pericles had to get buy-in and support for the project which he did by offering payment in return for some level of personal recognition.

- ○ The project was further popularized by making this a public works project were unemployed Athenians were given an opportunity to work. It also gave small businesses a boost in providing feeder industries. This raised the profile and support for the project very high amongst Athenians.

- ○ The project team used plans to communicate ideas. In particular, the craftsmen used templates and full scale models.

- Risk Management

  - ○ From a construction perspective the biggest risk was with a potential structural failure in the material affecting the 55 marble columns and their overall height of 10.4 meters (34.1 feet).

  - ○ This was greatly mitigated by assembling the columns from massive column drums, 0.7 meters (2.5 feet) thick, were it was far easier to spot defective (poor quality) stone, and swap out the drum. Careful tracking of where the stone was coming from within the quarry further helped in mitigating defects in the materials.

  - ○ The quarrying of oversized stone was widespread in ancient projects and in the medieval period. For example, with the Parthenon the Greeks cut oversized diameters of the column drums to protect them from injury in transit and during erection, a form of risk mitigation.

  - ○ Another risk was the use of very large cranes for lifting the drums and lintels to over 10.4 meters (34.1 feet).

  - ○ When it was discovered that parts of the substructure was higher in relation to the natural ground, the temple was made smaller all around in relation to the substructure to reduce the pressure on these parts.

- ○ To mitigate the risk of an earthquake the foundation was built partly on the bedrock and partly on a bedding of stone so that so it was more flexible.

- Procurement Management

  - ○ Plutarch wrote about the project as a "Public Works" and "services of which distributed plenty through every age and condition."

  - ○ The flourishing artistic talent of Athens was put into state-pay to keep unemployment down.

  - ○ The project depended on a number of feeder industries specifically the quarrying and transporting of marble.

  - ○ A vital part of the project was securing the procurements arrangements with the quarries to guarantee a continuous supply of marble.

  - ○ An important part of the project was the planning and administration of procurements and the incorporation of local small businesses into the overall supply chain.

Educators

- Discuss in the context of today's projects how quality management was incorporated into the Parthenon project?

- Discuss the challenges of public works projects and comment on the perceived differences of today's projects versus those of the past, like the Parthenon project?

- Discuss the impact of Pericles in sponsoring, initiating, and leading the project. Did a war time record help him?

CHAPTER 5

The Roman Empire (1ˢᵗ Century BCE - 500)

This period saw the Roman Empire reach its zenith were 25% of the world's population was part of it. From a project management perspective construction projects were driven primarily by the state of the Rome. New architectural forms like the arch reduced the volume of materials, and increased the strength of the structure. New materials primarily concrete, with a veneer of stone, provided flexibility. There was an increased usage of draft animals and more sophisticated equipment for lifting, and moving materials.

In this Historical Period

- Brief synopsis of trends/changes (natural disasters, wars, technology, epidemics).

  a. The Roman Empire grew to cover Europe, North Africa, and Middle East. Roman organizational abilities transfer from the military to projects.

- Impact of changes

  a. The principal monuments of Roman architecture belong chiefly to the period between 100 BCE and 300 as the Romans complete hundreds of high profile civic construction projects.

- Major events

  a. Birth of Jesus Christ in 0.

  b. Mount Vesuvius erupts in **79** and destroys the cities of Herculaneum, Stabiae, and Pompeii. It kills thousands of residents.

  c. Split of the Roman Empire (the collapse is covered in the next chapter).

- New tools, techniques and breakthroughs

  a. Roman Law and the legal system.

  b. Roman Financial System.

      i. Money economy emerges.

  c. Treatise on architecture[1] written by the Roman architect Vitruvius.

  d. Roman Army Specialists (Immunes).

  e. Roman Construction Materials and Techniques.

      i. Organizational ability based on the militar.

      ii. Scientific (concrete).

      iii. Roman roads.

      iv. Arch, barrel and tunnel vaults, and coffered roofs.

      v. Viaducts.

      vi. Baths.

  f. Chinese developments – paper, the wheelbarrow, the rotary fan, and the stern post rudder for junks.

  g. Stirrup.

  h. Plough.

- Regions

  a. Mediterranean, Western Europe.

  b. Middle East.

- Significant projects

  a. The Colosseum (70–82) in Rome.

    b.   The Pantheon (118–125) in Rome.

- Other Notable Projects

    a.   Bridge over the Rhine (army).

    b.   Hadrian's wall in Britain.

    c.   Aqueduct - Pont du Gard, France.

- Key players

    a.   Julius Caesar.

    b.   Vitruvius.

- Chapter Wrap-up

Trends and Changes

The world population in this period grew from 150 to 190 million. The population of Rome reached 1.25 million in 200. The Roman Empire grew to cover Europe, North Africa, and the Middle East. It is the largest empire in the world. Trade within the Roman Empire was substantial because it was one big trading organization with low tax rates, a common currency, and custom fees. The Romans transferred organizational abilities from the military to government and civilian institutions, and into project.

Impact of Changes

As the Roman Empire reached its zenith the Romans undertook a massive construction program of infrastructure (roads, towns, fortifications, civic buildings) to underpin the Empire.

Major Events

Birth of Jesus Christ (0)

The influence of Christianity and its impact on the Roman Empire, and on civilization for the next 2,000 years.

Mount Vesuvius erupts (79)

Mount Vesuvius erupts in 79 and destroys the cities of Herculaneum, Stabiae, and Pompeii, and kills thousands of residents.

Split of the Roman Empire (324)

When Constantine became the Emperor of Rome he realized the empire was getting unmanageable. So he split it into two halves, the Western and Eastern, and ruled the Eastern part.

Figure 5.1: Map of Territorial Expansion of Rome.[2]

New Tools Techniques and Breakthroughs

Roman Law and the Legal System

At the core of Roman Empire was Roman law which brought order and a level of societal stability. This in turn, became the foundation of law in all civil law jurisdictions across Europe and the West.

Roman Financial System

The Romans had a tradition of accounting which started in the home. Accounts began "as elaborations of the records traditionally kept by the heads of families." These records included daily entries of household receipts and payments in a day book called an adversaria. There were also monthly posting's made to *a codex accepti et expensi* which served as a cashbook for the families. The keeping of household accounts was very important because taxpayers were mandated by law to prepare statements of all their property and debts outstanding. Another reason for the compilation of these statements was the dependency of a citizen's civil rights on the amount of property that a citizen declared.[3] The sophisticated tax system (with different rates) was core to the Roman Empire along with a common currency, and fees for customs.

If we believe Caesar's Commentaries on the Gallic War, the Gauls were groaning in his time under the pressure of taxation, and struggled hard to remove it. Rome lightened their burden; but the fiscal system of the metropolis imperceptibly took root in all the Roman provinces. There was an arbitrary personal tax, called the poll tax, and a land tax which was named cens, calculated according to the area of the holding. Besides these, there were taxes on articles of consumption, on salt, on the import and export of all articles of merchandise, on sales by auction; also on marriages, on burials, and on houses. There were also legacy and succession duties, and taxes on slaves, according to their number. Tolls on highways were also created; and the treasury went so far as to tax the hearth. Hence the origin of the name, feu, which was afterwards applied to each household or family group assembled in the same house or sitting before the same fire. A number of other taxes sprung up, called sordides, from which the nobility and the government functionaries were exempt.[4]

In 48 BCE Julius Caesar took back from money changers the power to coin money and began minting coins on behalf of the state. With this plentiful supply of money he established the ability to complete many massive construction projects.

As in Greece, the Roman banking system was highly developed in that "bankers maintained three books: an adversaria, in which transactions were noted as they occurred; a *codex accepti et expensi* or cashbook; and a *liber rationum*, literally personal ledger or book of accounts, where data from the *adversaria* were classified".[5] In the Roman financial world the bankers were an imperial elite of senators and knights, who made loans either directly or through intermediaries. These loans financed a variety of operations: conspicuous consumption, tax obligations of provincial cities, and, to a lesser extent, also some production and trade (pp.28, 148). These elite financiers did not call themselves bankers. They were basically wealthy landowners, not entrepreneurs, who cherished a strategy of security, not of profit (p.24). This mentality imposed limits on the Roman economy (p.28). This elite thought in terms of networks, not of commercial companies; and this complex is supposed to explain the absence of a bill of exchange (p.26).[6]

The Romans had heaped up at home the silver of the conquered countries, and so silver was very abundant in Rome and scarce in the provinces. Rome was the money centre of the world, where the great companies were organized to lend money, construct public works projects, collect taxes, and engage in the shipping trade. With their central offices in the capital they sent out their representatives to all parts of the Roman world. In Rome, the borrowing rate was 4% to 5%, in the provinces not less than 12%. The row of banking offices which ran along one side of the Forum made it an ancient Wall Street or Lombard Street.

In 116 the Emperor Trajan in his eastern campaign reached Ctesiphon, the capital of Parthia (modern Iran), the supposed limit of the Roman world. Yet he found Roman merchants already settled there. Besides the merchants and capitalists who were engaged in business on their own account in the provinces, there were thousands of agents from the great Roman corporations scattered throughout the Empire.

Treatise on Architecture

This treatise on architecture[7] was written as a guide for building projects by the Roman architect Vitruvius and dedicated to his patron, the Emperor Caesar Augustus. The work is one of the most important sources of modern knowledge of Roman building methods as well as the planning and design of structures, both large (aqueducts, buildings, baths, harbors) and small (machines, measuring devices, instruments).

Roman architects were significantly different from their modern counterparts, acting as engineers, architects, artists, and craftsmen combined. Vitruvius was very much of this type, a fact reflected in De Architectura. He covered a wide variety of subjects which he saw as touching on architecture. This included many aspects which may seem irrelevant today ranging from mathematics to astronomy, meteorology and medicine. In the Roman concept of life, architecture needed to take into account everything touching on the physical and intellectual life of man and his surroundings. Vitruvius specified an architect to be:

> "A literate, skilled draughts man, and good at geometry, well versed in history and philosophy, knowledgeable about music, medicine, law, with experience in astronomy."

Roman master carpenters played a vital role equal to the architects. They built the centering and frameworks for erecting arches, vaults, and domes. They created scale models for centering's requiring a continuous surface.

Roman Army Specialists (Immunes)

Building work played an important part in the Roman army and the legionaries built roads, bridges, fortresses, canals, river widening's, and mines. Entry soldiers joined the Roman army as labor-working soldiers, the lowest class. After a period some soldiers achieved through promotion immune status. The immune's were legionary soldiers who possessed specialized skills but were exempt from tedious and dangerous tasks like guard duties, and latrine cleaning. The skills included the drill and weapons instructors, military police, musicians, engineers, carpenters, hunters and medical staff. They also included surveyors and ditchers (to mark out camps), transporters of materials, and builders including roof-makers, stone cutters, woodworkers, metal cutters, glass workers, plumbers, water pipe makers, cartwrights, blacksmiths, coppersmiths, and lime-burners. The clerical staff included a range of assistants for keeping records (of stores), and financial accounts.[8]

The Roman genius for organizing and controlling huge numbers of men in the army was transferred to large scale projects. Most teams of workers were small although the total workforce ran to tens of thousands.

Figure 5.2: Fresco of Roman builders (Tomb of Trebius Justus in Ostia) showing the simplicity of techniques – Source: National Museum of Rome.

Roman Construction Materials and Techniques

One of the most significant Roman Construction Materials was concrete (200 BCE). The Romans improved Greek Cement by using finely ground volcanic lava (pozzolana) in place of clay, and this was the strongest mortar in history until the development of Portland cement. The addition of small fragments of volcanic rubble resulted in concrete which made possible the arch.

With these materials it was possible to employ unskilled labor for the great bulk of this massive masonry, and to erect with the greatest rapidity and in the most economical manner those stupendous piles which, even in their ruin, excite the admiration of every beholder.

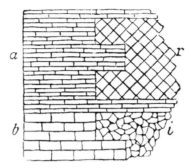

Figure 5.3a: Roman Wall Masonry with exterior of a), Brickwork; b), Tufa ashlar (blocks); Ordinary concrete walls were frequently faced with small blocks of tufa, called, according to the manner of its application; r), Opus reticulatum; i), Opus incertum.[9]

The arch made up of concrete and stone voussoirs (wedge-shaped stones in a semi-circle with a keystone in the center) was used in all aspects of Roman architecture and was pushed forward in various deviations like the vaulted ceilings, barrel vaults, bridges, and aqueducts.

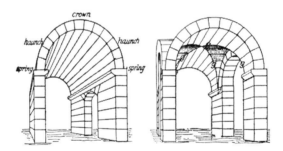

Figure 5.3b: Barrel and Groined Vaults extensively in the Colosseum.[10]

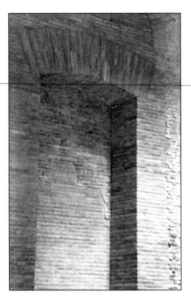

Figure 5.3c: Arch within the Pantheon made of up Tufa ashlar blocks.

Ancient Rome had 350 kilometers (220 miles) of ducts that supplied 1 million cubic meters of water per day to 11 public baths, 856 smaller baths, and 1,352 fountains and cisterns.

The Romans used a very wide variety of construction equipment in the form of various types of massive cranes and lifting devices, using complex configurations of pulleys. This helped the Romans establish over 80,000 kilometers (50,000 miles) of stone-paved roads including permanent stone bridges spanning at least the smallest rivers, primarily for military and governmental purpose.

Some of the construction equipment was massive and included large lifting towers whose four masts were arranged in the shape of a quadrangle with parallel sides, not unlike a siege tower, but with the column in the middle of the structure.

Travertine or lava, a heavy rock, was used as an aggregate for foundations. Stone of this class stands impervious to damage whether from a heavy load laid upon it, or from the weather, or exposure to fire. Travertine limestone was quarried in Tivoli, 32 kilometers (20 miles) from Rome.

Figure 5.4: Travertine quarries on the Roman campagna. The tracks or rails on which the carts ran were cut into the stone itself.[11]

Quarried stones were transported by ox and cart, as there were no rivers. Imported stones from around the Empire were transported by boat and stock piled in warehouses along the Tiber River.

Chinese Developments

The Chinese developed the wheelbarrow, the rotary fan, the stern post rudder for junks. They also created gunpowder.

The discovery of paper in 105 was attributed to a eunuch Cai Lun at the Chinese imperial court who presented the emperor with a report on the new material paper. In the 2ⁿᵈ Century it was produced in quantity in China from rags, the fibers of mulberry, laurel and Chinese grass. Its significance was its relatively low cost as a writing material. Overtime paper spread reaching Europe in the 13ᵗʰ Century.

Stirrup

The stirrup was developed as early as the 2ⁿᵈ Century BCE and was widely spread until the 7ᵗʰ Century. It was probable that early nomadic horsemen, such as the Scythians, used some form of looped fabric to support their feet. But the first direct evidence of a stirrup was a loop for the big toe as used by Indian cavalry from the 2ⁿᵈ Century BCE. Suitable only for use by barefoot warriors in warm climates, this device spreads gradually through

Southeast Asia. At some time before the 5th Century the Chinese, who needed to keep their boots on, transformed the toe loop into a metal stirrup for the whole foot. From China this crucial device moved westwards, through Iran to the Muslim world in the 7th Century, and then through the Byzantine empire to Western Europe. Coupled with the saddle in 365, the two devices were instrumental in harnessing animal power for projects.

Plough

In northern Europe a machine was developed, probably by the Celts in the 1st Century BCE, in which a sharp steel blade cut into the wet earth and an angled board turned it over to form a furrow. With heavier soil in Northern Europe, this type of plough was effective and was a major factor in the agricultural revolution.

Regions

The Roman Empire dominated the Mediterranean, Western Europe, and the Middle East.

Significant Projects

In this period there was a quantum leap in the number of large scale construction projects based on Julius Caesar's financial initiatives. The projects took a very systematic approach, through better organization of teams (based on the military) and a wider spread of knowledge, and specialists (Immune's). This led to two projects that defined Roman project management:

- The Roman Colosseum project (70–80) in Rome.
- The Pantheon project (118–125) in Rome.

Colosseum Project (70–80)

Background

The project was initiated in the aftermath of Nero's rule who after the great fire of Rome in 64 had built himself a huge pleasure palace. A much hated figure for his excesses Nero was forced to commit suicide in 68 when faced with military uprisings and civil wars engulfing the empire.

The next Emperor Flavian Vespasian (69-79), a very successful general, was under pressure to restore confidence and wipe away the memory of Nero. As Rome grew, swelled by immigrants, the state policy of civic control was through free benefits (bread) and entertainment (circuses).[12] Up to that point gladiatorial events were held in temporary structures.

Initiating

The Emperor Vespasian conceived the idea of a Colosseum on the site of the lake in the gardens of Nero's palace. He needed a pleasure palace for the people, capable of holding some 50,000 spectators, to host gladiator contests and other public events. It was a political project, a gift to principal stakeholders to highlight the generosity of the government and to give back something to the people. The project shored up the emperor's shaky regime as the Empire was close to ruin. So, there was a strong business case and rationale for moving forward. The project funding came from booty captured in the Jewish War by Vespasian. Roman political propaganda required these buildings to be showcases and also to emphasize that Rome was at the center of a vast empire.

The project was part of a continuum of projects delivering these types of structures that was a gradual development, and not merely one original design. The Romans had mastered several technologies that made the project possible. First, the use of concrete and bricks: both of these were strong and flexible. They used concrete for the core with a veneer of costly materials applied to the outside of the core. This material technology had been perfected over 300 years. Second, the extensive use of arches and vaults which allowed for vast structures to be built with only a fraction of the materials. Third, the increased use of sophisticated mathematics and geometry in the design process. With this combination it was possible to deliver a stadium capable of seating around 50,000 spectators in a relatively short time frame.

Planning

For the project vast quantities of essential equipment had to be planned for including capstans, windlasses, gins, cranes, as well as sleds and wagons. Some equipment had to be manufactured at the site like the centering for large arches or heavy shoring for a structure.[13]

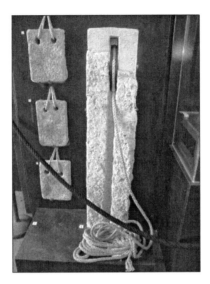

Figure 5.5: On the left three counter weights. On the right a simple pulling device. Source: Colosseum Museum of Rome.

The project was supported by a well organized work-yard.

The Colosseum consisted of 240 arches built throughout the complex to reduce the total volume of material required.

The project was delivered by four contractors.[14] Their contracts detailed the specifications of the work, the requirements for contractor guarantees, and the methods of payment and the schedule as well.

As to construction contracts the Romans employed highly sophisticated oral and written contracts that "were explicit in assigning detailed responsibilities" for all parts of the job, including labor and materials—and there were even arbitration clauses... they possessed the same basic function of our modern contracts.[15]

The government went through a rigorous procurement process. Much of this was lost with the fall of the Roman Empire and did not re-emerge until the Renaissance.

Construction had been one of the most important industries in the city. Any emperor who could not find work for this large body of men created a significant man management problem; Vespasian's awareness of this is demonstrated by his refusal to use certain labour-saving devices commenting "you must let me feed my poor commons". We should therefore also see the Colosseum in terms of the diversion of labour which was required to build it.

For the workforce the contractors used the guilds, the employment of soldiers (Immunes) from the Legion, and some government slaves. The skilled Roman workers were highly organized through the guilds (trade union or *collegium fabrum tignuariorum),* and this is referred to by the Theodosian Code. In the city of Rome there were more than 100 guilds.[16] The guilds were social clubs and burial associations, and often they took an active part in politics. The Theodosian Code also refers to unions of painters, ship masters (sailors), bread makers, pack animal drivers, armament makers, boatmen, collectors of purple dye, weavers, minters, herders, lime burners, grain measurers, porters, physicians, professors (teachers), public service guilds, ragmen and transporters of wood. There existed other unions that were not regulated by the Theodosian Code. They included unions of builders, sculptors, metal workers, leather workers, butchers and jewellery workers. Years later Justinian amended the Theodosian regulations to add several provisions. The most interesting amendment was the recognition that unions were given the legal right to manumit[17] slaves and accept them into their union.[18]

The planning detail for the project was worked out before the construction started as the building was built according to a set of architectural principles, developed through other amphitheaters constructions.[19] To date the project manager and principal architect are unknown.

In 2008 an analysis of the Colosseum Project Walkthrough[20] was completed by Karen Frecker. It summarized from the Colosseum calculations (dimensions, volume, and weight) the overall likely human/animal energy requirements (778,325 Labor days) based on the effort required to create the structure. From this the number of humans required per day for the project was calculated at 2,135. In all, the project workforce was between 20,000 to 30,000 which solved many of the unemployment problems of Rome.[21]

The design incorporated a three-storied arcade with rows of arches at each of the levels. The first level consisted of Doric "order" or style arches that were 7 meters (23 feet) high and 4.2 meters (14 feet) wide. The second level was designed with Ionic arches 6.4 meters (21 feet) high and 14 feet wide, and the third level with Corinthian style arches 21 feet high and 14 feet wide. This was a natural progression from the sturdiest and plainest arches (Doric) at the bottom, to slenderest and richest arches (Corinthian) at the top.[22]

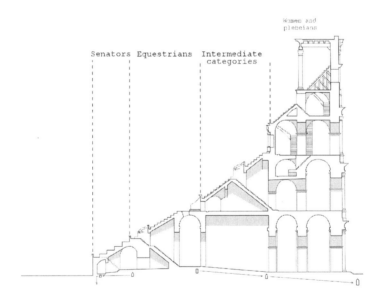

Figure 5.6: Colosseum side view profile highlighting the three-storied arcade surmounted by a fourth story pierced with window like openings. Each of the three arcaded stories originally had 80 arches. Behind each of the outer arcades, two corridors circled the Colosseum, and beyond these were two other smaller concentric corridors.[23] The seating reflected the hierarchy of Roman society with the most important sections at the bottom.

Figure 5.7: Colosseum front profile showing what it would have looked like. Note the masts for the awning or velum. Source: Colosseum Museum of Rome.

Executing

Construction work on the Colosseum began with a complex drainage network, the ring drain running 8m (28 feet) below the ground into Nero's lake, and then into the Tiber river. A heavy storm could deliver 175 liters of water a second.

The next activity was to build an enormous 14 meter (45 foot) foundation extending beyond the perimeter of the building by 6 meters (20 feet). This was in the shape of a donut, 32 meters (105 feet) wide, or 200 by 168 by 6.5 meters (656 by 553 by 22 feet). This hole of 100,000 cubic meters, 220,000 tons, was excavated using ox and cart (half ton load). Two great perimeter retaining walls were built (inner and outer) within the oval hole, 3 meters (10 feet) thick and to a height of 12.5 meters (40 feet). The inner gap (a volume of 250,000 cubic meters (850,000 cubic feet)) was then filled with concrete, lime, mortar, and sand mixed with volcanic rock and water.[24] A concrete foundation ring was capped with travertine (heavy limestone), upon which was built a skeleton of travertine concentric rings. These were in-filled with tufa (blocks), and roofed with opus caementicium (roofing concrete). The use of tufa, recommended by Vitruvius,

was durable under cover as screen walling in the non load-
bearing sections.

The site was not large enough to accommodate the whole
workforce. So, the outer perimeter was divided into four quarters
and independent teams worked in parallel. The outer wall, the
main pillars, the ground floor, and the skeleton of the interior,
up to the second story, were constructed of large blocks of
travertine bonded without mortar and 300 tons of iron clamps.
Travertine which is relatively easily quarried but hardens
on exposure to air was used on all the load-bearing piers. An
estimated 100,000 cubic meters (131,000 Cu. yd.) of travertine
were required. This was covered with a roof.

*Figure 5.8: Roman construction techniques show the use of winching
cranes, to move materials off ox pulled wagon. In a large project
over 100,000 poles would have been used for the scaffolding. The
Colosseum construction is to the far right of the image. Source:
Colosseum Museum of Rome.*

Inside the Colosseum there were four tiers of seating. Each
subsequent tier or layer was made lighter. The ceilings of the
passages and corridors which circled the arena on each tier
consisted of vaulted arches made of concrete supported by the

travertine. This added strength to the building without adding excessive weight. The vaulted arches made the ceilings much stronger than a flat ceiling would have been. The mixture of stone and roofing concrete on such a large scale solved the structural problems, height to weight ratios, inherent in such a design. Over 6,000 tons of concrete were poured.

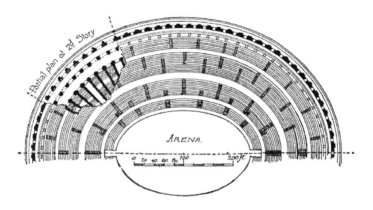

Figure 5.9: Colosseum Half Plan of design. This massive stadium was 550 meters (1,800 feet) in diameter, 195 meters (638 feet) long, 163 meters (535 feet) wide, and sits on a 6 acre lot. The outer walls reached 48 meters (160 feet) high.[25]

The aqueducts of Ancient Rome were run by a Water Commission which was the first in the world to "standardize" parts to supply water. These included settling tanks, ducts, mains, and lead home-delivery pipes. The Colosseum was constructed in the same way around the arch and vault. Outside the Colosseum there were three tiers of 240 arches. Each arch was assembled by a work gang or team of low skilled workers. They could then rapidly replicate the arch and mass produce these as they were standardized throughout the structure. In fact, the simplicity of the design meant minimum project supervision was required.

As is done today skilled work was done in the stone yards for pieces brought already cut and finished to be installed. Marble was used to cover the outside walls (veneer) and for some of the seating (elite classes), statues and ornaments, and drinking fountains. Tiles were used for the floors and the walls.

*Figure 5.10: The Colosseum in the 20ᵗʰ Century. The outer wall is estimated to have required over **1,000,000** cubic meters (**3,531,466** cubic feet) of travertine stone which were set, and held together by **300** tons of iron clamps.*[26]

Similarly, the stairs and seats were standardized, all built to the same size, and were interchangeable. They were constructed off-site in workshops and then brought to site and fitted by teams. They were filled in on two levels at the same time.

Lead and terra-cotta pipes were installed in the walls for carrying water throughout the complex. Rainwater was collected in the Cavea by concentric ducts and poured into vertical pipes leading to the ground floor. Because of a double incline in the floor the water flowed both towards the arena and the outside where it was drained into a brick conduit which surrounded the arena 2 meters (6 feet) below ground. The Aqua Claudia aqueduct was used to supply water to the area of the Colosseum.

In addition, a system of drinking water and toilets (latrines) was incorporated through the stadium. Over 100 drinking fountains were installed to discourage the consumption of wine in large quantities. Communal toilets, or latrines, consisted of a row of holed seats. A flow of water circulated under the seats. A system of small sewers led from all parts of the Colosseum to one great circular drain which surrounded the amphitheater which connected to the Cloaca Maxima, the main sewerage system of Rome.

The structure was covered with a retractable roof called a Velarium based on the principle of mast and sails.

It was the tallest Roman structure ever built. The Colosseum was the pinnacle of the Roman Empire when it was completed in 80.

Controlling and Monitoring

Over time the modifications continued and improvements were made. Further work was carried out at the top of the building, hence the markedly different styles of Corinthian capital around the interior. In effect, the building's style evolved over the reigns of three emperors.

Figure 5.11: The Colosseum in the 19ᵗʰ Century.[27]

The Colosseum was a pleasure palace for the people. It was capable of regularly holding some 50,000 spectators, and 87,000 when really pushed. The network of entrances through 76 numbered (and 4 unnumbered) arches ensured that all classes, citizens or non-citizen, would be guided directly by stairways and ramps. Each had a ticket with an entrance and their seat assignment.

Figure 5.12: The Colosseum today, most of the outer wall is gone due to earthquakes over the years, but much of the inner walls still remain.[28]

Closing

The Colosseum opened with 100 days of festivities. Some of the features that are worth mentioning include:

- The sand-covered floor was built of wood supported by deep substructures, from which elevators brought up animals from dens using elevators to the trap doors.

- Aqueducts supplied water, stored in three reservoirs, to flood the floor of the Colosseum for the extravagant naval fights using large galleys.

144

- Awnings, designed to shade about one third of the spectators, were supported by masts (Mediterranean pines) held by stone socles (plinths).

- Drinking fountains and toilets (latrines) throughout the structure.

The Colosseum was a continuum of stadiums and was followed by the great amphitheater at Capua.

The whole decade witnessed an unparalleled programme of sustained building activity demonstrating both his munificence and strengthening his grip on power.[29]

In later years Antoninus Pius is said to have restored the Colosseum.

The Pantheon Project (118–125)

Background

In 120, Hadrian was looking to rebuild the Pantheon which was burned in 110. He initiated a project to create a building on the lines of a Greek temple.

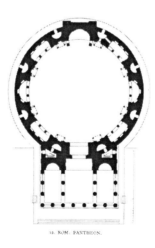

12. ROM. PANTHEON.

Figure 5.13: Plan of the Pantheon showing the drum and the ground entrance portico.[30]

Planning

The most striking design feature was the dome on the drum. The architect designed the structure so that it would fully enclose an imaginary sphere, 43.3 meters (143 feet) in diameter. The mathematics that were required to locate the construction points on the inside of this ball-like structure were a challenge. To date the architect is still unknown.

The practice of giving large sums of money to embellish the city, and to please the public, had grown up under the Republic. The people of Rome had come to regard it as the duty of their distinguished fellow citizens to beautify the city and minister to their needs and pleasures by generous private contributions. It was common for all the Roman emperors in varying degrees to follow this, as well as gifts from generals, from distinguished citizens, and from candidates for office.[31] The project funding in this case would come from the Emperor Hadrian.

> My intentions had been that this sanctuary of All Gods should reproduce the likeness of the terrestrial globe and of the stellar sphere... The cupola revealed the sky through a great hole at the center, showing alternately dark and blue. This temple, both open and mysteriously enclosed, was conceived as a solar quadrant. The hours would make their round on that caissoned ceiling so carefully polished by Greek artisans; the disk of daylight would rest suspended there like a shield of gold; rain would form its clear pool on the pavement below, prayers would rise like smoke toward that void where we place the gods.
>
> —Emperor Hadrian

The project charter entailed a vision where the principal stakeholder Emperor Hadrian would be enthroned directly under the Pantheon's oculus (opening or eye). He was a near-deity around whom not only the Roman Empire but the universe, the sun, and the heavens obediently revolved. The oculus was a key requirement of the structure.

The sequence of major activities included creating the foundation, the drum, and a dome over the drum and finally a ground entrance portico with massive granite columns.

Executing

Hadrian's project workforce cleared the site and prepared the foundations by digging a circular trench 8 meters (26 feet) wide and 4.5 meters (15 feet) deep for the rotunda's foundation and rectangular trenches for the pronaos and the connector. The trenches were lined with timber forms and layered with pozzolana cement. The Romans had been building with concrete and brick since about 200 BCE.

The work on the Pantheon was difficult and graduated. Because other buildings surrounded the site, the laborers lacked space in which to work. They also lacked machinery. Vitruvius (20 BCE), a noted Roman architect (see page 136), recorded the process followed by the Pantheon's project. Wet lime and volcanic ash were hand mixed in a mortar box with very little water so that the composition was nearly dry. The mixture was carried to the job site in baskets and poured over a prepared layer of rock pieces. The mortar was then tamped into the rock layer reducing the need for excess water and stimulating the bonding.

Eventually, work began on the drum. The workforce built extremely thick 20 foot (6 meters) footings and drum walls. The load-bearing walls of the drum, without reinforcing rods, had to be built upward with progressively changing concrete. The workforce reduced the weight of the concrete by using aggregate of different weights. The foundation had an aggregate of lava or travertine, a heavy rock. In the higher parts of the walls, the aggregate was lighter with. The drum itself was strengthened by huge brick arches and piers set above one another inside the masonry walls. Hidden voids and the interior recesses hollowed out this construction, so that it worked less as a solid mass and more like three continuous arcades which corresponded to the 3 tiers of relieving arches visible on the building exterior (see figure 5.10). Originally, these exterior walls were faced with colored marbles.

Figure 5.14: Outside the Pantheon and the entrance portico visible with massive granite columns, it is the most intact Roman building to have survived.[32]

The dome was the most complex part of the construction as it was unsupported with a large opening or oculus. The dome and its internal geometry would create a perfect sphere, since the height of the drum to the top of its dome would match its diameter: 43.30 meters (142 feet). Roman building techniques were sophisticated enough to vary the weight of concrete by controlling the *caementa,* the careful selection and grading of the aggregate material. This ranged from heavy basalt in the foundations of the building and the lower part of the walls, through brick and tufa (a stone formed from volcanic dust), to the lightest of pumice toward the centre of the vault.[33]

To prevent the risk of collapse lightweight concrete was used in the dome thus reducing the outward thrust on the walls. Empty clay jugs were roped together and embedded into the dome's upper courses to further lighten the structure and facilitate the concrete's curing. These ropes of jugs were used for reinforcement and compensated for the weakness and weight of the concrete. Quality standards required the tapering layers of concrete to be the thickest at the base 6 meters (20 feet), and thinnest at the oculus 2.3 meters (7.5 feet). The quality of the concrete composition was very carefully controlled. The builders also used stepped rings (see figure 5.11) because they understood that this would strengthen the dome.[34]

Figure 5.15: Inside the entrance of the Pantheon showing the dome and the opening or oculus.

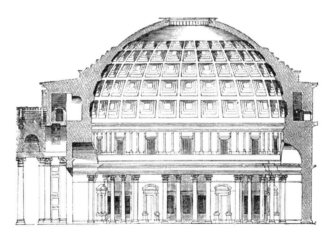

Figure 5.16: Inside the Pantheon showing the stepped rings of the dome, and the cornice lines around the dome. The roof was recessed with panels or "coffers."³⁵

The most likely approach to constructing the dome and the oculus was with heavy wooden scaffolding, from the floor to the highest point. This would have reduced the risk of collapse during the construction itself. Another approach speculates that centering was not required for the lower third of the dome, so the workforce used a lighter centering system supported from the dome's interior second cornice line.

The dome's oculus acted as a compression ring and the workforce built two circles of bipedales using handmade bricks that were 23.4 inches^2 (0.60 meters^2) and 1.56 inches (0.04 meters) thick. They laid the bipedales edgewise in three vertical courses and then circled the oculus with a bronze cornice. The oculus is open completely to the elements and rain water can pour through onto the Pantheon floor. There are rain water drains, holes in groups of four, in the floor.

Figure 5.17: The interior of the Pantheon in the 18th Century, painted by Giovanni Paolo Panini, showing its usage. The height of the drum to the top of its dome matched the diameter of 43.30 meters (142 feet) enclosing a perfect sphere.[36]

Figure 5.18: Inside the Pantheon showing the modern looking stepped rings of the dome, and the cornice lines around the dome. The weight of the dome decreased with each level of rings.

The interior of the building is lined with colored marble, and the walls are marked by seven deep recesses and screened by pairs of columns.

Transportation presented the project another challenge. All material had to come via the Tiber by boat. Hadrian ordered for the Pantheon's Pronaos (front façade), 16 gray granite columns (see figure 5.20 below), each 11.8 meters (39 feet) tall, 1.5 meters (5 feet) in diameter, and 60 tons in weight. These were quarried in Egypt's eastern mountains, dragged on wooden sledges to the Nile, floated by barge to Alexandria, and put on vessels for a trip across the Mediterranean to the Roman port of Ostia. From there the columns were barged up the Tiber.

*Figure 5.19: Example of Roman transportation techniques of an obelisk. The **16** granite columns (**60** tons each) were shipped this way from Egypt. Source: Vatican Museum of Rome.*

The columns would have added considerable cost to the project. But it did significantly enhance the beauty of the building giving a level of grandeur on the Parthenon.

Figure 5.20: The Pantheon and the 16 granite columns in the grand entrance portico shipped from Egypt.

The great domed temple remains today and is the most complete and best-preserved monumental interior to survive from Roman times. The dome has a span of 43.2 meters (142 feet), the largest dome until Brunelleschi's dome at the Florence Cathedral of 1420-36 (see page 250). Light floods through the 8 meter (27 foot) eye, oculus at the centre of the dome, and this was revolutionary as it was the only light source.

Figure 5.21: The Pantheon in the 20ᵗʰ Century, with its grand entrance portico.[37]

In later years Agrippa's Pantheon was rebuilt by Hadrian.

Other Notable Projects

Across the Roman Empire there were many notable construction projects that were used to highlight the power and might of Rome in its conquered territories.

Pont du Gard Aqueduct Project (124-122 BCE)

Built by the Romans through the 1ˢᵗ Century, the aqueduct route went around the east side of the higher Massif Central in France, following a total gradient of about 17 meters (58 feet) through a series of some 35 kilometers (22 miles) of tunnels. The Gardon valley was spanned with aqueduct bridges with the most intact remaining today at Pont du Gard. Rome is fed by 7 aqueducts. This fresh water allows the city to grow to 1 million.

Figure 5.22: Pont du Gard, Roman aqueduct in France, image taken in 1850s.[38]

At least 20,000 cubic meters (44 million gallons) of water were brought in daily to the Roman town of Nemausus (Nîmes) via an impressive aqueduct system of 50 kilometers (31 miles) from springs at the Fountaine d'Eure in Uzès.

Caesar's Rhine Bridge Projects

The Romans chose not to conquer some of the lands beyond great rivers like the Rhine, but used it as a natural barrier. However, they wanted to keep in check the populous on the opposite bank by the ability to build enormous bridges with great speed. Julius Caesar famously constructed a bridge across the Rhine in only 10 days and demonstrated the might and power of Rome.

Technical Details of Wooden Beam Bridge

Double timber pilings were rammed into the bottom of the river by winching up a large stone and releasing it, thereby driving the beam into the riverbed. The most upstream and downstream pilings were slanted and secured by a beam, and multiple segments of these then linked up to form the basis of the bridge. Conflicting models have been presented based on his description. Separate upstream pilings were used as protective barriers against flotsam and possible attacks while guard towers protected the entries. The length of the bridge has been estimated to be 140 to 400 meters (160 to 1,300 feet), and its width 7 to 9 meters (23 to 29 feet). The depth of the river can reach up to 9.1 meters (30 feet).[39]

The ability to build a bridge of this scale in such a short time frame required incredible organizational ability, and engineering skill as only the most basic tools were used. The barbarians soon learned to see the bridges themselves as Roman weapons.

Figure 5.23: Caesar's Rhine Bridge, by John Soane (1814). The supporting pilings were driven in at angles to strengthen the core.[40]

Hadrian's Wall Project (122-130)

The northern frontier of Britain was under threat from unconquered tribes. Under the Emperor Hadrian in 117 the Roman Empire had ceased to expand. Hadrian was concerned to consolidate his boundaries and had no desire to capture further territory. He visited northern Britain in 122, and ordered a wall to be built between the Solway Firth in the west and the River Tyne in the east a distance of 117 kilometers (73 miles) "to separate the Romans from Barbarians".

The wall was built by Roman legionaries, men from all over the Empire, and at one of loneliest outposts at the farthest edge of the Roman world. The Roman army had within its ranks highly skilled architects, mason builders, surveyors and carpenters as well as soldiers. Some local people may have willingly helped, as they would have undoubtedly benefitted from trade in goods and services. In all, the project workforce was made up of 3 legions (15,000 to 25,000 men).

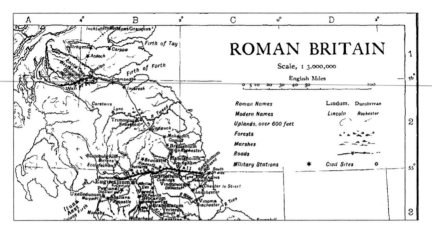

Figure 5.24: Hadrian's Wall (19th Century map) built for 117 kilometers (73 miles) across the width of northern England between two river estuaries.[41]

The construction was broken into lengths of eight kilometers (five miles). One group of each legion would excavate the foundations and build the mile castles and turrets whilst other cohorts would follow with the wall construction.

Figure 5.25: Painting by William Bell Scott.[42]

The actual sequence of activities was twofold. First, construction of the wall of stone itself, 2 meters (6-8 feet) thick, and originally perhaps 3.5 meters (14 feet) high, with a deep ditch in front, and a connecting road behind it in an open area which allowed easy access to all parts of the wall all along its length. This wall was the most striking feature. Second, there was the so-called "Vallum," a broad flat-bottomed ditch out of which the earth was cast up on either side (North and South) into regular and continuous mounds that resemble ramparts. Typically, the "turf wall," was constructed of sods first laid in regular courses, and then reconstructed in stone. Any perilous precipices and natural features were incorporated into the wall. Local limestone was used unless there were no useful outcrops nearby then turf was used instead. Eighty forts and mile castles and turrets were built along the connecting road where each could garrison between 8 and 32 men. During the six years of building the wall reached this final basic form see figure 5.26 below.

Figure 5.26: Outline Hadrian's Wall defense system (19ᵗʰ Century print).[43]

As Hadrian's project evolved, more legionaries were moved up to the wall and within a few years a further (14) full-sized forts were added along the connecting road length. A total of 17 forts each housed between 500 to 1,000 auxiliary troops. These forts had gates to allow traffic to pass north and south through the wall.

For the native inhabitants the ditch and mounds to the south may have signified the start of a sort of reserved military zone. From the point of view of a "barbarian" from the North, the wall might have seemed a psychologically daunting symbol of power.[44] The wall survived the Roman Empire and a few centuries beyond.

Key Players

Julius Caesar (100-44 BCE)

Caesar's financial changes were significant in funding a wave of construction projects in the Roman world which continued for many centuries and built a series of thousands of civic buildings.

Vitruvius (80/70 -15 BCE)

The Roman architect Vitruvius wrote the treatise on architecture, and defined the role of the architect in one role as a combination of the engineer, artist, and craftsman. He also defined the required education for an architect, and the required skills and traits.

On the assessment of project contractors he concluded:

"...things of this sort should be known to architects, so that, before they begin upon buildings, they may be careful not to leave disputed points for the householders to settle after the works are finished, and so that in drawing up contracts the interests of both employer and contractor may be wisely safe-guarded."[45]

Chapter Wrap-up

Conclusions

The Roman projects were remarkable because the Romans were able to deliver solidly built structures in a very short time frame (decades), when compared to the medieval cathedral builders that required centuries. The interior of the Pantheon is a testament to the quality of the project, and the fact it is the most intact structure of that age. The Romans also built with limited mechanization in the ox and cart, and treadmills, winches, and capstans.

The Colosseum project established a public works program to employ a large unskilled project workforce that has the characteristics of a modern project. When the Emperor Vespasian was presented a labor saving device, he rewarded the inventor but rejected the device, so he did not deprive laborers from their work. The availability of capital, through financial reforms, gave the project impetus. The project took a systematic approach, borrowed from the military, with a high level of organization of teams. It incorporated a wide spread knowledge, specialist skills, and far reaching construction techniques using the arch and concrete that allowed for a very large unskilled workforce.

The technical complexities of the Colosseum included:

- The aqueducts which supplied water, stored in three reservoirs, to flood the floor for naval battles.

- The awnings, to shade about one third of the stadium, supported by masts held in by stone plinths.

- The extensive drinking fountains and toilets (latrines) throughout the stadium.

- The wooden floor supported by deep substructures, with elevators and trap doors built in, used for bringing up animals, props, and performers.

The Pantheon project is somewhat similar in its vast use of concrete but more aligned to esthetics of the Greek style. The project incorporated the sophisticated use of concrete, where the weight of concrete varied through the aggregate material which ranged from heavy basalt in the foundations to the lightest of pumice toward the center of the dome.

The Pantheon was one of the most elegant buildings since the Parthenon, hence its style was copied not just for centuries but for millennia to come (the Pantheon in Paris had its façade modeled on the Pantheon in Rome).

Through this period the Roman's initiated a massive construction program of thousands of building projects including cities, civic institutions, forums, markets, religious temples, public baths, libraries, roads, bridges, amphitheaters, aqueducts, and sewers, military fortifications. For the Romans land surveying was fundamental to the creation of this vast program (see the ability to measure page 36). Toward the end of this period the program slowed down as funding dried up.

PMBoK® Guide Knowledge Areas

- Integration Management

  - As in many projects the Colosseum project charter was based on maintaining the principal stakeholder's (the Emperor Vespasian) power and position, by placating and gratifying the public.

  - The costs of the Colosseum project were anticipated in advance as much of the detail was worked out before the project started. The costs were funded through taxes as the business case was driven by political motivation. The principal benefit was to satisfy the Roman public with a place for lavish entertainment.

  - The business justification for these public projects was helped by the increased supply of money created by Julius Caesar's when he took back the power to coin money and cheap loans.

  - Booty from the military campaigns went towards funding the project.

- Scope Management

  - The principal requirements for the Colosseum were based on a rich history of experience for example the amphitheater completed in Pompeii in 70 BCE. In Rome a standard practice was to complete a wooden structure where it would be dissembled after the gladiatorial show.

    - Any further requirements were to come from the emperor and his entourage.

  - By the time of the Colosseum, construction time lines were mature, the Romans were very savvy when it came to scope. The site was in a valley where there was previously a lake so it had to be planned carefully.

  - In the ancient world, the design process would have defined the scope and involved floor plans (preserved on inscriptions) drawn to scale, a handful of three-dimensional scale models, perspective drawings, and for the artisans, some full-size design sketches.

○ The Romans worked with standard templates and made use of repeatable processes. The arch is a prime example of this. This approach simplified determining the scope.

○ The scope in terms of materials, and equipment was dictated by:

- The ground, set on the site of an artificial lake.

- The length, height, diameter of the oval and breadth.

- The availability of materials, quarrying, and transporting of traventine blocks and marble, and the production of concrete.

- The ability to fill and empty the building in a short time with 40,000 spectators.

- The ability to flood the arena for naval battles.

○ In this building of huge scale and complexity, much of the detail was worked out before the building started. The work breakdown structure laid out the project with the following:

- Site preparation installation of circular drainage system, deep foundations, plumbing, and leveling.

- Assembly of scaffolding, delivery of materials (within site quadrants).

- Construction of the structure, both on and off site workshops, blocks to a precise position in the structure.

- Removal of scaffolding and site clean up.

○ A further example of the work breakdown structure can be found in Appendix A.

○ The Pantheon's oculus increased the complexity and scope of the project but it proved to be the most significant esthetic feature that differentiates the structure.

- ○ The varying thickness and weight of the concrete roof (dome) towards the oculus increased the scope.

- Time Management

  - ○ Most ancient projects had to keep to a deadline. The Roman Colosseum was built by the government as a political project to highlight its generosity, and to give back something to the people. Any delays would have been problematic. Time cost money then as it does today.

  - ○ Further into the Roman Colosseum project cycle, activities related to the assembly and coordination of construction equipment was almost as significant as the erection activities. For example, vast quantities of essential equipment were required including capstans, windlasses, gins, cranes, sleds and wagons. Some equipment had to be manufactured at the site, like the centering for a large arch or heavy shoring for a structure.[46]

  - ○ In addition, the principal sponsor Emperor Vespasian was sixty years old and was very anxious to see the project completed. This put the project manager under tremendous pressure to finish quickly.

  - ○ To help keep their building projects on track and to schedule the Romans used a building technique with concrete for the core and a veneer of costly materials applied to the outside. The structural core was constructed in a single building campaign. Thus the structure could be built without any work contingent or delays because of uncertain shipments from afar of costly imported finishing materials. Both the Pantheon and Colosseum exemplify this approach.

  - ○ The use of standardized parts where pieces were made to be interchangeable, like the same width stairs, simplified installation and saved time.

  - ○ With the Colosseum project Roman concrete was a relatively new material and the project required efficiencies when using it, because of the speed at which it was laid.

- Cost Management

  - A history of amphitheater project experience was critical to the estimates although this was a considerably larger project.

  - The arch was so fundamental to projects the Romans were able to estimate it parametrically.

  - Managing a project budget came from a Roman tradition and requirement of managing household accounts, and preparing statements of property and debt for taxes.

  - The Romans, in their public building projects like the Pantheon and Colosseum, controlled costs by using concrete for the core with only a veneer of costly materials applied to the outside. The structural core was constructed in a single building campaign, thus lowering the costs. For the Romans the raw materials for concrete were in abundance where volcanic sand lay just below the surface, and great quantities of large aggregate came from the stonecutter's yards, and discarded pottery from community waste heaps.

  - Costs were also reduced by the use of:

    - Unskilled labor enabled, with the use of concrete, and the arch.

    - Construction equipment like lifting towers reduced costs by reducing the project workforce.

    - Standardized interchangeable pieces which were made in workshops and then brought to site for assembly.

- Quality Management

  - Construction blocks were imported in great quantities from marble quarries right across the empire and from distant provinces. For example, marble slabs in the Colosseum were transported by sea from the Greek island of Paros in the Cyclades. Each block was carefully marked with the quarry, the section it came from (locus), the owner (Emperor Vespasian), and

163

with the freeman who excavated it. The blocks were catalogued according to their quality and provenance.

- ○ The Colosseum was built in an identical way around the structure using the arch and vault. Up to 240 arches were replicated in three tiers. Such standardization allowed tight quality control at a local or micro level. Because the arches were standardized they could be mass produced quickly and consistently by relatively unskilled laborers.

- ○ Quality control was applied where it really mattered and the degree of accuracy was very high. For example, the Voussoirs (wedge-shaped stones) in the arches, crucial to the stability of the structure, were close to identical. The outermost annular corridor on the ground floor was 5 meters (17 feet) wide and varied along it entire length by less than one per cent. Where it mattered less, like the stairways or some arches, there was a variation of several centimeters.[47] Inspections were widely used in these areas.

- ○ The Pantheon project controlled the careful selection and grading of the aggregate material to get a grading of weight with progressively changing concrete. Each concrete batch was hand-carried in baskets to the placement, a very slow process, and with very strict quality control.

- ○ Lime was one of the first man-made products that relied on a chemical reaction of burning limestone pieces in a crude kiln. Again strict quality control was required for the processing temperature and hydraulic reaction.

- ○ Roman legions built roads and had milestones that carried the names of the troops' commanders, acknowledging them as the road builder. These were simple rewards and recognitions of the achievement put in place which instilled a sense of pride in the builder and helped establish quality standards and a level of quality control. It made sure that quality improvements spread throughout the organization.

The Roman army was based on a regime of harsh discipline.

- Human Resource Management

    ○ Roman architects individually had a broad training and experience which incorporated the skills of an engineer, artist and craftsman.

    ○ In the Colosseum project the main construction force was made up of contractors, who used the guilds for the workforce. Aside from these the employment of soldiers from the Legion, some government slaves (Servi Caesaris were the most renowned and on the path to freemen) were used.

    ○ Concrete allowed for the increased use of unskilled labor.

    ○ Simplicity of design with the arches allowed for minimal supervision of the teams from the overall designer.

    ○ Under-architects led teams for less critical parts of the structure where high levels of quality mattered less.

    ○ The Colosseum project was designed from two semi-amphitheaters joined back to back. This allowed the workforce to be split because of the restricted size of the site. The use of standardized interchangeable pieces allowed a large off site workforce to complete these in workshops.

    ○ Caesar's Rhine Bridges were estimated to be up to 400 meters (1300 feet) long, 9 meters (29 feet) wide, in a river depth of 9.1 meters (30 feet). The ability to deliver a bridge of this scale in only 10 days required incredible project management acumen, organizational ability, and engineering skill as only the most basic tools were used.

    ○ Even though slaves existed in these times, with 21 million in the Roman Empire[48], these were mainly owned by private families and individuals.

- ○ Major projects tended to use citizens or contracted companies rather than the mass use of slave labor. Construction became the single largest industry in Rome creating much needed employment.

- • Communication Management

  - ○ For the Romans transport and communications were at the heart of the Roman phenomenon of order, form and stability. Roman roads helped the Roman army make much use of dispatchers with written or oral communications. Roads also greatly assisted the operation of the Imperial postal service, the *cursus publicus*,[49] and good communications across the empire.

  - ○ The Colosseum project was popularized with the public through constant communications. These reiterated it was a public works project that helped keep the unemployed off the streets.

  - ○ There are five different representations of the Colosseum on coins and medals of the Emperors, one of Vespasian, two of Alexander Severus, and one of Gordianus. These all appear to have been made from the architect's designs before the work was completed. This was a clever way of promoting the project, and building awareness and interest. The project was also popularized with the public by communicating it as a public works project providing employment (stakeholder management).

  - ○ For the project team the plans (ground and floor), perspective drawings, and models were absolutely critical in communicating spatial ideas non-linguistically. In particular, to guide the builders, perspective drawings were very analytical and communicated a feel for the building in space.[50]

- • Risk Management

  - ○ Roman architect Vitruvius laid out in his treatise specific dangers; everything from erecting roof vaulting, to digging wells (working in confined spaces), to working with large construction equipment, and to the dangers of fire to certain materials.[51] Overtime these types of best practices were incorporated into the building codes of municipalities like Rome and medieval London that assured good practices in building construction.[52]

○ Risks during the construction of the Pantheon's dome roof and oculus were somewhat mitigated by a system of heavy wooden scaffolding.

○ The construction of the structure of the Pantheon's concrete dome carried significant risk, in its ability to hold together with a large oculus at the center. The risk was reduced by lightening the weight of the dome as much as possible, progressively getting lighter towards the centre.

○ The Colosseum had to deal with a public safety issues and was able to safely accommodate 50,000 people entering and exiting a 48 meter (157 foot) high structure.

○ The Romans could only estimate the strength of concrete a relatively recent invention. With the Colosseum project they didn't want to take risks with such a prominent building. So, they combined concrete with stone, a tried and tested material. The corridor ceilings were vaulted arches made of concrete but, the supports were of heavy limestone.

• Procurement Management

○ *As to construction contracts the Romans employed highly sophisticated oral and written contracts that "were explicit in assigning detailed responsibilities" for all parts of the job, including labor and materials—and there were even arbitration clauses... they possessed the same basic function of our modern contractors.*[53]

○ The Colosseum project was very much in line with the practice of contracting out public works projects. The project was delivered by four contractors whose contracts detailed specifications of the work, requirements for guarantees, and the methods of payment and time. Thousands of small building firms were engaged through this process.

○ The architect's plans were very much principal deliverables in the contract.[54] The Roman architect Vitruvius provided procurement guidelines in his landmark work 'De architectura.'

○ The volume of materials that was procured for the project was considerable. An extensive supply chain of raw materials included:[55]

- Travertine limestone from the quarries in Tivoli, 32 kilometers (20 miles) from Rome. The quarried stones were transported by ox and cart.

  - Lime was produced from the limestone.

- Pozzolano was available within 3 kilometers (2 miles) of Rome.

- Water was transported 300 meters (1000 feet).

○ The raw materials were combined and gave:

- Tuff which was used in the preparation of special cements, produced within a range of 1 kilometer (0.4 miles).

- Tiles and bricks which were produced from clay mixed with water and often with sand, straw and finely ground pozzolana, within a range of 1 kilometer (0.4 miles).

○ Roman Cement which was produced by mixing finely ground limestone and clay. Concrete was produced by adding sand, small stones or pebbles. The percentage mass by material was rubble 35%, travertine blocks 32%, mortar 21%, tufa 6%, brick 6%, pozzolano 14%, lime 4%, and water 3%.

Educators

- Discuss in the context of today's projects the importance of understanding the properties of building materials, like the Romans did with concrete?

- Discuss the influence of the Roman military on the organization of project teams? In particular, Caesar's Rhine Bridges.

- Discuss which of nine *PMBoK® Guide* Knowledge Areas were very significant in the Colosseum project?

- Discuss whether the overall Roman construction program contributed to the fall of the empire?

The Middle Ages

This period is comprised of:

- The Early Middle Ages, lasting from about 350 to about 1050.

- The Central Middle Ages, lasting from about 1050 to about 1300.

- The Late Middle Ages, lasting from about 1300 to about 1450.

CHAPTER 6

The Early Middle Ages (350 - 1050)

The last two centuries of the Roman Empire were spent in a losing battle, where the will for supporting the behemoth empire was waning with the Roman citizens that were burdened with its upkeep. From 450 to 476 the Roman Empire gradually collapsed. By 750 the Roman world had given way to three heirs: the Byzantine Empire (Eastern Roman Empire, see page 181), Islam, and the West. In European historiography (the West), the period 350 to 1050 (4th to 11th Century) was known as the Dark Ages, a period of stifled growth. From a project management perspective, the Byzantine Empire and Islam are where construction projects continued driven primarily by religion, but nowhere close to the construction boom under the Roman's at the height of their Empire. There was an increased in the use of natural materials like stone and wood.

In this Historical Period

- Brief synopsis of trends/changes (natural disasters, wars, technology, epidemics).
    a. Collapse of the Roman Empire (single largest in the world).
- Impact of changes
    a. Growth of Byzantine Empire.

       b.   Dark Ages in Europe.

- Major events

       a.   455 - Vandals sack Rome.

       b.   476 - Fall of the Roman Empire.

       c.   481 - Clovis became King of the Franks.

       d.   732 - Battle of Tours.

- New tools, techniques and breakthroughs.

       a.   Paper bank notes in China.

- Regions

       a.   Europe.

       b.   Byzantine Empire.

       c.   Islam.

- Significant projects

       a.   Cathedral at Hagia Sophia.

- Other Notable Projects

       a.   Toltec Pyramid.

- Key players

       a.   Anthemius of Tralles (Aydin).

       b.   Isidorus of Miletus.

- Chapter Wrap-up

Trends and Changes

The world population in this period modestly grew from 190 to 310 million.

Collapse of the Roman Empire

The bureaucracy of the Roman Empire had stifled progress. When Gaul became part of the Roman Empire there was a net positive effect. By 450, Gaul was requiring an influx of cash from Rome. The collapse of the Roman Empire was dramatic, as it left a disparate set of rulers incapable of supporting the Roman civic infrastructure in place. In the cities merchants lost the security of safe conditions for manufacture, trade and commerce. Traveling or carrying goods over distance became unsafe and there was a collapse in exports. The loss of educational institutions and of a unified culture saw intellectual development suffer.

Impact of Changes

Growth of the Eastern Roman (Byzantine) Empire

Before the collapse of the Roman Empire in 476, the Eastern Roman Empire broke away in 330. The root cause for both was the same: the Roman Empire was too expensive and difficult to administrate. The Emperor Constantine I transferred the capital from Nicomedia in Anatolia to the city of Byzantium, and renamed it New Rome or Constantinople. As a second Rome, it was well-positioned astride the trade routes between east and west. The Emperor Constantine recovered much of the Roman Empire's military strength and enjoyed a period of stability and prosperity.

By the 7th Century the Byzantine Empire had taken on a distinct new character, under the reign of Emperor Heraclius. His reforms changed the Empire's military, and recognized Greek as the official language.

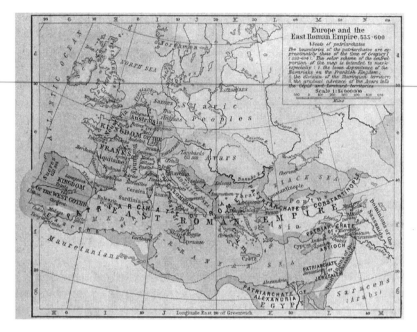

Figure 6.1: The Byzantine Empire (Eastern Roman Empire) during its greatest territorial extent under Justinian. c. 550.[1]

Dark Ages in Europe

In Europe there was general demographic decline and political fragmentation, a shift of power from kings to feudal lords, and limited building activity and material cultural achievements. There was little social or economic interchange and any news relayed to villages was communicated verbally by "village criers." Many changes took place in Europe, like:

- Fragmentation of countries into feudal states, that were typically run by warring tribes.

- Political fragmentation where kings had less power because countries were divided into areas controlled by feudal lords.

- The decline of urban life as towns were the prey of continual violence and depredation, and towns people were exposed to merciless extortion and plundering at the hands of their feudal lords. As a result, trade collapsed and municipal organizations ceased to function.

- The Christian Church helped unify the people of Europe because of its centralized authority and hierarchical nature and it was one of the only constants left. The Church preserved ancient and biblical writings.

- Far less writing, less education, most of it was ecclesiastical based.

- The reversion to an agricultural economy where trade and national revenue reached its lowest point.

Major Events

Vandals sack Rome 455

High taxes, Roman prejudice, and government corruption turned the tribes against the Roman Empire. The Visigoths sacked Rome in 415 but spent only three days in the city. The Vandals plundered Rome for fourteen days which marks the fall of the Roman Empire.

Fall of the Roman Empire 476

In Europe there was a collapse of trade after the fall of the Roman Empire which made bankers less necessary than before, and their demise was hastened by the hostility of the Christian church to the charging of interest.

Clovis becomes King of the Franks 481

The inhabitants of the Roman conquered territories began to assert their independence. The Franks, in northern Gaul, came under Clovis who expanded his rule by wiping out the remnants of Roman control in Gaul and fought other tribes who bordered his territory. The Frankish Kingdom included nearly all of France, the northeastern part of Spain, and a great part of Germany. Clovis embraced Christianity and it thrived and flourished with the Franks throughout the Dark Ages.

Birth of the Prophet Muhammad 570

The central figure of the Islam religion and regarded by Muslims as a messenger and prophet of God (Arabic: Allāh), the last and the greatest law-bearer in a series of prophets of Islam.

Battle of Tours 732

A Muslim army crusading in search for land and the ending of Christianity, after the conquest of Syria, Egypt, and North Africa, and Spain crossed the Western Pyrenees and marched towards the Loire River. They were met just outside the city of Tours by Charles Martel, and the Frankish Army. This was an extremely decisive battle for Christians as it stemmed the Muslim invasion of Western Europe.

New Tools Techniques and Breakthroughs

As Western Europe fell into the Dark Ages, technologies thrived in the East. In India in the 4th Century iron production was well advanced and exemplified when Indian foundry men built an iron pillar 7 meters (24 feet) high at Delhi. Even today it shows no signs of rust as it was protected from rust by a thin coat of manganese dioxide.

Paper-making evolved in China in the 8th Century and factories spread to Samarkand in 751, Baghdad in 793, Cairo in 900, Morocco in 1100 and to Spain in 1150.

The first paper bank notes appeared in China in 806. In all, China experienced over 500 years of early paper money, spanning from the 9th through the 15th Century. Over this period, paper notes grew in production to the point that their value rapidly depreciated and inflation soared.[2]

During this period there were virtually no standard measures in Europe. As a result, pupils were taught to gauge scales and sizes so they could estimate sizes. Reckoning was done by eye and on the spot also known as the "rule of proportion."

Regions

- China continues to innovate and come up with new inventions.

- The Byzantine Empire grew from the Roman Empire and adopts its mantle.

- Islamic culture continues to thrive and evolve unlike Western Europe that stagnates under the collapse of the Roman Empire.

Significant projects

The collapse of the Roman Empire (single largest in the world) in the West saw a demise in large scale projects.

Cathedral at Hagia Sophia Project (532-537)

Background

Under the stability of the Eastern Roman Empire, under the Emperor Constantine, large scale projects continued. In January 532, there was a revolt by partisans of the chariot racing factions, known as the Nika riots, against Justinian in Constantinople. He was forced to dismiss some of his ministers, but then it turned against him. He ordered the brutal suppression of the riots which resulted in the death of 30,000 unarmed civilians. During the riots many fires were started which destroyed large parts of the city including a predominant church, the Church of the Holy Wisdom (Hagia Sophia). Its destruction provided Justinian an opportunity to create spectacular new buildings.

Initiating

The Emperor Justinian ordered the construction of a cathedral at Hagia Sophia. As the principal stakeholder he set the project charter that the cathedral was to be a showpiece and surpass in magnificence all earlier cathedrals. It was to be a spectacle of marvelous beauty. The building heavily drew on the Pantheon in its design. He made full use of all his Empire could offer as his ambition was to make the cathedral unique which spurred him on to an unremitting effort. No expenses were spared as artisans were gathered from around the world.

Justinian's ambitions were to re-establish the Roman Empire from the east. During his reign he conquered territories to the east, and in North Africa. He also brought Italy under the empire.

Planning

Justinian as the patron personally took care of the project funding and supervised the construction. He became very involved directly. The project would establish himself and the empire as the successor to the might of Rome and showcase its capability.

The time line for the project was set to complete in an extremely aggressive five year period to gratify the Emperor Justinian so he could realize the benefits in his reign. This could only be achieved with experienced architects and a very large workforce that was very well organized.

The two most famous architects of the age; Anthemius of Tralles (Aydin) and Isidorus of Miletus, were entrusted with the construction of the building. Anthemius was the best military engineer that Justinian had and was not merely a master-builder but a geometrician of the first order. They were both academics of considerable status, equivalent of university professors with strong theoretical backgrounds. The architects located it on a hill it so it would adorn and dominate the City of Constantinople like a watchtower.

Justinian wanted a very symbolic building that would incorporate the round dome of state power on the square base of a Christian church. The Romans had tried this before and failed, believing only a cylinder could support a dome like the Pantheon.

The main ground plan consisted of a rectangle, 70 meters (230 feet) in width and 75 meters (246 feet) in length. This shape was carefully proportioned so it was harmonious in length and width.

This was the first structure to combine the rectangular plan of traditional basilicas with the central dome of imperial buildings such as the Pantheon in Rome.[3]

The architects designed an unobstructed hall of unrivalled proportions. This was to be covered by a combination of half-domes, increasing in span and height, as they successively led up to the stupendous central dome, which was to rise 55 meters (180 feet) into the air and fitly crown the whole. The design was complex in its geometry and without a precedent. The central dome was flanked by half domes of the same diameter which created the illusion of a very large dome.

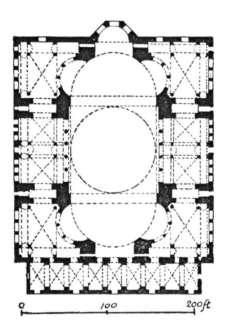

*Figure 6.2: Hagia Sophia plan shows how the dome rests upon four arches bounding a square, into two of which open the half-domes of semicircular apses that are equal in diameter to the central dome. The whole vast nave, measuring **60 x 30** meters (**200 x 100** feet), is flanked by **128** enormously wide aisles.[4]*

The imposing effect of this low-curved but loftily-poised central dome, resting on a crown of windows, would make its summit visible from every point of the nave (as seen in figure 6.3 below). The interior was paneled with costly colored marbles and ornamental stone inlays. Its interior design surpassed anything ever created.

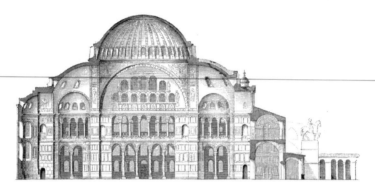

Figure 6.3: The interior of Hagia Sophia highlights how it was designed (19ᵗʰ Century print). The central dome (a height of 54 meters (177 feet)), seems rendered weightless by the unbroken arcade of arched windows under it, with no supporting walls.[5]

What is surprising with this design, of exceeding complexity, is the level of geometry used was no more sophisticated to what Archimedes had outlined several centuries earlier.

Executing

The architects had direct access to the Emperor Justinian,[6] the project stakeholder, at any time. The project hierarchy was run on the lines of the Roman guilds and included one hundred master-builders supervised by the architects, and ten thousand laborers. Either of the architects could have directed the project. The large workforce also presented a challenge in that it had to work in a relatively confined environment.

An extensive supply chain was set up as materials used in the construction were brought in from Rome and Delphi. The whole interior was built of brick and lined with costly marble. The design allowed a flood of light into building which was reflected off the marble interior to create a very well illuminated building.

To add to its splendor columns were plundered from the temples of the ancient gods (at Heliopolis, Ephesus, Delos, Baalbec, Athens, and Cyzicus). This further underlined the importance this project carried at the time for the Emperor Justinian.

To render the building as light as possible, it was constructed of pumice stone and Rhodian bricks (hollowed and porous bricks, made from lightweight clay imported from Rhodes). This added to the cost as these bricks had to be shipped in versus being produced locally. The bricks were light enough that they could float. The bricks were held by much thicker mortar, almost as thick as the brick. The arched walls reduced the volume of material, and were considerably lighter, and less expensive. The floor of the main nave was covered with grey and white marble.

The workforce discovered a cement with earthquake-resistant properties withstanding earthquakes of up to 7.5 on the Richter scale. The risk of fire through the building was mitigated in the project by not using any wood in the construction, except for the doors

Figure 6.4: Interior of Hagia Sophia shows an unobstructed hall of unrivalled proportions with a combination of half-domes increasing in span and height as they lead up successively to the central vault 54 meters (177 feet) into the air.[7] Two of the pendentives (concave triangular sections) can be seen clearly above the arches. The interior glistens with fragile golden mosaic tiles illustrating Christian figurines and scenes.

The cathedral was covered by a central dome with a diameter of 31 meters (102 feet). This was slightly smaller than the Pantheon's. It was shaped like a scalloped shell or the inside of an umbrella with ribs that extended from its top down to its base. The dome sat on four arches making a square. The pendentives were four concave triangular sections of masonry (brick and mortar) sitting between the four arches that made a smooth structural transitions between the curved tops of the four arches and the bottom of the dome. These pendentives carried the circular dome on a square base and then transitioned its weight downward to the massive marble columns that formed the corners of a square. The columns were buttressed by massive piers to restrain horizontal forces in the arches. The design was an absolute first and the geometrical complexity of this was remarkable.

A marble sheathing was useful in camouflaging the large pillars, as well as to give it a brighter look.

The dome was further complicated by forty arched windows which did reduce the overall weight. The dome was made of lighter bricks, fired at lower temperatures, making them lighter. The domed roof was made of heavy black lead. The roof stretched 54 meters (177 feet) into the air. Even the Gothic cathedrals of France did not reach these heights. The construction complexities required massive scaffolding to support this. The weight of the circular dome was too heavy for the base and it pushed it out. The change had to be adjusted from a square to a rectangle. The shape of the dome was changed to an elliptical dome. The weight of the building was so great that flakes of marble came off the columns. The architects created a huge space in the center of the building for worship.

Controlling and Monitoring

The project ran five years, ten months and four days, from February 23rd 532 to December 27th 537 and cost 20,000 pounds of gold which was a fortune at the time. This equates to $3 billion today. The speed of construction was remarkable considering the cathedrals proportions, size and complexity of the structure. It was a rush job and today this is visible in the overlapping and ill fitting pieces of marble in some areas.

Figure 6.5: Hagia Sophia it is internally one of the most perfectly composed and beautifully decorated halls of worship ever erected (19th Century print).[8]

Figure 6.6: Hagia Sophia modern image. Source: British Museum London.

Closing

The forty arched windows around the base of the dome formed a crown of light, an effect almost as striking as the Pantheon oculus. The dome seems rendered weightless by the unbroken arcade of windows, with no supporting walls.

The cathedral was shook by an earthquake in 558 and major cracks appeared in the dome and a portion of it collapsed. It was repaired by the son of Isidorus over a four year period who rebuilt the dome, changing its shape to an elliptical dome so its weight was more evenly distributed down. He also strengthened the interior walls so that they were vertical changing the shape from a square to a rectangle.

The cathedral became a prototype for many churches, for example, domed architecture spread throughout the Byzantine Empire. The most famous copy was St. Mark's, Venice (1063-1071). The cathedral is considered today to be an embodiment of Byzantine architecture and of great beauty. It also had the distinction of remaining the largest cathedral in the world until 1520. The Cathedral was damaged many times by earthquakes and fires, and had to be repaired and reinforced. In 989, workers replaced the great dome damaged through earthquakes.

In 1453 when Constantinople was conquered by the Ottoman Turks the Sultan Mehmed II ordered the building to be changed into the Ayasofya Mosque.

Other Notable Projects

The 10th Century Grand Mosque of Cordoba was the second-largest mosque in the world. Hagia Sophia influenced the Sultan Ahmed Mosque or Blue Mosque of Istanbul and the Mosque of Mohammed' Ali, Cairo, Egypt. The "Alabaster Mosque," whose lofty and graceful minarets are so conspicuous from the distance that they form one of the major landmarks of Cairo.

Toltec Pyramid at Cholula Project

The Toltec, ancient Native American polytheistic culture, were the dominant people in the region of Mexico from 900 to 1200 CE. The city of Tula (capital Tollán) at its height housed a population of around 40,000-60,000.

The Toltecs built step pyramids, with temples on top, more akin to the ziggurats of Mesopotamia than to the pyramids of ancient Egypt.

Between 900-1000 the Toltecs built the largest pyramid in the world by volume, a man-made mountain of 133 million cubic feet (4.45 million meters), a base of 450 x 450 meters (1476 x 1476 feet) and a height of 66 meters (217 feet), 300 meters (1,000 feet) on a side and 52 meters (170 feet) long covering 45 acres at Cholula. The total volume is almost one third larger than the Great Pyramid of Giza. The workforce created a radial pyramid with stairs covering all four sides so that the summit could be approached from any direction. Each group that dominated Cholula added to the pyramid (technically called a temple, not a pyramid). In the 12th Century, the civilization was gradually supplanted by the Aztec.

Piramide di Cholula

Figure 6.7: Toltecs Pyramid at Cholula.[9]

Example of a Master Plan

In a well documented example from the 9th Century, a plot plan for a large monastery was created. This was an "authorized standard" or an "ideal plan" that was then repeatedly used as a master, where it would be subject to modification as local requirements dictated. The plan was created by a council of churchmen-administrators who had to carefully think it out because of certain constraints.

As part of the requirements the monks had to be isolated from the lay workers (craftsmen, gardeners) and the surrounding public who could use some of the facilities. The plan is preserved in the chapter library of St. Gall in Switzerland[10] (see Figure 6.8). The monastic builders, with little technical training, but with plenty of willing hands, sought out new architectural paths to meet their special needs. Remote from classic and Byzantine models, and mainly dependent on their own resources, they often failed to realize the intended results.

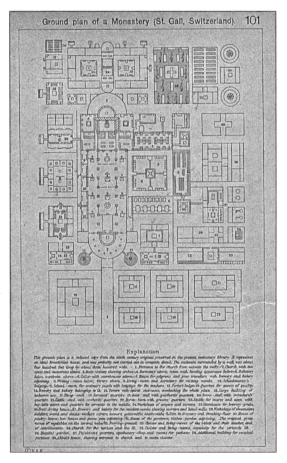

Figure 6.8: Ground Plan of a 9th Century Monastery (St Gall, Switzerland) that represents the ideal Benedictine house. It shows the detailed level of planning required to isolate the monks from lay workers (craftsmen, gardeners) and the surrounding public.[11]

Key Players

Anthemius of Tralles (Aydin) (474 – before 558 BCE)

He was a Greek professor of Geometry in Constantinople who described the string construction of an ellipse and wrote a book on conic sections. This proved to be of immense value in the construction of Hagia Sophia.

Isidorus of Miletus

Isidorus was a Greek professor of physics in Alexandria and then Constantinople. He wrote a commentary on earlier published building books. Furthermore, he may have been responsible for the T-square and string construction of a parabola.[12]

Chapter Wrap-up

Conclusions

Emperor Justinian was determined to see the success of the cathedral project. It was a project of personal prestige as the leader of the remnants of the Roman Empire. He set up the core project team with two renowned architects of the age, and a massive skilled workforce with a disproportionally high ratio of master craftsmen.

The complexities of the cathedral included the floating domes pushing the architecture and geometry to its limits. The use of materials as light as possible, pumice stone and Rhodian hollow bricks, indicate the architects fully understood the risks of the design. The use of a cement with earthquake-resistant properties and no wood to prevent fires indicate how cautious and risk adverse the workforce was. The speed of construction, just under six years, was remarkable when compared with the elongated time-lines of the Gothic cathedrals to follow in the 11th and 12th Centuries.

The project design goal of Emperor Justinian of surpassing earlier cathedrals was reached as it certainly was a larger structure than the Pantheon, and the interior had a greater space, although the dome was not as large.

Key Lessons

PMBoK® Guide Knowledge Areas:

- Integration Management

  - With a high-profile stakeholder like the Emperor Justinian, taking a hands-on-approach in the project, there was a high chance of the project scope changing. Therefore, the project required tight change control. However, the aggressive schedule constrained change and the Emperor's role may have helped to keep the project on track.

  - The project was politically motivated and under-pinned the Emperor's personal prestige. Even though the project's business justification was a minor factor the cathedral was significant to the city. Like future medieval cathedrals, it would attract visitors and pilgrims.

  - The project had to find a balance where the design was focused on the esthetics, beauty, and quality, versus the schedule and delivery (within a 6 year period), and to a lesser extent costs.

- Scope Management

  - The direct involvement of the emperor and the desire for the project to finish in his life time better controlled the scope of the project.

  - The scope in terms of materials, and equipment was dictated by:

    - The location in the middle of an earthquake zone.

    - The height, diameter and thickness of the dome structure.

    - The availability of materials, mortar and brick production (imported from Rhodes).

  - The work breakdown structure laid out the project with the following:

    - Site preparation.

    - Assembly of scaffolding, delivery of bricks.

- Construction of the structure in particular the pillars (four arches) to support the dome.
- Removal of scaffolding and site clean-up.

- Time Management
  - The aggressive time-line of five years required a massive on-site workforce throughout the project.
  - With such a large workforce the project scope had to be decomposed and the sequence of activities had to be carefully managed and scheduled because of the limited size of the construction site, and to avoid conflicts and collisions.
  - To meet the schedule the dome had to be built within the first three years of the five year project.

- Cost Management
  - The most significant cost was the project workforce.
  - Costs were controlled in several ways.
    - First, by following the traditional Roman techniques of building with a low cost core of bricks then clad with an expensive veneer of marble, and tiles.
    - Second, like the Colosseum the project made the use of bricks and arches to keep the costs down. Bricks were versatile, practical, easy to handle, and very durable.

- Quality Management
  - Like the Parthenon project quality was ingrained into all areas of the project to ensure it would meet the high standards of esthetics required.
  - The project took into consideration that the final deliverable was in an earthquake zone. The fact it still stands today is a testament to the design and the quality of the construction, and much to the team's foresight despite many earthquakes over the 1500 years.
  - Bricks were stamped with a mark for provenance.

Figure 6.9: Roman brick stamps – National Museum of Rome.

- Human Resources Management
  - The project selected two renowned architects when one would have done. Anthemius died four years into the project.
  - A workforce of ten thousand laborers for a project of this size was very significant and would have been a challenge to manage. They were divided into two teams of 5,000.
  - The architects designed spiral ramps that allowed the workforce to carry materials to the top most heights of the structure, removing the need for cranes.
- Communication Management
  - A large workforce of ten thousand laborers very much depended on communication through the hierarchy of one hundred master-builders that were supervised by the architects.

- During the European Dark Ages the Roman system collapsed and many communities were fragmented and lived in isolation. Without communication the little news that would reach a village from outside would take weeks or months, and be cried by a town crier.

- Risk Management

  - With Hagia Sophia risk was somewhat mitigated by appointing the two most famous architects of the age and a very large workforce.

  - The design pushed the limits of geometrical knowledge.

  - The weight of the circular dome was so heavy it pushed out and distorted the base. The repairs made by the son of Isidorus in 558 changed it from a square to a rectangle and the circular shape of the dome to an elliptical dome.

  - The architects understood the impact of earthquakes on large buildings and incorporated the following into the design to mitigate the risk[13]:

    - They made the building light and flexible, a modern concept. They realized that forces in a dynamic system are proportional to mass.

    - They utilized cement with earthquake-resistant properties with calcium silicate that could reform as cracks appeared. This is similar to that used in today's Portland cement.

    - They used light Rhodian hollowed bricks made from lightweight clay.

    - They used crushed brick in the mortar that gave it a high tensile strength.

    - They added 40 windows into the dome to avoid cracking.

    - They used thick mortar joints thicker than the bricks which made the material more like reinforced concrete.

- They secured the building from fire as no wood was used in its construction except for the doors.

- They created shock absorbers using lead at the foundation of the major columns carrying the dome.

- Procurement Management

  - Certain materials that could not be secured locally and had to be procured from far and wide. For example, Rhodian bricks were selected for their lightness and imported from the island of Rhodes by boat. They were light enough to float.

  - The total number of bricks used for the project kept the Rhodian brick makers busy for years. It was essential that the project had a priority to this supply of bricks.

  - Other material that had to be procured in quantity include:

    - Volumes of copper and iron. Recent conservation work of the dome mosaics has found quite a considerable number of copper and iron nails and cramps used in the project.

    - Estimates for the weight of glass used in the tesserae (mosaics) is well over 400 tons.[14]

Educators

- Discuss in the context of today's projects the impact of the fall of the Roman Empire on the ability to continue projects of scale?

- How well was risk managed with the Hagia Sophia project?

- Discuss the pros and cons of having two renowned architects.

- Discuss why the pressure of the project sponsor, Emperor Justine, actually helped the project.

CHAPTER 7

The Central Middle Ages (1050 - 1300)

T his historical period was known as the Early Gothic Era and brought to an end the Dark Ages in Europe which was a period of stifled growth. The period saw the urbanization of Europe, military expansion (Crusades), and intellectual revival. China was undoubtedly the most technically advanced region in the world, particularly with regard to the use of coke in iron smelting, canal transportation, and farm implements. Bridge design and textile machinery had also been developing rapidly.[1] From a project management perspective in Europe, there was a rediscovery of lost ancient knowledge that inspired a period of learning and growth. It also provided a new confidence for the master-builders when examining the past, for example, the architectural achievements of ancient Rome. The period was dominated by the construction projects for the Gothic cathedrals driven by the Christian church and the towns' vision for economic prosperity. There was also a boom in castle building that extended through the period.

In this Historical Period

- Brief synopsis of trends/changes (natural disasters, wars, technology, epidemics).

    a. Most projects were controlled by the church but the state starts to emerge from its predominance.

  b. Climate changed with a warm up.

  c. Large scale urbanization in China due in part to the Grand Canal.

- Impact of changes

  a. The population in China started to boom with large scale urbanization.

  a. Lower level of economic activity in Europe by 1250.

- Major events

  a. Leif Ericsson reached Newfoundland, 1000.

  b. Norman Conquest of England, 1066.

  c. First Crusade to the Holy Land, 1096.

  d. The Grand Canal opened in China in the 11th Century.

  e. European rediscovery of lost Greek knowledge, 1100.

  f. Genghis Khan founded the Mongol Empire, 1206 and China fell under the Mongol Empire.

- New tools, techniques and breakthroughs

  a. Viking Technology.

  b. Chinese Technology.

  c. The Agricultural Revolution of the Middle Ages.

  d. Water driven bellows lead to hotter and larger furnaces.

  e. New building techniques developed for cathedrals.

- Regions

  a. Europe.

  b. China.

- Significant projects

  a. Great Gothic cathedral projects (Chartres, Durham).

- Other Notable Projects

  a. Great castles in Europe.

    b. Angkor Wat.

    c. Grand Canal of China.

    d. Viking voyages to Greenland and Vineland.

- Key players

    a. William of Normandy.

    b. Villard de Honnecourt.

- Chapter Wrap-up

Trends and Changes

The world population in this period grew from 310 to 360 million. In 1000 there were 22 million people in Europe, 60 million people in China, and 79 million people in India, and 40 million under Islamic governance. Baghdad, the largest city in the world, grew to a population of nearly 1 million. It was the centre of the Islamic Empire after 750.

Church versus State

During the Dark Ages in Europe the church was the centre of each community. Men lived in daily contact with religion. The church was the centre piece for a population that relied on the spoken word (99% illiteracy). The church was a conduit of information for villages. The church provided a support system that could influence, through parish priests and the pulpit, people's daily lives and intellectual horizons. In these fragmented communities most people lived within an 11 kilometers (7 miles) radius of where they were born, for most of their lives. The Papacy reigned supreme and Papal taxation on land was very high. For example, by 1279, the Catholic Church in England had amassed enormous wealth through taxes and land.

Merchants were also a key factor in the growth of the medieval towns. Guild merchants organized themselves to fight for the rights of the city, and their leadership was rarely challenged. Gradually, new forms of government evolved in towns with a mayor and council. Merchants demanded a communal freedom of the towns from the feudal lords. In

England monarchs granted town charters of limited self-government in return for their support.

Climate Change Warming up

In Europe during the eleventh and twelfth centuries a slight warming of the climate and improved agricultural techniques allowed lands that had previously been marginal or even infertile to become fully productive. Europe enjoyed an economic and agricultural boom. Dry and warm summers prevailed in Northern Europe in the first half of the 13th Century but in 1250 the climate cooled off again.

Large Scale Urbanization in China

In the 11th Century the population in China started to boom due in part to the development of the Grand Canal (see page 228) and the ability to transport quantities of rice, the population spiked to an estimated 115-123 million. Urbanization increased to 20% of population, and shifted south, with the growth of the leisured middle classes.

Government Run Projects in China

In some societies a centralization of power controlled projects and the availability of human labor. For example, in China the population started to boom in the 11th Century with large scale urbanization due in part to the Grand Canal.

Impact of Changes

Chinese Population Boom and Large Scale Urbanization

In parallel, new technologies like moveable printing evolved in China (in 1045), made from an amalgam of clay and glue by a Chinese alchemist named Pi Sheng.

Under the Sung dynasty centralized power in the hands of the Emperor through a huge and monolithic a governing bureaucracy which was huge and monolithic with over 100,000 civil servants. This gave the Emperor control down to the village level, and there was no authority to challenge this control. The omnipresence of the Chinese Mandarinate (civil service) restricted any projects outside of official channels.

Lower Level of Economic Activity

In Europe by 1250 the climate cooled off again and the decreased agricultural output could no longer support the same level of economic activity. As early as the middle of the 13th Century the economy began to weaken. The 14th Century saw a declining population, shrinking markets, a decrease in arable land, and a general mood of pessimism as economic conditions deteriorated.

Major Events

Discovery of Vineland (1000)

Leif Ericsson reached the shores of Vineland (modern day Newfoundland) and created a Viking settlement at L'Anse aux Meadows which is somewhat successful. However, their technology advantages were not significant enough for them to hold on for more than several years (see page 205) before being ousted by the native population.

Norman Conquest (1066)

William of Normandy (the Conqueror) conquers England. This event was significant as it shifted power in England and was a catalyst for change. The Domesday Book was a record of the great survey of England completed in 1086, executed for William the Conqueror. It was similar to a government census today. Its purpose was to determine who held what, and what taxes had been liable under the defeated King, Edward the Confessor. This influx of tax revenues allowed William to initiate a building boom of the great Gothic castles (see page 225). He built more than 40 castles scattered throughout England and a ring of castles around Wales (see page 228) to keep the Celtic nation at bay. His successors continued what he started. Military architecture developed into a new science of building and in the 11/12/13th Century, as hundreds of imposing castles were built all around Europe often artistic as far as military priorities would allow.

Fig. 35.—William, Duke of Normandy, accompanied by Eustatius, Count of Boulogne, and followed by his Knights in arms.—Military Dress of the Eleventh Century, from Bayeux Tapestry said to have been worked by Queen Matilda.

Figure 7.1: William of Normandy (the Conqueror) conquered England 1066 and the event was a catalyst for a massive building boom for more than 40 castles.[2]

First Crusade to the Holy Land (1096)

As Europe came out of Dark Ages there was a movement that united European countries morally. This was driven by religious impulse.

China Opens Grand Canal Project

(See page 220.)

Knowledge Refound in Europe

In the 12th Century the Europeans in Spain rediscovered ancient Greek and Muslim knowledge from the Muslim civilization. It included the major scientific disciplines (Logic, Philosophy, Engineering, Architecture, Maths, Geometry, Astronomy, Medicine, Pharmacology, Botany, Anatomy, Zoology, Physics, Chemistry, Optics, and Meteorology) and the experimental sciences. This acted as a catalyst and the first European university offering a degree (6 years of study) was set up in Bolognia in 1088. Paris University was set up in 1160 under a new rationalism where logic and free thinking were accepted. Paper was first manufactured in Europe (Spain) in 1200, using techniques from the Islam world, and the cost of parchments plummeted. In 1220-1244 Vincent of Beauvais created the Speculum Majus, the most comprehensive general encyclopedia of knowledge until the 18th Century.

Through the 12th Century the universities revived Roman law which in turn, became the foundation of law in all civil law jurisdictions.

Genghis Khan Conquers China

In 1206 Genghis Khan founded the Mongol Empire and started the Mongol invasions and raids, to eventually occupy a greater part of Central Asia and China.

New Tools Techniques and Breakthroughs

Viking Technology (900-1000)

The Viking's developed seaworthy ships and reliable navigational techniques that allowed them to take extensive sea journeys. The use of a broad ax over a saw enabled an oak tree trunk to be split into long, thin planks. These were then fastened with iron nails to a single keel and each other where each plank overlapped the next. This differed from the more conventional method of building an inner skeleton for the hull. Evenly spaced floor timbers were affixed to the keel and not the hull to provide flexibility. They then added crossbeams to provide a deck and rowing benches, and secured a massive beam along the keel to support the mast.

This technique contributed to a very unique construction where the longships were light, and achieved speeds of up to 14 knots. They were highly maneuverable and could easily navigate shallow river estuaries. They could reef square sails in strong winds and adjust these for rapid tacking. The larger of the longships measured up to 28 meters (95 feet) in length.[3]

Indian Technologies

In the 12th Century Indian blacksmiths created iron girders and beams of a scale not seen anywhere else in the world at the time. The temple of Puri contained 239 iron beams and the temple of Konarak had one beam that was 14 meters (35 feet) long. Similar to the Delhi iron pillar, the beams were 99.64% iron.

Chinese Technologies

In about 1040 a Chinese manual on warfare was issued under the title Compendium of Military Technology. It was the first document to describe gunpowder. The earliest reference to the compass is in a Chinese manuscript of the late 11th Century; within the next 150 years it was referenced in Arabic and European texts. Around the 10th Century paper money (a promise to pay gold or silver or bronze in exchange for this piece of paper) was first used in China during the Sung Dynasty. Moveable printing using wooden and ceramic moveable type was developed in China in 1045. Another remarkable technology was Su Song's astronomical clock tower completed in 1088. It had a clepsydra tank (measured time by marking regulated flow of water), waterwheel, escapement mechanism, and a chain driven armillary sphere with 113 striking clock hammers for sound and displaying plaques.

The centre of technological innovation was China where iron production became a state enterprise under the guidance of metallurgists. Production rose from 13,500 tons in 806 to 125,000 in 1078 because of limited resources of copper and tin for bronze. In contrast England produced 68,000 in 1788. Chinese iron industry used a water powered bellows to provide a blast and smelted the ore with coke.[4]

Figure 7.2: This medieval printed illustration from China depicts waterwheels powering the bellows of a blast furnace in creating cast iron.[5]

The abacus arrived in Europe in 1200 and provided a tool for making rapid and accurate financial (mathematical) calculations, useful in managing and controlling project budgets.

The Agricultural Revolution of the Middle Ages

In Europe the introduction of the horse shoe, modern harness and collar (making specially bred farm horses more effective in ploughing and pulling loads than oxen), the three year fallow system, the heavy wheeled plough and other innovations contributed to a large increase in food production. The effects of this on the diet and living standards of the people were considerable, when subjected to modern nutritional analysis.[6] By 1300 the population of Europe trebled from 26 million to 79 million. The introduction of the farm horse greatly reduced the manual labor involved, and freed it up for other labors.

Fig. 18.—Ploughmen.—Fac-simile of a Miniature in a very ancient Anglo-Saxon Manuscript published by Shaw, with legend "God Spede ye Plough, and send us Korne enow."

Figure 7.3: The heavy plough created a more cooperative peasant society and caused small hamlets to combine into larger villages in order to share ox teams.[7]

Water and Wind Technology

In this period waterwheels arrived in Europe and by 1086 William the Conqueror recorded some 5,624 water mills in England. Over a third of manor houses owned one.[8] The technology spread rapidly across Europe and by the 13th Century there were some tens of thousands in France. As the technology evolved with the use of uniform machine parts and gears, the first clock towers appeared (First Public Striking Clock is in Milan in 1335). The technology increased the need for more precise measurements.

In 1200, windmills arrived in Europe from Islam. Consider the assessment provided by the historian Jean Gimpel:

> "There were so many windmills, bringing in such high profits, that Pope Celestine HI (1191-1198) imposed a tax on them. (Gimpel 1976, 25)." [9]

Keeping Time

During the Early Medieval Era or the Dark Ages in Europe had virtually no calendars, clocks, or written records. The world moved at the pace of nature. Units of less than one hour were rarely used. In some towns, where sundials were available, watchmen called the time on the hour. It was the monasteries that drove the mechanization of time in the 10th Century, using water clocks to support the routine of daily monastic life. The monks adhered to a rigorous schedule of prayer and work. Time was divided into hours and minutes[10] for the first time in 1345.

New Building Techniques for Cathedrals

A number of innovative building techniques were a contributing factor behind a building boom in Cathedrals across Western Europe.

Pointed Arch

An improvement over the Roman arch was the Pointed Arch which was introduced into Europe from Moorish Spain. The Islamic pointed arch was made by forming each side of the arch from a different centre point, the greater the distance between the two points the sharper the point. The center line was more closely aligned to the forces of compression and much stronger.[11]

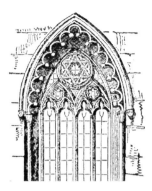

Figure 7.4: Early Gothic Pointed Arch.[12]

Flying Buttress

Many of the medieval cathedrals did not have solid stone walls.
Even though the walls appeared to be 2 meters (6-8 feet) thick
of solid stone, they were actually constructed in 3 parts. The
inner and outer surfaces were of stone blocks, while the centre
cavity, using wood and chain as reinforcement, was filled with a
mixture of mortar and small stones. The Romans had used arch
buttresses, at the Baths of Diocletian[13], to control the outward
thrust of a wall of a building. These evolved into Flying Buttress
which supported thinner walls and allowed for more stained
glass in the walls.

Figure 7.5: Early Gothic Flying Buttress.[14]

Status Quo in Construction Techniques

For 250 years (from the end of the 13th Century to the beginning of the 16th) there was little technical progress made in the construction techniques of Gothic cathedrals.[15]

Regions

China was the predominant world power in terms of technology. Islam's empire stretched from China across India and the Middle East, into Africa and Europe. Europe comes out of the Dark Ages.

Significant projects

This medieval period in Europe was dominated by Gothic cathedral projects. From 1050 to 1350 in France alone, 80 cathedrals and 500 churches were built.

Gothic Cathedral Projects

Background

Through a period of several hundred years cities in Western Europe strived to out do each other in creating the ultimate cathedral in splendor and height. In these years more stone was excavated than at any time in ancient Egypt.[16]

Initiating

The cathedral boom was spurred by the competition for pilgrims between cities. The cathedrals had a major impact on the prosperity and importance of the city as they attracted thousands of pilgrims. The merchants in a city recognized the value a cathedral could bring. In this period world record fever gripped the cities as they poured in resources for their own cathedral projects. It was broken five times within 62 years.

Start	Finish	Height	Cathedral
1152	1240	32.8m	Notre-Dame Paris
1145	1220	36.55m	Chartres
1209	1290	37.95m	Rheims
1219	1269	42.3m	Amiens
1247	Not completed	48m	Beauvais (collapsed in 1284)

Table 7.1: World Record Fever Led to a Race Between Cities.

The cathedrals were a showcase for the church, reaffirming its power and authority. The towering walls of stained glass windows poured colored light in, an amazing spectacle in the medieval world, and used to inspire the congregation. These windows narrated stories and transmitted information to all inside. Chartres had over 3,000 meters^2 (32,292 feet^2) of stained glass.

The Gothic cathedral projects elapsed long time spans, over generations, (some over a hundred years, very few were completely finished), and required long-range planning as the output had to be built with painstaking precision and incredibly complex detail. Cutting corners carried great risk and could lead to calamitous consequences, in lives and having to redo years of work. A cautious and safe approach was required daily.

Governance and Project Team

The project came under the direction of a Bishop and a group of clergymen known as a Chapter who controlled the finances. The chapter included a number of dignitaries a chancellor, who acted as secretary, a treasurer, and a cantor who organized religious services. Because of the time scales the costs had to be tightly controlled, and the project budget was closely managed. The chapter's role was also similar to a director of urban redevelopment and responsible for compulsory purchasing, and adjudication. In addition, the chapter appointed a director of the site, or an overseer who would be a works accountant. Chosen for his business acumen and knowledge of architecture he was responsible for the accounts the income and expenses (see table 7.2). The chapter searched for and appointed a master-builder (an architect) to design and supervise the project. Sometimes the search would take years. The master-builder hired the master craftsmen including master quarry men, master stone cutter, master sculptor, master mortar maker, master mason (free masons), master carpenter, master blacksmith, master roofer, and master glass maker. Each master-craftsmen ran a workshop for their trade and had many assistants and apprentices. The on-site workshops were known as lodges and were used as the living quarters of the masons, where they would "lodge."

#	Description – Income	Sum
1	Taxies levied on the chapter at Autun	
2	Income from vacant benefices in the city and diocese of Autun which, by the authority of the Holy See, has been assigned to the cathedral works	
3	Income from indulgences granted to benefactors to the fabric	
4	Income from the collection and from the Saint-Lazare brotherhood at the Pentecostal Synod	
5	Casual offering not accounted for by the works fund	34 pds 19 sous
6	Income from collection boxes designated for the fabric	10 pds 17 sous
7	Additional item deducted from the collections at Autun Cathedral	42 pds 13 sous
	Total =	400 pds 9 sous
#	Description – Expenses	Sum
1	To the quarries for the excavation of stone	8 pds 10 sous
2	To the same for a year's supply of lime	9 pds 8 sous
3	For the cutting and transportation of cask wood for vaulting in the church (carpenters and laborers)	17 pds 2 sous
4	To the forge at Autun for the year	42 pds 10 sous
5	To the quarry forge	
6	To the laborers for opening the said quarry	4 pds 15 sous
7	For investigations into the situation at the quarry atMarmountain	1 pds 10 sous
8	To the laborers who laid the tiles of the roof of the church	9 pds 11 sous
9	For the poles for the rafters	5 sous
10	For the making of 12 carts including the ironwork	1 pds 15 sous

11	To the carpenters for the cask wood cut in the chapter's forest	8 pds 16 sous
12	For repairing the roof of the church	9 pds 11 sous
13	To the carpenters for lathing at the church	10 pds 8 sous

Table 7.2: Robert Clavel's statement of accounts for the cathedral at Autun, gives a very good indication of the scale of the project budget and the on-going project procurements. As an overseer he was responsible for providing the working site with raw materials, manufactured goods, and the transporting of goods, and the organization of workmen.

Guilds

A very important feature of medieval society were the trade-guilds which embraced both the guilds-merchant and the craft-guilds. The latter were associations of workmen, for maintaining the customs of their craft, each with a master, or alderman, and other officers. The trade-guilds had their provisions for mutual help for themselves and for their widows and orphans, and they had their religious observances. Each had its patron saint, its festivals, and its treasury. They kept in their hands the monopoly of the branch of industry which belonged to them. They had their rules in respect to apprenticeship, etc. Almost all professions and occupations were fenced in by guilds.[17]

Each craft was a separate guild where the skills were secret and closely guarded, a "closed shop", which you could marry into or entered through an apprenticeship under a Guild Master (Craftsman). The apprentice would learn the craft in one place and then later spend time as a "Journeyman" gaining experience from job to job. They were highly mobile traveling from town to town. Eventually, they became a guild master member when their work was considered to be the equal of a master craftsman and they could train their own craft apprentices. Master craftsmen had no technical education but were of wide experience. Knowledge that was gained was not written down but passed aurally. Each workshop had specific tools. All metal tools were made by the blacksmith and wooden pieces by the carpenter.

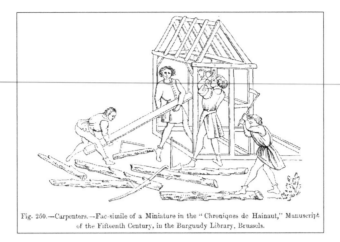

Fig. 250.—Carpenters.—Fac-simile of a Miniature in the " Chroniques de Hainaut," Manuscript
of the Fifteenth Century, in the Burgundy Library, Brussels.

*Figure 7.6: Medieval Projects were wholly dependent on the
organization of guilds - the companions here (carpenters) belonged to
a community of their own special craft.*[18]

*Figure 7.7: Wood engravers in Germany during the latter half of the
16th Century.*[19]

The heavy work was carried out by laborers with no particular trade or skill. They were at the bottom of the medieval ladder, but had the opportunity to better themselves. They could become a specialized craftsman or save money and set themselves up as a contractor. Often, they were recruited from the rootless like serfs fleeing their feudal lords. If they were not found within a year and a day, they became freemen and citizens of the town. Records show the work given to the laborers varied: they transported cask wood for carpenters, dug the quarries or the foundations, carried a variety of materials in baskets (panniers) on their backs and took the tiles to the roofs. Stonecutters and masons (specialized workers) usually had a number of laborers to help them called servants or assistants, for example, making of mortar and plaster.[20]

Planning

The master-builder would draw out the design on a parchment or two pieces of plaster that would then be approved by the bishop and chapter. One would have the floor plan and the other the wall elevation. The master worked in the tracing house set up with trestle tables and a slab of plaster of Paris on the floor on which the large drawings would be scratched out.

At the core of the design the master-builder would use three shapes namely, a circle, square, and equilateral triangle. Every element of the design incorporated these shapes. Starting with a base line a small scale design would be set up. Using the methods of rule of proportions and counting units the rest of the plan would follow.[21]

One of the main challenges in planning the Gothic cathedral was to consider how the slender walls with their massive stained glass windows, would support the weight of the massive roof structure.

Executing

Trenches were dug for the wall lines and pits for the columns by the laborers. Rough stone would be tipped in and flooded with lime mortar to make a concrete foundation. A ceremony marked the laying of the "foundation stone" usually by the bishop or notable person, or major contributor to the project. Hewn stone was supplied from local quarries (within an 8 kilometer (5 mile) radius) after it had been inspected by the master. Sometimes it would be procured overseas (Caen) if unavailable locally. Quarry men worked in groups of eight under the direction of a master quarry man. Typically, a cart of stone would leave the quarry every 15 minutes, and weigh about 1 ton.[22] The high cost of transport made it more cost effective to send the stone cutters to the quarry to square off stones according to certain measurements. Working masons would shape these stones cutting a true flat face on the front and beds on both ends - top and bottom. The front face was tooled with a regular pattern on the surface of small pits or grooves from side to side.

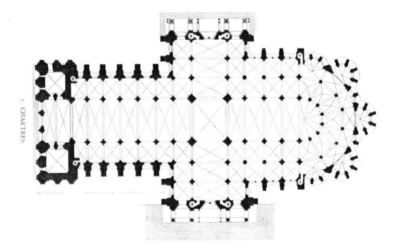

Figure 7.8a: Chartres floor plan built over a 75 year period.[23]

Gothic cathedral projects were planned and executed in logical segments, that defined an overall scope. These consisted of numerous phases that built the cathedral in sections to an order in the erection procedures (see figure 7.8b). In any of the intermediate phases, the sequence of the construction operations, like the dispositions of scaffoldings, and materials, had to assure a safe state of equilibrium. So, for example the lateral naves had to be built before the main (central) nave.[24] As sections of the cathedral were completed, they could be put into use for church services.

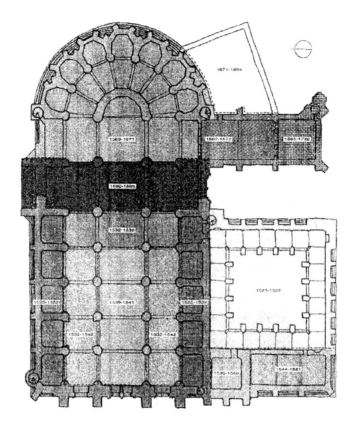

Figure 7.8b: Phases of construction delivery of Segovia Cathedral that show sections with the various completion dates (1525-1607).[25] These became functional as they were built.

As the walls rose to five feet much scaffolding had to be built by the carpenters, made of poles lashed together with rope, and then hoists were attached. The scaffolds also held work platforms for the masons made of mats of woven twigs called "Hurdles," which could be easily moved. Small stones were carried by basket. Larger ones were hoisted using rope and pulley. The carpenters also built wooden centers for the arches based on the masters drawings. Sometimes, where wood was more predominant than stone as a building material, the master carpenter was required and took a more prominent role than the master mason in the construction project.

Other significant contributors to the project were the blacksmiths who made the iron bars to strengthen pinnacles, and to tie together the sides of arches. Also plumbers laid lead gutters and roofs of case lead sheets.

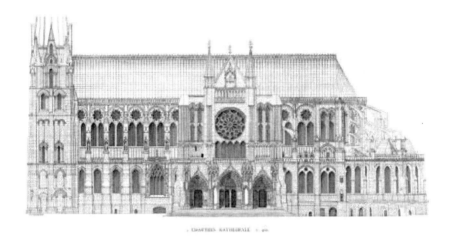

Figure 7.9: Chartres, South elevation.[26]

Skills of a Master-builder

One of the most important traits of the master-builders was that they had to have a comprehensive grasp of the totality of building operations, and all their ramifying complexities, so that they could sequence the activities most efficiently. They needed a thorough knowledge of all related building trades.[27] They carried the simplest of tools a measuring stick or rod, a straight-edge and ruler, a pair of calipers, compass, square, angles, proportional divider and string. What they lacked in technology they made up in ingenuity and personal skill. Overall they had to read and write, understand building operations and geometry, and manage the project workforce on site. At Amiens there were over 50 sculptors working at one time.[28]

The breaks in construction due to a lack of funds meant that the master-builder had no option but to move to another project and then be replaced when construction resumed. Without one single master-builder responsible for a project, from beginning to end, new masters would have to adapt their plans to what was already constructed. On the Chartres project, up to nine different master-builders[29] have been identified. These nine did not all exert the same influence on the final project and this change of leadership affected the project. These masters did not all use the same standard of measurement and had the added complication of continually making adjustments in their calculations based on what was done before. The process was done over a thirty year period and the final result stands as a magnificently constructed building that is unified through a variety of themes and plans.

Figure 7.10: Carvings around choir ambulatory, Chartres (there are eighteen hundred statues, and almost as many delivered with the projects at Amiens and at Rheims and Paris).[30]

One reason for the superiority of French figure sculpture in the 13th Century is that the French used models, as preserved in the sketch book of mediæval architect Villard de Honnecourt with studies from life.

Sophisticated Technologies

The master-builder of Chartres outlined new principles, through sophisticated technologies like flying buttresses that supported the slender walls and allowed more stained glass to be incorporated. The walls appeared to be made of glass and were the star attraction, magnificent in the brilliance of the holy light. The massive stained glass windows (up to 10 meter (32 feet) in diameter) were very labor intensive from the creation of the stone tracery to the manufacture of colored glass, which was cut to a pattern and then set into a lead frame.

Figure 7.11: Medieval Glass Makers of [31] Chartres produced over 3,000 meters² (32,292 feet²) of stained glass.

The Pointed arch also evolved from the Roman arch. The point on the arch further helped distribute the weight through the sides of the arch making it stronger and better able to take more weight. The cathedral at Chartres would inspire all the great architects of the 13th Century.

The workforce also used sophisticated technologies for a wide range of equipment like a human powered (treadmill) crane as a hoisting device that could raise loads of considerable weight of stone masonry to heights of 60 meters (200 feet).

The use of beasts of burden was common were the medieval workforce would have used an ox and cart to transport stone to the cathedral site. In Laon (in the Picardy region of France) oxen were considered so important they were immortalized as sculptures in the towers.

Weeks	Stone-cutters	Monumental Masons	Masons	Carpenters	Sanders	Smiths	Glass-makers	Roofers	Labourers	Total
Feb 1 – April 18	74	45	24	4	13	20	15		131	326
April 28 – May 4	29	14	20	32	13	19	14	4	150	295
May 5 – 12	39	15	26	32	13	19	14	4	176	338
May 12 – 18	39	15	26	32	15	17	14	6	200	364
May 19 – 25	41	16	31	32	15	17	13	6	213	384
May 26 – June 1	41	16	31	32	15	17	13	6	213	384
June 2 – 8	41	16	35	33	15	18	13	6	213	390
June 9 – 15										
June 16 - 22	42	18	28	33	15	17	14	4	220	391
June 16 - 22	42	18	28	33	15	17	14	4	220	391

Table 7.3[32]: Men Employed at Westminster Abbey Resource Calendar shows how the project was affected by the seasons.

Similarly, in the same era Chinese author Li Jie (1065-1110), for the Directorate of Buildings and Construction, created The Yingzao Fashi[33] a Treatise on Architectural Methods or State Building Standards. He estimated the monetary costs of hiring laborers of different skill levels and types of expertise in crafts.

His *estimates* are on the basis of a day's work and include the materials needed, taking into account the season in which the work is done.

Closing

The end results of the project were stunning. The spire at Strasbourg Cathedral reached 142 meters (almost 500 feet). The nave at Beauvais Cathedral reached 48 meters (150 feet) and could hold a fourteen floor building. When a Gothic cathedral was completed it was by far the largest indoor space in the town. The buildings were vast caverns of space, through pointed arches and vaulting for a high roof. At Amiens the cathedral floor space of 7,700 meters^2 (8,500 yards^2) could hold the entire population of the city some 10,000 people. The buildings were used for numerous purposes not just religious, for example, public or civic meetings, magistrates or heads of guilds meetings, or fairs with stalls. The Gothic cathedral allowed men from all social classes to meet from kings, to noblemen, to peasants.

Figure 7.12: Interior of Notre Dame highlights the ratio of glass to stone over the height and length of the building.[34]

Figure 7.13: Cologne Cathedral, Germany is probably the most magnificent Gothic edifice in the world. The foundation-stone was laid 1248. It was finally completed in 1880.[35]

Other Notable Projects

Great Castle Projects (in Europe 12th/13th Century)

William of Normandy conquered England in 1066, and this event was the catalyst for a massive building boom of castles across England including in a ring of castles around Wales. Hundreds were built across Europe and the Middle East. The most notable were Conway Castle in Wales, Castillo de Loarre, Aragon, Spain, and Krak des Chevaliers, Homs, Syria.

A construction project for a castle was not that different to a Gothic cathedral in terms of approach, although it required fewer skilled men but a larger workforce because the time span was much shorter. When Beaumaris Castle was built in Wales between 1278 and 1280, 1,630 workmen were employed including: 400 masons, 30 smiths and carpenters, 1,000 laborers and 200 carters. The number of specialized workmen on this site was comparatively low (25%) whereas on a church site, such as Westminster Abbey, this would be as high as 50% among the workforce. These projects provided much needed employment.

217

However, conscription was used, particularly in England, where the King had the power to recruit 25 to 40 men for the site of a fortified castle which might be several hundred miles away.[36] Obviously there was a lot of resentment and it didn't do much for project morale, but these projects were deemed as high priority to national security.

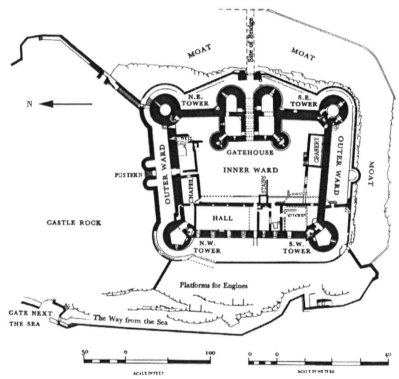

Figure 7.14: Layout of a Gothic castle Harlech, Wales.[37]

In 1270 concentric castles appeared that were more effective against well-elaborated sieges and could resist a besiegement for more than a month. Concentric castles were built with a well for water and carried a large amount of provisions. They had up to four lines of defense which incorporated farmland, fruit trees and wells. A breach in the wall could be held off by archers firing from the top of a keep. They were also located near the sea or lakes so provisions could be brought in even when under siege.

Angkor Wat Project (1113-1145)

A temple at Angkor, Cambodia, was built for King Suryavarman II in the early 12th Century as his state temple and capital city. It was completed in 1145 in only 32 years by a project workforce of up to 50,000. The world's largest religious site has a moat and an outer wall of 3.6 kilometers (2.2 miles) in length. It was built from sandstone (over 5 million tons) that was transported from a quarry approximately 40 kilometers (25 miles) away by raft along the Siem Reap River. The project time scale was significantly shorter than that of the French Gothic cathedral projects. It was at the center to one of the largest cities in the pre-industrialized world that was approximately 700 kilometers^2 (400 miles^2). It carried a complex network of canals, 1,000 man-made ponds, and roughly 70 long-lost temples. Even though the rainfall was ample to support agriculture for part of the year the water engineering works were required as there was a long dry season. By building reservoirs to retain floodwater, irrigation of an extra rice crop became possible by utilizing a complex network of canals.[38]

Figure 7.15: Angkor Wat, the highest tower (in shape of the Lotus flower) is taller 43 meters (141 feet) or 65 meters (213 feet) above the ground) than the tallest Gothic cathedral in Europe.[39]

Grand Canal of China Project

In China, paper money was first issued in the 10th Century to become the dominant currency in the 12th Century which fueled economic growth and a rise of major projects. Boats on the canals were used for transporting grain to the South. Canal traffic increased with grain and iron barges.

The Grand Canal was a building project of ancient Emperors similar in scope to the Great Wall. It's the world's longest man-made waterway which spans over 1,800 kilometers (1,150 miles) long connecting the cities of Beijing in the North and Hangzhou in the South. The Canal connects the Yangtze and the Yellow River valleys that flow from west to east, containing 24 locks and 60 bridges.

Figure 7.16: Shows the world's longest man-made waterway, over 1,800 kilometers long, connecting the cities of Beijing in the North and Hangzhou in the South. The Canal connects the Yangtze and the Yellow River valleys.

The waterways of eastern China went through three building periods which began in 486 BCE during the Zhou Dynasty,

220

extended during the Qi Dynasty (dates unknown), and later by Emperor Yangdi of the Sui Dynasty during six years of furious construction from 605-610. The final period was prompted when China's "breadbasket" shifted from the wheat and millet producing regions of the north to the rice fields of the south. This final canal section was designed to supply the armies protecting the northern frontiers, improve the administration, and increase the economic interdependence of the north and south.

With primitive building techniques (digging with simple tools), conditions were harsh for the workforce. It is estimated that half of the 6 million men (peasants) died building the Canal. However, the Canal accomplished for China what the real Nile had done for Egypt thousands of years ago. It integrated the economies of the north and south, and China emerged as the most powerful state in the world.

Figure 7.17: Grand Canal of China (19ᵗʰ Century photo).[40]

In 1327 the Grand Canal stretching 1800 kilometers (1,150 miles) was completed.[41] The canal's main purpose was moving rice to the empire's wheat-growing north.

Viking Voyages to Greenland and Vineland (circa 1000)

The Viking's relied on the sea for their raiding, attack and escape. This was a motivation for them to develop seaworthy ships. The Knar was an ocean-going cargo vessel, higher and wider in relation to its length, with a limited numbers of oars, and cargo decks installed fore and aft. With experience and more reliable navigational techniques they were able to travel increasingly longer distances over open sea and ocean up to 800 kilometers (500 miles). Through a string of settlements and outposts in Iceland and Greenland they were eventually able to reach the shores of North America almost 500 years before Columbus.

Eric the Red, having been banished from Iceland for a series murders, was the first to venture and sail west in 986 to Greenland. Eric's son, Leif Ericsson, continued exploring and in 1000 sailed southwest from Greenland to the islands off northern Canada and the shores of Newfoundland.

Figure 7.18: L'Anse aux Meadows the site of a Viking settlement in Vineland (Newfoundland).[42]

The Vikings' visits to Vinland ("wine land" for its wild grapes) are recorded in the Norse sagas. They found the land inviting, so they stayed through the winter before returning to Greenland. Leif Eriksson's brother Thorvald led a second expedition in 1003, and the following year a colonizing expedition of 130 Vikings was abandoned after warfare with the native Indians. The final expedition was led 1013 by Erik the Red's daughter Freydis.

The Vikings didn't stay longer because their technology advantage was not significant enough against a hostile local population for them to hold on more than a few years. These expeditions (projects), were well planned and executed and not just meandering. The Vikings could support their far-flung trade networks and the colonization of Iceland, Greenland, and North America.

Key Players

William of Normandy (1027-1087)

His invasion of England sparked off a building boom in castles and churches. He built more than 40 castles scattered throughout England and a ring of castles around Wales. His successors continued what he started which led to hundreds of imposing castles built all around Europe.

Villard de Honnecourt (1200s)

His notebook/model-book[13] provided insights on a wide range of topics that were part of a tool kit of a cathedral master-builder. These included architectural plans, elevations and details, ecclesiastical objects and mechanical devices like a mill-driven saw, a number of automata, and lifting devices. Villard's detailed sketchbook of construction work was based on his travels through many cathedral building-sites. He likely visited in France the cathedrals of Cambrai, Chartres, Laon, Meaux, Rheims, the abbey of Vaucelles, and the cathedral of Lausanne in Switzerland. He lived in Picardy in northern France.

Chapter Wrap-up

Conclusions

The most striking thing about the Gothic cathedral project was the commitment to keep the project going and finishing it off, even if it ran into decades, and centuries. Many projects were built in parallel competitively across cities and towns in specific regions. The projects did not have the resources of a nation, or a city-state but those of a small town and community, for which they were incredibly important. The scale of the building at 13 stories dwarfed the surrounding 2 story town. The engineering complexities of the cathedrals were in several areas like for example in the ratio of glass to stone over the height and length of the building, for the painstaking precision required, and the incredibly complex detail. The construction sites were dangerous places and the workforce had to be very adept in such an environment.

This period also saw a boom in building projects as Great castles were built across Europe, similar projects to the Gothic cathedrals but more labor intensive. They did not have the same constraints of man power availability as the master-builders, under the king's authority particularly in England, had the power to recruit or conscript up to 40 men within a radius of several hundred miles.

In Asia the Grand Canal of China was very significant to economic development but, tragically it was built at a huge cost in human lives. Angkor Wat was a stunning construction but little is known about the project itself.

The Viking Voyages to Vineland were incredible achievements, yet they were unsustainable, and didn't have the lasting impact they should have had. As projects they were somewhat opportunistic in their approach, land hoping, and their technology advantage was not significant enough against a hostile local population.

Key Lessons

Lessons Learned

Many of the Gothic cathedral projects took up to one hundred years to complete:

- They depended on a governance framework to carry them through this period. The project funding was levied through taxes, collections and indulgences.

- Typically, each time a master-builder took over they would use different standards of measurement, and therefore had to continually adjust calculations, based on what was done before. For the chapter a sequence of different master-builders was inevitable and acceptable.

- The chapter (project team) depended on "true blinding faith" to carry the project through.

PMBoK® Guide Knowledge Areas:

- Integration Management

  - In medieval construction projects the master-builder needed the most complete knowledge and grasp of every phase of the project, and a structural and constructional understanding from the planning to the execution of the project. A great number of trade guilds and master craftsmen had to be integrated into the project. At Chartres there were over 43 trades represented.

  - The projects had chapters who oversaw the project, and an overseer of works who monitored the project on a daily basis and managed change control tightly. Significant changes had to be submitted to the chapter for approval. The chapter controlled the finances.

  - With Gothic cathedral projects the issue of losing key staff due to breaks in construction was real (most often owing to a lack of funds). It almost certainly meant the master-builder had no option but to move to another project. When construction resumed he had to be replaced. This risk was very real but there is not much evidence that contingencies were put in place when this happened. There was an expectation that this would happen and that no one single master-builder could finish the project off.

- ○ The Church monopolistically initiated the finance of the projects from its own revenues derived from church owned land and also from the community. Collectors would be sent out to raise money and appeals went out. The church could also send its holy relics on tour, which would raise donations. The challenge was to financially sustain the project over several generations where the original participants would never see it to completion.

- ○ Very few of these projects were achieved in one continuous campaign of building.[44] The business justification was remarkable as the full benefits often only materialized beyond-the-life of many of the patrons. This required tremendous assurances in the ability to continue the project to completion, through a "Transition Plan."

- ○ A strong motivation for completion was the fact that each cathedral was the largest and most important building in its town, and dominating it in size (40m or 130 feet). Most of the surrounding buildings were a mere couple of stories high. So, the cathedral had to be completed otherwise it could end up as a redundant structure.

- ○ Most cathedral projects were influenced by the merchants of the town or city. The economic benefits of having a landmark cathedral, to attract pilgrims, motivated the merchant classes into sponsoring the project. Typically, prime spot stained glass windows, for advertising, were reserved for them.

- ○ The projects had significant impact on the employment of a town.

- Scope management

  - ○ A couple of driving requirements for the project were related to:

- The esthetics of the building, and creating "the most radiant windows." These stained glass windows were central and communicated messages, stories, and advertising for local business, traders, and guilds.

- The competition between the cathedrals projects led to a race to beat the world height record.

○ With the Gothic cathedral projects the unparalleled commitment and time frames required a very well defined scope of what was being built as it was passed along through the ever changing project team.

○ Cathedrals were built in sections so they could be put to use right away (see figure 6.9).

○ The scope in terms of materials, and equipment was dictated by:

- The location within the center of a town.

- The height of the towers, transepts, and nave.

- The size of the windows and the ratio of glass to stone in the walls.

- The availability of materials, mortar and stone (quarried locally if possible).

○ The work breakdown structure laid out the project with the following:

- Site preparation.

- Assembly of scaffolding, delivery of bricks.

- Construction of the structure in logical segments.

- Removal of scaffolding and site clean-up.

- Time Management

○ The order and sequence of activities in a substantial building like the Gothic cathedrals were significant. The erection sequence had to be worked out and adhered to. Constraints like access to materials and work-in-hand inventories had to be carefully thought through. For example, the close distance between

227

pillars made it possible for only handcarts to be moved along prepared roadways into the interior of the building.

○ Labor schedules were created for the project to manage over seasonal variations.

○ Gothic cathedral projects are a very good example of long-range planning and scheduling. The final project output was built with painstaking precision and incredibly complex detail. In some cases the project took over a hundred years to complete, Notre Dame in Paris took 88 years.

• Cost management

○ Gothic cathedral projects predominantly used cost management throughout the project and kept a projection of costs. The church authorities used a system of accounts for managing the project budgets in a very similar way to what the Romans used (see page 213), tracking income versus spending.

○ The costs were enormous and the project required the whole town to get behind it, so the idea had to be heavily promoted and sold. One approach was to raise funds through marketing.

• The stained glass windows incorporated and promoted in their layout local merchants and professional groups (stonecutter, carpenter, clothier, cartwright, etc) to advertise their trades. This was an early example of self-advertising. These windows were located closest to the public in the side aisles, and low in the window frame. Similar to selling advertising-space today.

○ The project was rarely one continuous process and was often halted for one reason or another. There were periods when the funds to pay for the project workforce, or supplies, or materials, became exhausted. Work ceased until the funds became available. At this point canvassing began, for example the sale of indulgences like the Butter Tower of Rouen Cathedral.[45]

○ The building materials (stone, lime, sand, and wood) were sourced from local quarries, where possible, which significantly reduced costs.

○ Cost control was practiced as techniques evolved. For example, Chartres (completed 1220) was one of the first cathedrals that had massive buttresses. These weighed 2,000 tons and supported flying buttresses. At Bourges (completed 1324), the flying buttresses were longer and more slender, blocking less light, a third of the weight and most importantly a lower cost.

- Economy of design was essential but it was possible to take it too far as with Beauvais cathedral which collapsed in 1284. The nave was taken to a height of 150 feet (48 meters). The transept (crossing the nave at right angles), and choir are all that remain today.

○ Labor was sourced locally where possible although certain skills were traveling like the master-builder and craftsmen.

○ The general laborers mostly lived near the site of the cathedral and received cash in hand at the end of each working day.

- Read more at *Building a Medieval Cathedral: Construction of Places of Worship in the Middle Ages.* http://www.suite101.com/content/building-a-medieval-cathedral-a44972#ixzz10nHLP457

○ The projects were seasonally driven and had a seasonal workforce. As the project teams shrank in the winter, only the skilled workforce was kept busy.

○ At Chartres, the sculptures were only 3% of the building expense and the windows were a similar cost (James, "What Price the Cathedrals?" 1972: 53). The structure was the bulk of the cost and this utilized local unskilled labor for quarrying, cutting, hauling, and lifting, etc. (James 1972: 63).

- Quality management

  - The mason's marks were used in the Gothic cathedral projects, with requirements for accurately cut and closely fitting blocks. These marks helped the stone setters more accurately set individually "hewn to shape" blocks into place without errors. The marks also introduced "traceability" so if a block didn't pass an inspection these would identify the stonecutter so he could be held accountable for its craftsmanship.[46]

  - The guilds brought their own organizational standards for work completed and templates.

  - The responsibility of the master-builder was to ensure the designed structure would work and not collapse of its own weight or from inadequate bracing.[47] The master-builder was aware of jerry building and shoddy practices.

  - Even though a few projects failed, notably the collapse of Beauvais in 1284, most cathedrals have lasted until today. This is a testament to how the approach to quality was built right into the project plan and quality control was used continually in the project. The quality assignments were part of the schedule so as these cathedrals were built in sections the choir, transept, and nave were rigorously inspected by the master-builder.

  - As the ratio of glass to stone increased in the walls and the choir rose to stunning heights (48 meters (160 feet)), the threat of building collapse was a continuous and real risk. To transfer the weight off the walls support was provided by flying buttresses. Quality control became a priority as these supports had to be built with great accuracy.

- Human Resource Management

  - There was an expectation that the project would have several generations of architect/master-builders (PMs).

  - There was a constant acquisition of apprentices into the medieval trade-guilds to be developed and maintained into a skilled and qualified workforce. The training and development of apprentices followed an oral and not written tradition. Skills were learned by visual example and by demonstration at the site. There is a virtual absence of written records that explain comprehensively and in detail the methods followed in the building project. This knowledge, the secrets of the craftsmen, was jealously guarded and never committed to writing. Even the design was not disclosed until the end of the medieval period.[48]

- The detailed records of Walter of Hereford, the master builder of the Cistercian Abbey of Vale Royal in Cheshire, England from the years 1278-1281 suggest that 85% of the quarry men for the project were drawn from the local population, while only 5-10% of the highly skilled masons and stone-cutters were hired locally.[49]

- For the medieval castle projects conscription was widely used, particularly in England, where the King had the power to recruit 25 to 40 men for the site of a fortified castle which might be several hundred miles away. This caused a lot of resentment and it didn't do much for project morale but, these projects were deemed as high priority of national security.

- The workforce was well looked after. The on-site workshops were more than a workplace. They provided living quarters in lodges for the workmen. This helped keep the team together for an extended period, a job for life.

- The project workforce worked for a salary, and there are extensive tax records, that exist today from the municipality of Paris in the 13th Century tax register, that reveal the tax by profession.[50]

- To better support the workforce there was a substantial effort to acquire and manage essential equipment. Even today there is evidence that the great wheel device for lifting heavy loads was left in place in the vaults so that, it could be used for repairs.

- ○ The size of the team was seasonally driven. During the winter the team shrank to 30 from a summer peak of 230 as men returned home. All masonry work proceeded inside workshops as all external work stopped. Because of the seasonal break the project workforce would work a 14 hour day in the summer, versus a 9 hour day in the winter. The ratio of skilled men to laborers was about 2:1 at peak and 3:1 off-peak periods. Typically, the masons (skilled) would have working for them the quarry men, mortar mixers and carriers.

- ○ The medieval project workforce on average worked a 4 to 5 day week when the 30 feast days are included. The medieval working population was not overburdened and the project workforce was certainly not to be pitied.[51]

- ○ The laborers were paid very little but the projects gave laborers an opportunity to better themselves.

- ○ An early example of a reward system is with the Gothic cathedral projects where the architect (master-builder) would have his portrait inlaid in white marble with that of the bishop. The honor of having your name engraved was an encouragement and inspiration. Similarly, the stained glass windows closest to the side aisles incorporated and promoted in their layout professional groups and local merchants (carpenter, stonecutter, cartwright, clothier).[52]

- Communication Management

  - ○ The Gothic cathedral project was driven by a governance framework that included a Bishop and Chapter. The initial communications to establish buy-in could be done within the existing hierarchy of the church using the local diocese to communicate and organize from the pulpit. This was an obvious advantage to the project.

  - ○ There are many aspects of modern communications management in the approach used in the Gothic cathedral projects. For example, the chapter (at the centre of the project) had to determine stakeholders, plan communications, set expectations, distribute information, report performance, and manage stakeholders.

○ On site communications management within the Gothic cathedral project would have relied on the master-builder and trade guilds, who had very clear lines of responsibility. For example, if the structure was built more in wood than stone then the master carpenter had more influence than the master mason. Each master-craftsman ran a workshop for their trade guild with assistants and apprentices, so communications was very tight within this hierarchy. Communication on construction sites today is not much different with the foreman and supervisors for a work team. The Gothic trade guild also readily provided a level of technical expertise, guidance, and information, to solve the problems with materials, equipment, and also to make sure that good practices were followed.

○ The Gothic cathedral projects are a good example of the importance of communication management in long scale projects. They were wholly dependent on communication management. There were two aspects to this the first was to get buy in to the project. The idea had to be sold to the town and the inhabitants had to get behind it. The structure was going to be a major imposition on the town in terms of scale. With a long-range vision and plan established the second aspect was to continuously communicate through the life time of an incredibly long project that could easily take up to a hundred years to complete. New generations had to be introduced and be groomed ready to take over.

○ The town of Chartres had a population of 9,000 which seems miniscule in today's terms for a cathedral project of such as size. Rallying the town behind the project was essential.

- Another common form of written communication was through the mason's marks. These ciphers communicated a record of the work done, the identity of the stonecutter so he could be credited for the work, and held accountable for its craftsmanship. The stone-setters could then take the individually "hewn to shape" blocks and place these without delays, ambiguities or errors of placement. Cyphers were also widely used in timber frame construction projects.[53]

- Risk Management
  - There was a financial risk where insufficient funding was available to carry the project through to completion.

  - The construction sites were very dangerous places carrying many risks. Not surprising when considering the heights of the structure and the small spaces available for maneuverability.

  - A cautious and safe approach was required daily. As the ratio of glass to stone increased the project workforce had to be sensitive to the degrees of stress within these walls as they worked within stringent structural tolerances.

  - To mitigate the risk of collapse new support techniques were developed like flying buttresses that transferred the weight off the walls. Pressures were directed and channeled to areas where they could most effectively and securely be received and grounded.

  - The guilds took a responsibility in safeguarding their members in such an environment, similar to an insurance.

  - The competition that grew between the various cathedral projects to be the tallest, and the increase in window to stone ratios, both greatly increased the risks. This was further complicated by the trial and error approach, rather than empirically based, which increased the risk by the time Beauvais Cathedral project was underway. The experience of structural

failures was no longer within living memory, so
assumptions made went untested.

- Procurement Management

  ○ To keep costs down materials were procured locally
  when the quality of the stone met the master-
  builder's standards.

  ○ The immense scope of the projects kept the local
  guilds fully employed for several generations, and the
  local economy stimulated.

  ○ The Gothic cathedral projects are a good example
  of the importance of procurement management in
  long scale projects. The schedule was affected by the
  seasons where the project slowed down in the winter.
  The project workforce shrank from a high of 230 to
  30. The services of the workers were procured. In
  France they came from all over Paris and had to bid
  for the work.

  ○ The Gothic cathedral projects in Southern England
  were built largely with a fine grained Oolitic
  limestone, desirable as an unstratified stone that was
  quarried near Caan in Normandy and shipped across
  the channel.[54]

  ○ Another critical aspect to these projects was
  compulsory purchasing, and adjudication. Handled
  by the chapter, and similar to urban redevelopment
  today, the chapter dealt with both public and private
  conflict. For example, in 1240 the obstinate monks of
  the Hotel-Dieu in Amiens had to be given a hundred
  pounds a year for five years and their hospital had to
  be rebuilt near a large waterway.[55]

  ○ Further evidence of procurement in the medieval era
  exists in the Fabric Rolls, early public records that
  give accounts of expenditure on buildings like English
  Cathedrals. These list the nature and quantities
  of materials purchased, who supplied them, what
  was paid for them, as well as the names and wages
  assigned to each category and grade of the labor
  force. The long Fabric Roll of 1253 shows that in

that year a large quantity of iron was brought from Gloucester for the works, and in this year there were no less than nineteen smiths at one time engaged on the building.[56]

Educators

- Discuss in the context of today's projects the importance of the role of the Gothic cathedral project chapter?

- Discuss the continuum of the Gothic cathedral project over decades and the challenges involved?

- Discuss the voyages to Vineland and whether the North American settlements had a chance of success?

CHAPTER 8

The Late Middle Ages (1300 - 1450)

The historical period was dominated by the cataclysmic epidemic of the Black Death which had a massive impact on China, Europe in terms of the population, economy, and trade. The latter part was dominated by the start of Renaissance period which changed many aspects of medieval life. From a project management perspective the cathedral boom continued because of its economic impact on towns. The Black Death had an enormous negative impact by stopping and closing projects. However, at the start of Renaissance, there was a resurgence in construction.

In this Historical Period

- Brief synopsis of trends/changes (natural disasters, wars, technology, epidemics).

    a. Changes in climate Little Ice Age, 1300-1460.

    b. Black Death (plague) in China 1331 and in Europe 1347-1351.

    c. Projects controlled by the church and state but commercially sponsored projects start to become more predominant.

- Impact of changes

    a. The first half of the 14ᵗʰ Century saw European overpopulation.

    b. Financial Revolution and the emergence of the money economy and banks.

    c. Change in economy after the Black Death.

    d. Growing influence of the architect/engineer.

- Major events

    a. 1356 Ming dynasty coalesced a coherent campaign that expelled the Mongol Emperor in 1368 and unified China.

    b. Start of Renaissance period, 1420, in Europe.

    c. Portuguese sail beyond equator, 1434.

- New tools, techniques and breakthroughs

    a. Industrial revolution (water power).

    b. First European canon cast, 1330.

    c. The Mechanical Clock.

    d. Medicine.

    e. Perspective Drawings.

    f. Cartography.

    g. Finance.

- Regions

    a. Europe.

    b. China.

- Significant projects

    a. Florence Cathedral Duomo.

- Other Notable Projects

    a. Voyage of the Chinese Fleet.

- Key players

    a. Filippo Brunelleschi.

- Chapter Wrap-up

Trends and Changes

The world population in this period dropped from 360 to 350 million.

Climate Change Little Ice Age (1300-1460)

The climate in medieval Europe began to cool and agricultural innovations could not maintain the productivity of frontier lands that became marginal again. This lasted until 1460 when the climate warmed for the next century until 1560.

Black Death or Plague (1331-1351)

It struck China in 1331 and then again in 1353. A census in China in 1200 recorded a population of 120 million, which by 1393 had dropped to 65 million through a combination of predominantly plague and war.[1] The plague struck Europe in 1347 - 1351 and had a significant effect on labor pools. It killed an estimated 75–200 million people in the 14th Century.[2] In less than 20 years half of Europe's population had been killed.[3] Livestock died, and land went uncultivated (up to 60% in Germany). The cost of projects got more expensive as there was a dramatic increase in the cost of labor, and money was the only form of payment accepted. It took three centuries for the towns to recover populations prior to the plague.

Projects

In Europe major projects were still controlled by the church and state. The state (kings) could second key master-builders and architects to their projects.

Impact of Changes

Northern European over Population

With the Little Ice Age the first half of the 14th Century saw Europe burdened by overpopulation as the agricultural output of northern Europe fell. This resulted in lower living standards for the peasantry as their land holdings were subdivided or they were forced into unproductive areas. In some areas, poor weather created meager harvests which led to mass starvation reducing the population by 15%.

Financial Revolution

The period saw the emergence of the money economy and banks in parts of Italy. The first income taxes were levied in Europe since Roman times. In Florence, the revival of guilds and extensive commerce and trade gave rise to institutions for commercial credit and banking. Over 80 banks were created in 1338 and by 1400 there were over a hundred. These banks kept detailed records of the debts owed as their business depended on the accuracy of records. The publication of Pacioli's 5th book in 1494, *Summa de Arithmetica, Geometria, Proporioni et Proportinalita* covers the topics of algebra and arithmetic, their application in business, bookkeeping, money, and exchange, and pure and applied geometry, and proportion. The Italians became the leading merchants of the Middle Ages and nearly monopolized international banking. They sent agents to other countries to trade but carried letters of credit instead of cash which allowed parties to transfer large amounts of money across Europe without carrying cash. Of course there was a charge fee for this service.

Change in Economy after the Black Death

In Europe after the Black Death (1351) there was a significant employment boom and a feeling of reckless joy of being alive. The survivors inherited what the dead had left behind so there was a massive spending spree and economic boom. In the Renaissance armies of Europe the ranks were now filled with mercenaries who fought until the king's money ran out.

Influence of Architect/Engineer

The growth in projects (commercial and military) saw the whole period come under the influence of the architect/engineer. They built cathedrals and other large buildings, whilst the military engineer built castles and other fortifications.

Commercially Sponsored Projects

In this period wealth was generated through increased commerce. Commercial projects were becoming more predominant with the growth of commerce and towns. Private capitalists promoted industry and technological innovation.

Major Events

Ming Dynasty 1356-1644

The dynasty united a coherent campaign that expelled the Mongol Emperor in China and ushered in a new period of peace and prosperity for China. The emperors of the dynasty tried to subdue the nomads and partially restored the Great Wall. Beijing was built into a magnificent city with 12 meter (40 feet) high walls around a perimeter of 23 kilometers (14 miles). Within the centre was the emperor's palace complex - the Forbidden City.

Renaissance Period 1420

The term means "rebirth" and was a major cultural movement that spanned the 14th to 17th Century encompassing a revival of classical knowledge in art, literature, philosophy, linear perspectives in painting, and widespread educational reform.

Portuguese Sail beyond Equator 1434

The European explorers lead the way in global exploration. At first timidly hugging the coasts they quickly reach the limitations of this travel.

One superstition related to the equator, and Cape Bojador, which many thought was the Gates of Hell. The North Star, the main navigational guide, disappeared south of the Equator. Sailors were reluctant to cross that line and for 15 voyages turned back before one expedition in 1434 finally braved its passage without being swallowed up.

New Tools Techniques and Breakthroughs

Industrial Revolution[4] of the Middle Ages

After the Black Death the shortage of human labor in Europe led
to the search for alternative sources of power. One source was
waterpower and many workshops and factories were relocated
into the hills were fast running water could be used to drive
machinery for many purposes, including the fulling of cloth and
the pressing of olives. This is contrary to the stereotypical view
of the Middle Ages as a backward, superstition-ridden, and
technologically primitive time. Organizations could achieve more
far ambitious projects as water and wind power was harnessed.

First Canon Casts 1330 in Europe

The first recordings in Europe are in Florence in a document
that shows the city acquiring 'metal canon' and iron shot.[5]
Although canon had evolved in China some two hundred years
earlier, the Europeans fully exploited and rapidly evolved this
into a very potent technology that was central rather than
peripheral to their arms and ensuing conflicts.

The Mechanical Clock

The European monasteries drove the concept of mechanization
of time to support the routine of daily monastic life. The monks
adhered to a rigorous schedule of prayer, and work. Between
1280 and 1320, there is an increase in the number of references
to clocks and horologes in church records, and this probably
indicates that a new type of clock mechanism had been devised.
Existing clock mechanisms that used water power were being
adapted to take their driving power from falling weights. This
power was controlled by some form of oscillating mechanism,
probably derived from existing bell-ringing or alarm devices.
This controlled release of power - the escapement - marks the
beginning of the true mechanical clock. In 1365 came the first
sub division of hours into sixty minutes.

Perspective Drawing Lines

The appearance of the cannon led to the drawing of trajectories of fired shot, which in turn saw the development of grids. Around 1410 Brunelleschi used grids to create perspective drawings, and view 3 dimensional objects in 2 dimensions.

...He had (Brunelleschi) made a hole in the panel...which was as small as a lentil on the painting side...and on the back it opened pyramidally, like a woman's straw hat, to the size of a ducat or a little more.[6]

Brunelleschi held a picture of the Baptistery painted on a panel, with its back to an observer, and had the observer squint through a small hole in the painting. Through the hole the observer saw a mirror reflecting the painting so the observer saw the front of the painting in the mirror. When the mirror was removed the real Baptistery was visible through the peephole, yet both views were so similar. But this was no ordinary painting it was the first accurate perspective picture. This breakthrough increased Brunelleschi's ability to measure the world around him.

Regions

China

At the beginning of this period China still led the world in technology but, by 1450 Europe had caught up. The Mongol invaders had been invading parts of China and subjugating it. It was not until 1356 under the Ming dynasty that a coherent campaign coalesced that succeeded expelling the Mongol Emperor in 1368.

Significant projects

Florence Cathedral Duomo Project (1417-1436)

Like many medieval cathedrals the overall project spanned 140 years. The project begun in 1296 and the major construction was completed by 1407 without the dome or cupola.

Background

The project was driven by a rapid population expansion, and a financial revolution which drove the emergence of the money economy and banks. The prosperous city of Florence wanted to match or exceed in size the much larger cathedrals that were being built. For Florence this project was all about prestige. The Consuls of the Guild of Wool made up the project sponsors who sat on the executive board.

Initiating (1417-1420)

In 1348 the project was halted for a number of years because of the Black Plague. By 1407 only the dome was uncompleted, although this was the most complex part of the project. A brick model of the planned octagonal dome had existed from 1367. The design was unprecedented for the time as it called for a massive dome, up to 30 meters (100 feet) in height, to be built above the 40 meters (133 feet) wide space above the church's chancel. It was the height of the Pantheon (43.30 meters (142 feet), matching the diameter) built 1,000 years earlier. It would weigh over 35,000 tons.

In 1407 the project sponsors and the Wardens of Works of the cathedral, summoned a congress of architects and engineers from around the country to discuss methods for constructing the dome and to find a candidate to lead it. The use of buttresses to support the dome, as in northern Gothic cathedrals, was ruled out by the building's design. Among the congress was Filippo Brunelleschi, a renowned master-builder. He advised not to build directly from the roof, according to the original design of the architect Arnolfo, but to build from the roof a frieze (drum) fifteen braccia[7] (9 meters or 30 feet) high (see figure 8.1). The drum would not only take the weight off the supports of the tribunes, but it would be easier to build the dome. Surprisingly, the congress did not end up appointing a project manager. Brunelleschi who was looking for this appointment made a number of models and machines in secret, and stayed on in Florence for many months to continue this work.

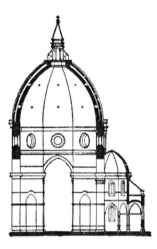

Figure 8.1: Filippo Brunelleschi's suggested incorporating a drum 9 meters (30 feet) in height with a large round window in the middle of each of its sides raising the dome 32 meters (108 feet) in height).[8]

On hearing talk that the project sponsors were looking at engineers to the raise the dome, Brunelleschi returned to Rome, believing that he would be sought out and be in greater demand than if he stayed in Florence. He had delivered his presentation with a confidence unique to the other master-builders. The project sponsors together with the master-builders thought the project unfeasible, and that there was no way to technically deliver the dome. They could not envisage a structure strong enough to sustain a framework and the weight of the dome. The diameter of the done was too broad for beams to cross it or to make a bridge.

Brunelleschi instinct was proved correct. The Wardens and project sponsors wrote to him in Rome, pleading for him to come back to Florence. He desired nothing more and readily returned to meet with them and the master-builders to walk through all the raised issues. Brunelleschi spelled out the challenges:

"My Lords the Wardens, there is no doubt that great projects always present difficulties in their execution, and if any ever did so, this of yours presents them, and even greater than perchance you are aware of, for the reason that I do not know whether even the ancients ever raised a vault so tremendous as this will be; I have often pondered over the framework necessary both within

*and without, and how it may be possible to work at it safely. I
have never been able to come to any resolution, and I am aghast
no less at the breadth than at the height of the construction, so
if the cupola could be made round, we might use the method
used by the Romans in raising the dome over the Pantheon in
Rome, whereas here we must follow the eight sides, and bind
the stones together with ties and by dove-tailing them, which will
be something very difficult..."*

Brunelleschi laid down his terms. He would show the method
if he was made the project manager:

*"But how can I help you since this project is not mine? If the work
fell to me, I would have resolution and courage enough to find
the method whereby the vault might be raised without so many
difficulties; but as yet I have given no thought to it, and you want
me to tell you the method!"*

Brunelleschi suggested the project manager should be of
proven ability and suggested a congress of architects to advise on
the project:

*"And when at last you determine to have it raised, you will be
forced not only to make trial of me, for I do not think myself able
to be the sole adviser, but also to spend money and bring within
a year many architects to Florence, not merely Tuscans and
Italians, but Germans, French, and of every other nation; and
explain this project to them, to the end so that, after discussing
and deciding among so many masters, a start is made, and the
commission given to him who has concrete proof of his ability or
demonstrated the best method and judgment. Nor could I give
you a better plan or advice than this."*

The Wardens and project sponsors pleaded with him to stay
but he returned to Rome.

Planning (1417-1420)

On May 26, 1417 the Wardens decreed Brunelleschi a financial
retainer, entered in the books of the Office of Works, and made
drawings for the dome. In 1420, as Brunelleschi suggested, the
most prominent architects and masters, from all over Europe
were finally assembled in Florence for a competition. They all

met, in the Office of Works, in the presence of project sponsors and Wardens, together with selected citizens, to hear each architect in turn present their method and plan. This varied, some suggested that piers be built from the ground up with arches upon them to hold up the wooden bridges used to prop the weight of the dome. Others suggested building the dome from pumice-stone so it would weigh less. Another opinion was to fill the centre with earth mingled with small coins so that it would encourage groups to find the coins, removing the expense of clearing these from the project. The consensus was for a pier built in the center, and the dome raised in the shape of a groined joint or vault (see Figure 5.3), like that of S. Giovanni in Florence.

Brunelleschi alone said that it could be raised very easily without so much wood-work, piers and expensive arches, framework, and earth. Everyone thought he spoke nonsense, something that was unfeasible. The masters asked Brunelleschi to provide details, and show his model, as they had shown theirs. But he refused and proposed instead that whoever could make an egg stand upright on a flat piece of marble should undertake the project. Taking an egg everyone tried to make it stand upright, but no one could find a way. When Brunelleschi was asked he gave one end a blow and made it stand upright. The craftsmen protested that they could have done the same but Brunelleschi answered, laughing, that they could also raise the dome, if they saw the model or the design.

Brunelleschi's persistence convinced many that he did in fact know what he was talking about. He laid out the design on paper and convinced the project sponsors to give him the commission, making him principal superintendent (or project manager).

But they did not contract with him for more than twelve braccia (24 feet, 8 m) of the whole height, saying to him that they wished to see how the work succeeded, and that if it succeeded as well as he promised they would not fail to commission him to do the rest. It appeared a strange thing to Brunelleschi to see so great obstinacy and distrust in the project sponsors and Wardens, and, if it had not been that he knew himself to be the only man capable of executing the work, he would not have put his hand to it.

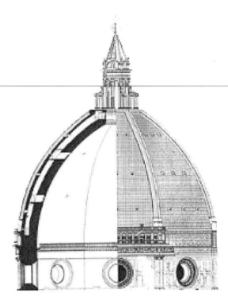

Figure 8.2: Florence Cathedral Duomo[9] section of the dome shows the distinctive octagonal design of the double-walled dome. It followed a rigorously geometrical design. The total height of the dome and lantern is 114.5 meters (375 feet) from the pavement. This was the first large, self-supporting dome ever to be built without centering.[10]

Executing (Construction and Techniques) (1420-1436)

Brunelleschi was given the job on a trial basis, but the matter did not end there. The craftsmen and the citizens heard about the commission and there were mixed reactions. As preparations for the construction were made, a faction appeared before the project sponsors and Wardens, complaining that the appointment was made to quickly, and that the project should not be carried out by one man alone. There was an abundance of excellent masters and that it didn't do the city any credit, because, if the project failed they might be blamed for giving too much control to one man, without considering the impact on the City of Florence. They suggested giving Brunelleschi an assistant Lorenzo Ghiberti to restrain his impulsiveness.

The problems did not end there as the assistant was given the same salary. Brunelleschi was dissuaded by his friends from leaving the project which he pursued with little enthusiasm as he had to divide the potential honor and fame equally with Ghiberti. He decided that he would have to find a way of removing Ghiberti from the project. Brunelleschi made a scale model with all the exact proportions, and complex parts of its design, such as the lighted and dark stairways, windows, doors, ties, and buttresses, together with a part of the gallery. Ghiberti, hearing of this model, wished to see it, but Brunelleschi refused. Ghiberti flew into a rage and ordered another model to be made for himself, so that he would not appear to be drawing his salary for nothing and not peripheral to the project. The chapter paid Brunelleschi 50 lire for his model, but Ghiberti was paid 300 lire for the labor and expense of his, more because of the friendship and favor that he enjoyed.[11] Brunelleschi's attempt had failed.

The situation continued for Brunelleschi for several years until 1426, when the friends of Ghiberti called him a master equal in stature to Brunelleschi. This infuriated Brunelleschi so much that he decided to try again and get rid of Ghiberti, recognizing that he was of little value to the project.

The project had completed the dome right round, at a height of twelve braccia (24 feet, 8 meters) or 2 vaults. The vaults (a few meters apart) had to be fixed with ties of both stone and wood. As this was a complex design, he discussed it with Ghiberti, to see if he had considered this. He found Ghiberti had not and referred it but to Brunelleschi the inventor. Ghiberti's answer pleased Brunelleschi, as it gave him another opportunity to remove Ghiberti from the project by exposing his limited value to the project and show that he did not possess the right skills to be in the position. Brunelleschi feigned an illness and removed himself from the project. The masons and master-builders engaged in the project were at a standstill, waiting to be shown how to start the next section, and how to continue the vaults up and bind them with ties. As they waited they became suspicious that neither Ghiberti nor Brunelleschi had enough conviction to carry on the work.

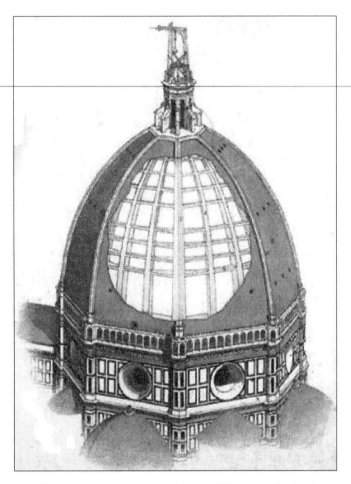

Figure 8.3: Cut away of the large dome of Florence Cathedral created by Brunelleschi.[12] "The dome itself consists of two octagonal vaults, one inside the other. Its shape was dictated by its structure. Brunelleschi made a design feature of the necessary eight ribs of the vault, carrying them over to the exterior of the dome, where they provide the framework for the dome's decorative elements, which also include architectural reliefs, circular windows, and a beautifully proportioned cupola."

Brunelleschi returned to the project, but, seeing that Ghiberti was still strongly favored, even though the project stopped, he made the following speech to the Wardens in the presence of Ghiberti:

"My Lords the Wardens of Works, if the time that is lent to us to live were as surely ours as the certainty of dying, there is no doubt whatsoever that many things which are begun would be completed instead of remaining unfinished. The illness which I have suffered might have cut short my life and put a stop to the work; therefore I have thought of a plan whereby, if I should ever fall sick again, or Lorenzo, which God forbid, one or the other may be able to pursue his part of the work. Even as your Lordships have divided the salary between us, let the work also be divided, to the end that each of us, being spurred to show his knowledge, may be confident of acquiring honour and profit from our Republic. Now there are two most difficult things which have to be put into execution at the present time: one is the making of the scaffoldings to enable the masons to do their work, which have to be used both within and without the building, where they must support men, stones, and lime, and sustain the crane for lifting weights, with other instruments of that kind; the other is the chain of ties which has to be placed above the twelve braccia, surrounding and binding together the eight sides of the cupola, and clamping the fabric together, so that it may bind and secure all the weight that is laid above, in such a manner that the weight may not force it out or stretch it, and that the whole structure may rest firmly on its own basis. Let Lorenzo take one of these two works, whichever he may think himself best able to execute; and I will take the other without difficulty, to the end that no more time may be lost."

Ghiberti was forced unwillingly for the sake of his honor to accept one of these tasks. He took the chain of ties, being the easier option. He relied on the advice of the masons and on remembering that in the vaulting of S. Giovanni in Florence there was a chain of stone ties, from where he might borrow the design. Brunelleschi took the other task and built the scaffoldings with great ingenuity, where the workforce stood on them, as securely as if they had been on the ground. The workforce was able to work safely and draw up heavy weights without any problem. This design swayed them and they threw their weight behind Brunelleschi.

Ghiberti completed the chain of ties on one of the eight sides with the greatest of difficulty. When it was finished, the Wardens had Brunelleschi look at it. He said nothing to them. He remarked to some of his friends that another kind of fastening was needed, and had to be placed differently because it was not strong enough to withstand the weight laid above it, and it did not bind the masonry firmly enough together. He added that the chain he had made had been discarded. When the Wardens heard about this he was asked to show the best way of making such a chain. At this, he immediately showed them the designs and models he had already made. They recognized the great mistake they had made in favoring Ghiberti. Wishing to atone for this they made Brunelleschi overseer and superintendent of the whole project for life, where nothing could be done without his command and gave him 100 florins. They also voted him an allowance of 100 florins a year for life.

Controlling and Monitoring

Brunelleschi, ordered the project to continue, and he pursued it with such scrupulous care and so great attention, that not a stone could be put into place without his blessing. For example, the curvature of the rising walls had to be very closely monitored. Ghiberti found himself defeated, and put to shame, but assisted by his powerful friends he went on drawing his salary, claiming that he could not be dismissed for 3 years.

There were many technical problems in building the dome. So, Brunelleschi was forever making designs and models of the stages for the workforce, and the lifting machines for heavy weights. He invented special hoisting machines and lewissons for large stones.

Brunelleschi built the dome out of bricks, as the Roman experience of building with concrete had been lost for over a 1,000 years, although he drew inspiration from the great dome of the Pantheon in Rome. Great quantities of brick had to be procured, in the region of 400,000 every year.

The diameter of the dome at 42 meters (130 feet) prevented the traditional use of wooden structuring (a framework) from the ground up through the center to support the construction of the two vaults. A circular platform was built around the drum to permit the movement of machinery and workers, and the temporary storage of materials. As the octagonal design of the double-walled dome rested on a drum and not on the roof itself, it allowed for the entire dome to be built without a framework.

The bricks were laid on sloping beds. Before closing each ring of bricks, the workmen placed a row of bricks whose longer sides protruded with respect to the bricks resting on the conic surface. This arrangement, known as a herring-bone, displayed a spiral profile. This also helped shift the weight through the bricks outwards to the dome's support. The inward slope increased continually from the base to the dome's closing stone ring, called the oculus or the eye. This ensured better stability of the brickwork courses. These two parallel shells were connected by brick spurs.

The ties were now finished right around the eight sides. The masons were laboring vigorously but were harried more than usual by Brunelleschi and were getting aggrieved by his daily reprimands with regard to the project to the point that they conceived a grievance against him. Moved by this and by envy, the foremen banded together and declared that the work was laborious and dangerous, and that they would not complete the dome without an increase in pay, even though their pay was higher than normal. This angered the Wardens and Brunelleschi, and they decided to dismiss them all. On the following Monday Brunelleschi set ten Lombards (apprentices) to work, and by standing over and directing them, he taught them so much in one day that they were able to work forward for many weeks. The masons could not find as well paid work and sent mediators to Brunelleschi, saying that they would willingly return. Brunelleschi kept them guessing for many days and then reinstated them at lower wages.

The execution of the project went so smoothly Brunelleschi was fully recognized and the project critics were silenced. Those not biased before maintained he had shown such boldness and daring in his design that it was unmatched by no other architect ancient or modern. When he brought out his model everyone could see how much thought he had put into the planning of the staircases and of the lights both inside and out so that no one would be injured, and how many iron railing he had placed for the stairways. He had even thought of the irons for fixing scaffoldings within.

Figure 8.4: The view from inside the dome provides a sense of the height and why the project workforce was worried about constructing the dome. Workers were paid according to the height they worked at.[13]

Figure 8.5: The stairs between the two vaults of the dome are 1 meter (3 feet) wide, highlighting the complexity of the structure. The walls slightly curve to the right. The internal wall is two meters (6 feet) thick and the external one just 80 centimeters (2.5 feet) thick.[14]

254

Everything was very carefully arranged. He also placed different kinds of water-gutters, some were covered and some were open, in the least dangerous positions, and along with these various holes and apertures, to break the force of the winds and to prevent movements of the earth from causing damage. He demonstrated how he had learned from his studies during the many years in Rome. He was widely recognized for his ingenuity in dove-tailing, joining, fixing, and binding together the stones. There was nothing, however difficult, that he could not personally tackle. For example, he showed this in the lifting of weights, by means of counterweights and wheels, that one ox could raise what required six pairs of oxen to have been done before.

The building was now at such a height that it was greatly inconvenient for anyone who had climbed to the top to descend to the ground. The workforce was losing much time in the day going for food and drink, as well as suffering a degree of discomfort in the heat of the day. Brunelleschi arranged for canteens with fully equipped kitchens to be opened on the dome, and for wine to be sold there, so that no one had to leave work until the evening. This was convenient for the men and very advantageous for the project.

Figure 8.6: The Cathedral, Florence, Italy.[15] *This enormous dome weighs 37,000 tons and contains over 4 million bricks. The drum sits over a 42 meter (139 feet) wide space above the church's chancel. The total height of the dome and lantern is 114.5 meters (375 feet) from the pavement. The drum starts at 50 meters (165 feet) above ground level and is 9 meters (30 feet) high. The dome atop of the drum is 30 meters (100 feet) in height. The height of the lantern atop the dome is 22 meters (71 feet).*

Closing

The project was completed in 1436 and the cathedral was consecrated by Pope Eugene IV on March 25, 1436. It was the first 'octagonal' dome in history to be built without a wooden supporting frame and was the largest dome built at the time. The Roman Pantheon, a circular dome, was built in 117–128 with support structures. Brunelleschi's solutions were ingenious and unprecedented. His ability to transcribe a circle on a cone face within the innermost double-vaulted wall makes the self-sustaining "horizontal" arch construction possible, since geometrically, a circular plan is needed for such an erection.

The work was completed in 16 years. It is still the largest masonry dome in the world and has been one of the most impressive projects of the Renaissance. For Brunelleschi this was very difficult because not only had he to keep proving himself but he had to also drive the project to completion. He was hampered by his project sponsors, the stakeholders (citizens) and his appointed assistant Ghiberti, which he skillfully got rid of. In the end Brunelleschi proved to be a great success. His ingenuity and skill resulted in the construction of the first dome since the time of the Romans, but much must be attributed to the way he managed the project overall, managing his stakeholders, and his communication management.

Other Notable Projects

Journey of Marco Polo (1271 to 1295)

Marco Polo is probably the most famous Westerner to have traveled on the Silk Road. The Polo brothers, Niccolò and his Maffeo, were prosperous merchants who traded with the East. They left Constantinople for the East, in 1259. They were convinced by the Mongol ambassador that the Great Kublai Khan wanted to receive them which he did very hospitably. He had set up his court at Beijing, built by Kublai as his new capital after the Mongols took over China in 1264.

Marco was 15 years old when his father and his uncle returned to Venice. Two years later the three of them embarked again on a journey to Cathay for the second time. They passed through Armenia, Persia, and Afghanistan, over the Pamirs, and all along the Silk Road to China. They traveled overland for 3.5 years and covered 9,000 kilometers (5,600 miles). The Polos spent the next 17 years in China and the Kublai Khan liked Marco that he sent him on many diplomatic missions throughout his empire. He also entertained the Khan with his stories and about the lands traveled.

The Travels of Marco Polo, written in 1298, were very important as they influenced both future explorers Columbus and Magellan in finding these same places but by sailing west rather than traveling east by foot.

Voyages of Chinese Fleet (1405 - 1433)

Admiral Zheng He (1371 - 1434) was given the project charter to fulfill a series of seven extensive expeditions into the Indian Ocean. The project objective was to establish diplomatic relations with other nations to encourage trade, and to collect information on navigation, geography, and natural history. The business justification lay in establishing profitable trade certainly through fine porcelain.

The planning and preparation was significant to the success of these voyages. A great shipbuilding effort took place between 1403 and 1419 with some exceptionally large vessels constructed of up to 100 meters (330 feet). These were designed for extended voyages and at a time where the largest European caravels were 30 meters (100 feet) in length. Between 1405 and 1433 up to 2,100 technically advanced vessels were constructed with compasses, bulkheads and water tight compartments, desalination. The ships had four decks with massive space for cargo, six masts and were up to 100 meters (330 feet) in length and 1500 ton displacement, carrying up to 1,000 sailors. The 1405 expedition consisted of a fleet of 62 treasure ships supported by approximately 190[16] smaller ships. Some ships were purported to have reached a length of 180 meters (600 feet). The fleet was manned by over 27,000 crew members, including navigators, explorers, sailors, doctors, workers, and soldiers.

The project succeeded in visiting practically every Asian country with a coastline. The large fleet reached the coast of East Africa between 1431 and 1433 a distance of 4,800 kilometers (3,000 miles). This proved that voyages of great distance could be completed. The project was shutdown for vague reasons. One was that the Chinese Government officials (mandarins) were resentful of the profits made by the middle class merchants running these expeditions, so they put pressure on the emperor to end them.

Figure 8.7: Early 17ᵗʰ Century Chinese woodblock print, thought to represent Zheng He's ships.[17]

Key Players

Filippo Brunelleschi (1377-1446)

He was a learned person and the son of a very famous physician, named Maestro Ventura Bacherini. His father placed him in the gild of goldsmiths, after observing his talent for all making sorts of mechanism. He quickly became a skilled workman, and perfected his knowledge of sculpture, perspective and geometry.

He designed some portions of houses in Florence, and in 1401 he was one of the competitors for the design of the gates of the baptistery of San Giovanni. Although he did not win the work he obtained praise and soon afterwards set out for Rome. He studied hard, and resolved to do what he could to revive the older classical style of Rome, which had died out in Italy. Moreover, he was one of the first to apply the scientific laws of perspective to his work. In 1407 he returned to Florence, just at the time when it was decided to complete Florence cathedral and the dome. When he died in Florence on the 16th of April 1446 he was buried in the cathedral church of his native city.

Chapter Wrap-up

Conclusions

Cathedral projects in this period continued in the style of the Gothic period as towns pushed for the world height record. With experience the designs became more elaborate and ambitious. As trade and commerce grew so did the banking system and more loans were available so it became easier to fund large projects. The approach to construction projects changed little only in that the designs got more ambition. Styles changed as Gothic became the Tudor style of architecture around 1485.

The most ambitious project of the time was the completion of the dome of Florence cathedral. The scale of the project was the largest since Roman times, a 42 meters diameter, close to the Pantheon 43.3 meters (143 feet). For the architect Brunelleschi this was a very difficult project were he had to keep proving himself to his project sponsors, and the stakeholders (citizens) who hampered him by their meddling. In the end Brunelleschi proved to be successful where his ingenuity and skill carried through the construction, but much must be attributed to the way he managed the project overall, managing his stakeholders, and his communication management. He was able to deliver a project against the limitations of the technology of his time. As an example, he was able to build a 29,000 ton structure without no centering, initially thought impossible and to radical an idea to base his appointment on. The solution was to create a self-supporting structure which gradually got acceptance with the sponsors.

This period also saw the voyages of Chinese Fleet which were a precursor to the European voyages of exploration and discovery, that started 70 years later.

Key Lessons

PMBoK® Guide Knowledge Areas:

- Integration Management

  - The design of the Florence Cathedral Duomo was unique, innovative, extremely ambitious, and pushing what was feasible in this era with the available materials and technologies.

  - The project had a very protective chapter over its interests. Brunelleschi ran into numerous conflicts with it as they were concerned over how the work was to be sequenced and delivered, so they could tightly manage it and any changes.

  - The project governance was very influenced by commercial interests (Consuls of the Guild of Wool) even though this project created a religious structure.

  - Brunelleschi had a very clear vision of how to complete the project. The complexity of the design required an almost total control over the delivery.

  - Brunelleschi assumed that without total control of the project he would not be able to supply such a complex final deliverable.

  - The business justification was driven by prestige. The cathedral, with a spectacular dome, would reinforce Florence as the main financial center in Europe.

  - Change control existed within the project and was used. For example, in March 1422 the project committee met and reduced the weight of the dome by replacing stonework by brick at a height of 12 braccia instead of 24.

- Scope Management

○ A driving requirement for the project was to match the scale (height and diameter) of the Pantheon built 1000 years earlier.

○ The scope in terms of materials, and equipment was dictated by:

  • The height and diameter of the structure.

  • The quarrying of stone and the production of bricks, and their transportation to site.

○ The work breakdown structure laid out the project with the following:

  • Creation of a solid foundation through the drum.

  • Construction of scaffolding, and delivery of materials to where needed.

  • Construction of the dome (internal and external)

  • Removal of scaffolding and site clean-up.

○ Through the Duomo project the scope was carefully managed. For example, scope increased with the double vault of the dome, but this had a positive impact on time and cost because it reduced the amount of scaffolding required, and the enormous challenges with that approach.

○ Brunelleschi constructed wooden and brick models to guide the craftsmen. This allowed discussions around the more abstract elements. The models were intentionally incomplete, as to ensure his control over the construction.

• Time Management

○ Brunelleschi had to get the sequence of activities right because of the risks in building at such a great height.

○ Scheduling of activities was important as the supply chain of brick had to continuously feed the brick laying teams. The rate of brick and mortar production had to work in lockstep with the project construction schedule.

- The pay clerk was appointed to check the daily production of each workman, to record overtime and the periods of leave which were granted only by special permission. Workers taking unauthorized leave were dismissed.

- An hour glass was kept on the wall to measure the breaks in the working day, and this was monitored by the pay clerk.

- The schedule was carefully adhered so as to ensure a perfect setting.[18]

- The pace of construction has been calculated at approximately 2,000 tons a year, based on a 29,000 ton structure. About 20 bricks per hour.

- Cost Management

  - As brick was an extremely common, tried and tested material, estimating the overall cost was relatively straight forward once the designs and models were completed.

  - Brunelleschi closely managed cost by taking a hands-on approach and working very closely with his laborers. When the project was slowing down he responded quickly to keep the momentum going.

  - The project expenses incurred were much lower that they would have been if the organization had been so efficient and tightly controlled by Brunelleschi.[19]

  - The project workforce wages varied accord to the number of working hours completed. This was unusual but on this project the skilled workforce enjoyed a rare degree of job security.

  - Brunelleschi was compassionate towards his workforce. When the dome reached a height where it was very inconvenient to descend to the ground he took action. Brunelleschi arranged for canteens with equipped kitchens to be opened on the dome, so that no one had to leave work until the evening, which was much more convenient for the workforce. It was very advantageous for the project from a cost perspective as morale improved (so did productivity), and time was not lost.

- ○ The special pay clerk kept a register and at the end of each working day payments were made and transcribed into the ledger.
- Quality Management
  - ○ The pay of the project workforce was adjusted on the basis of the quality of the work they had turned up in the previous semester.
  - ○ Brunelleschi was widely recognized for his ability to use grids to create perspective drawings, and view 3 dimensional objects in 2 dimensions. He created very detailed and accurate scale models of how a particular construction should look to help prevent mistakes.
  - ○ Brunelleschi took a very hands-on-approach related to the inspections of materials for quality, and guiding his workforce who were often irritated by his presence.
    - *...he was continually labouring, going in person to the furnaces where the bricks were being shaped and demanding to see the clay and to feel its consistency, and insisting on selecting them with his own hand when baked, with the greatest diligence. When the stonecutters were working at the stones, he would look at them to see if they showed flaws and if they were hard, and he would give the men models in wood or wax, or made simply out of turnips; and he would also make iron tools for the smiths. He invented hinges with heads, and hinge-hooks, and he did much to facilitate architecture, which was certainly brought by him to a perfection such as it probably had never enjoyed among the Tuscans.*[20]
  - ○ He checked the mixing of the lime He also inspected the construction itself of the walls and ties.
    - The curvature of the rising walls had to be very closely monitored.

- ○ Timber varied in quality and was carefully selected based on its use. For example, white fir was selected for its length, oak for its strength, beech and chestnut for carvings. Timber was inspected, cut, and graded with great care.

- ○ Contracts with local quarry men stipulated that white marble had to be unblemished

- ○ Bricks were carefully checked to see they were good, well baked and clean.

- ○ The curvature of the rising brick walls had to be very carefully monitored and controlled.

- Human Resource management

  - ○ The experience gained in organizing the project workforce for the cathedral project greatly helped Brunelleschi in the organization of the complex and vast Duomo project.

  - ○ The project ground to a halt as labor pools were affected by the Black Death.

  - ○ The skilled workforce enjoyed a rare degree of job security. Through the *ruoli* or a special list of workers, there was a great deal of continuity with experienced workers on the project. Workers were selected and paid according to performance.[21]

  - ○ He was continuously present on site.

  - ○ He appointed eight master masons, one for each side of the octagon, to carry out the actual building work.

  - ○ He used models to guide the craftsmen so they had a much better understanding.

  - ○ He had to change the standard work day because of the difficulties in getting to the site with the heights.

  - ○ Brunelleschi also had to manage a group of renegade foremen who were demanding increased pay.

    - *As the project progressed Brunelleschi pressed the workforce more than usual. They resented certain reprimands received with regard to the building*

> *and other things, they had conceived a grievance against him. Wherefore, moved by this and by envy, the foremen leagued themselves together into a faction and declared that the work was laborious and dangerous, and that they would not build the cupola without great payment—although their pay had been raised higher than usual— thinking in this way to take vengeance on Filippo and to gain profit for themselves.*

- ○ He understood the game they were playing and cleverly side stepped them by dismissing them and hiring his own workforce.

  - • *This affair displeased the Wardens and also Filippo, who, having pondered over it, made up his mind one Saturday evening to dismiss them all. They, seeing themselves dismissed and not knowing how the matter would end, were very evilly disposed; but on the following Monday Filippo set ten Lombards to work, and by standing ever over them and saying, "Do this here," and, "Do that there," he taught them so much in one day that they worked there for many weeks. The masons, on the other hand, seeing themselves dismissed, deprived of their work, and thus disgraced, and having no work as profitable as this, sent mediators to Filippo, saying that they would willingly return, and recommending themselves to him as much as they were able. Filippo kept them for many days in suspense as to his willingness to take them back; then he reinstated them at lower wages than they had before; and thus where they thought to gain they lost, and in taking vengeance on Filippo they brought harm and disgrace on themselves.[22]*

- ○ The project provided much employment for locals.

- ○ The project workforce was organized into teams each directed by a master mason.

○ The project workforce was bound by the rules of conduct and not adhering to these would exclude them for the list (*ruoli*).[23]

○ The non-skilled project workforce was not kept on the list (*ruoli*).[24] They were recruited through a different process.

- Communications Management

  ○ Brunelleschi had a very challenging time managing his stakeholders, the Wardens and project sponsors, who lacked the confidence that the project would be completed. This was even after he had presented his design. He initiated a conference that brought the most prominent architects and masters of design from all over Europe in Florence, to discuss the project and put forward ideas. Through the dialogue and communication of the conference the stakeholders were convinced that he had the best plan and method. It also initiated the project with Brunelleschi as the principal architect.

  ○ Brunelleschi was constantly managing the project sponsors and the Wardens of Works for the cathedral, through the life cycle of the project. Even though he was the only architect who articulated a solution for raising a dome they were constantly questioning his approach, pulling away commitment, and undermining him.

  ○ He was constantly managing and communicating with his workforce, his stakeholders (Public citizens), and his sponsors. He removed hurdles and built confidence that the project could be completed.

  ○ Even though he was the only architect who could consistently articulate a solution for raising the dome they were constantly questioning his approach, pulling away commitment, and undermining him. He refused to show his model of the work until they committed to a commission. When they finally agreed to a contract it was not for the complete work as they wished to see how the work succeeded, and that if it succeeded as well as he promised they would not fail to commission him to do the rest. They also assigned him an assistant who they gave the same salary to.

- In accordance with ancient custom the completion of each important stage was celebrated with the project workforce to boost morale. For example, an inaugural banquet, or the offering of wine. On the closing of the project there was great celebration.[25]

- Risk Management

  - Brunelleschi's Duomo was pushing the limits of known architecture, materials and technology. None of Brunelleschi's counterparts believed it could be readily built and the building risks were too great.

  - Brunelleschi had to work very hard to win the right to lead the project, and he very much put his reputation was on the line that he could successfully complete the project. As a result, he more readily accepted various project risks.

  - With the curving walls if the bricks were laid too quickly before the mortar set there was risk that the walls could distort.

  - The risks of constructing at the dome heights required Brunelleschi to pay great attention to the safety and protection of the project workforce, for example, through the use of stable scaffolding.

  - As the project proceeded the risks increased as did the safety measures. For example, as the platform was elevated more protective parapets were added and as well as screening boards to block the view. The consumption of wine was restricted when working at these heights.

  - Each worker could choose tasks of lower or greater risks which determined their pay. For example, they could work at lower elevations, with lesser risk and pay.[26]

- Procurement Management
  - The project had secured advantageous arrangements for obtaining materials, especially timber and marble. It even supplied these materials to other builders in public or private sectors as they had been granted rights to forests of the Casentino, and the quarries of Campilgia.[27]
  - The mainstay and core material was brick (over four million). Although bricks were supplied with a local brick maker the vast quantities required supplies from non local producers. Eventually, in 1418 a contract was signed with the local brick maker for the supply of 200,000 bricks over a two year period. The order was placed well in advance because of the limited capacity of this sector in the local economy. This kind of volume would have kept factories busy for years.[28]
  - Iron was needed in large quantities for the chains, bars, nails, brackets, etc. It was sourced from deposits around the Elba but had to be transferred to foundries in the Tuscan Appenines were there was plenty of wood fuel. So critical was iron to the project numerous contracts were set up with Appenine blacksmiths.[29]
  - The scale of the supply chain and the materials pouring into the site required a number of subsidiary building sites to be set up around the Cathedral. These had roofs to protect the materials from the elements, and were also used to house the tools and machinery.

Educators

- Discuss the impact of the Black Death on projects at the time?

- Discuss Brunelleschi's strategy with his project sponsors and stakeholders?

- Discuss whether Brunelleschi's desire to lead the project compromised his approach to risk management?

CHAPTER 9

The 15ᵗʰ Century and Renaissance (1450 - 1500)

The period was dominated by the Renaissance. The closure of the trade routes east by the fall of Constantinople spurred the age of exploration driven by Spain and Portugal. From a project management perspective the development of new tools and technologies facilitated these journeys of exploration. Large projects were sponsored by the crown and church.

In this Historical Period

- Brief synopsis of trends/changes (natural disasters, wars, technology, epidemics).

  a. Printing.

  b. Age of exploration driven by Spain, Portugal.

- Impact of changes

  a. European rediscovery of the Americas.

- Major events

  a. Fall of Constantinople to the Ottoman Empire in 1453.

  b. Columbus lands in the West Indies in 1492.

     a. Vasco da Gama (1460-1524): Round Africa to India, 1497-1498.

     b. John Cabot (1450-1499): Voyage to North America, 1497.

- New tools, techniques and breakthroughs

     a. Gridlines and cartography.

     b. Caravel.

     c. Longitude.

     d. Moveable Typeface Print (Guttenberg) in 1440.

     e. Donatello's bronze statue of David 1469.

- Regions

     a. Europe.

- Significant projects

     a. 1492 Columbus voyage of discovery.

- Key players

     a. Christopher Columbus.

     b. Leonardo da Vinci.

- Chapter Wrap-up

Trends and Changes

The world population in this period grew from 350 to 420 million. The Renaissance brought a different view about the world. One that was more humanistic and naturalistic as opposed to a supernatural one.

In Europe the growth of the nation-state lessened the importance of feudalism and stabilized the monarchical authority.

Printing

Printing was a catalyst for projects as it spread information about new technology and ideas. The Koreans had metal type printing since 1234, and printing was available in the Islamic countries as early as 1294 when Chinese printers were employed in Iran to produce notes for an experimental issue of paper money.[1] But it was not widely adopted probably because the demand for books was still by copying by hand.

Exploration

The age of exploration was driven by principally the Portuguese who were sailing further south down the coast of Africa. They were soon followed by the Spanish when they saw the riches coming back.

Impact of Changes

Columbus's discovery of the Americas spurred the Europeans into an age of exploration.

Major Events

Fall of Constantinople

For Europe the most significant event was the fall of the city of Constantinople (modern Istanbul, Turkey) to the Muslim Ottoman Empire in 1453. Constantinople was the capital of the Orthodox Christian Byzantine Empire for centuries. It was the most important center for trade between Europe and Asia. Its capture closed an important trade route off for European merchants although they could still buy Asian goods in Alexandria, Egypt. But there was no longer a direct route to Asia to purchase Asian products directly.

Columbus Lands in the West Indies (1492)

See significant projects below and voyages of exploration and discovery.

Vasco da Gama Voyage to India (1497)

From the early 1400s, the nautical school of Henry the Navigator had been following the African coastline and extending Portuguese knowledge of it. The goal in the 1460s was to round Africa to gain easier access to India for its spices through a reliable sea route. In 1497 a fleet left Lisbon consisting of four ships and a crew of 170 men to complete the voyage.

John Cabot Voyage to North America (1497)

This voyage was significant as it highlighted the extent of the continent of Americas to the North. It also brought the English into the race for new lands as national imperialism and economic competition between nation states took root.

New Tools Techniques and Breakthroughs

The European push to find a sea route to Asia was enabled by the availability of certain technologies namely, the Astrolabe, the Caravel, and advances in cartography.

Astrolabe

The first astrolabes were invented by the Greeks in the first/second centuries BCE (often attributed to Hipparchus). An early astrolabe was a rudimentary analog calculator capable of working out several different kinds of problems in spherical astronomy. Brass astrolabes were developed in the medieval Islamic world, chiefly as an aid to navigation and for finding the direction of Mecca.

In the 15th Century metal astrolabes improved on the accuracy of their wooden precursors and the first known European metal astrolabe was developed by Rabbi Abraham Zacuto in Lisbon. Along with portable sundials these became popular scientific gadgets of the day.

The metal astrolabe had a map of the major celestial bodies inscribed on it. This allowed a mariner to tell the location by positioning the stars on the astrolabe to match the stars in the sky. But the astrolabe worked only when the skies were clear and the positions of the stars were known. On cloudy days or when the stars in the sky were unfamiliar, celestial navigation and the astrolabe were ineffective.

Figure 9.1: Example of a 15th Century Astrolabe.

Advances in Cartography

Although cartography had evolved overtime in the 15th Century the discovery of Ptolemy's *Geographica,* a copy of the greatest cartographical text, by Florentine business men, sparked a renewed interest in maps. Coupled with the new perspective drawing which used gridlines, the new grids were applied to maps for the first time. This important breakthrough improved the ability to predict distances. Maps influenced Columbus who concluded that if he sailed west from Spain he would reach Japan in approximately 2000 miles.

Other Navigational Aids

These included assorted time pieces like an hour glass and sundial, compasses, cross-staffs, nocturnal, and traverse boards.

Caravel

Developed in the Mediterranean the caravel was the preferred vessel for Portuguese explorers. The small caravel (50 to 200 tons) was very fast, well balanced, agile, and economic. It could sail upriver in shallow coastal waters for inshore surveying of unknown waters, a clear advantage over the larger, full-rigged square-sailed ships which could not always be sailed with such precision. It was also easy to repair without a dockyard. The exploration done with caravels opened up the Spice Trade for

the Portuguese and Spanish. Eventually, the caravel was later
replaced by the larger carrack or nau which was much more
profitable for trading.

The caravel generally carried two or three masts and had a
large number of small lateen and square sails. The number of
and positions of the sails could be varied to different weather
conditions. Early caravels had a weight of 50 tons, a length of 20
to 30 meters, a high length-to-beam ratio of around 3.5:1, and
narrow ellipsoidal frame (unlike the circular frame of the nau),
making them very fast and maneuverable but with somewhat
low capacity. Towards the end of the 15th Century, the caravel
was occasionally modified by giving it the same rig as a nau with
a foresail, square mainsail and lateen mizzen, but not the high
forecastle or sterncastle, which would make it unweatherly.
In this form it was sometimes known as caravela redonda (a
bulging square sail is said to be round, redonda, in the Iberian
tradition). It was in such ships that Christopher Columbus set
out on his expedition in 1492.

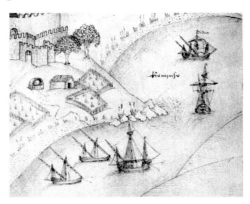

Figure 9.2: Examples of 15th Century Caravels.[2]

Longitude

The problem of determining longitude was not adequately mastered until the 18ᵗʰ Century by John Harrison with the invention of the marine chronometer. Up to then sailors relied on dead reckoning were the pilot estimated the ship's speed with a log-line, a knotted line with a weighted wooden float, thrown from the stern where the number of knots pulled off the reel determined the ship's speed. Time was measured with one-minute glasses and combined with the known direction of the compass determined progress along longitudinal lines. Each time the ship changed tack the time, distance, and direction were measured.

Moveable Typeface Print

Written communications forever changed when the movable type printing press arrived. In 1440 it arrived in Europe (originally established in Korea in 1234, page 274) when Guttenberg developed a printing press near Mainz, Germany. It used standardized, movable type and could print 300-500 sheets per day. In 1455 Guttenberg published his first Bible. By 1490 Guttenberg's moveable typeface printing was common in major cities across Europe.

Broadsheets and Newspapers

Printing with the availability of low cost paper was a catalyst for a communication's revolution. In 1450 newspapers appeared in Europe in the form of Broadsheets[3] which radically changed the ability to communicate stimulating a renewed demand for literacy and education. Around 13,000 books were in circulation in Europe by 1500 and this started to breakdown the monopoly of learning in universities. Books were printed on every subject available, and this replaced the reliance on experts which was useful for projects. The modern contract also evolved at this time with the overall increase in literacy.

Donatello's Bronze Statue of David 1469

This was the first significant statue since the Roman era. It was important as it marked Europe's awakening through the Renaissance.

Regions

Europe

Significant projects

Columbus Voyage of Exploration and Discovery (1492-1493)

With the fall of Constantinople a race developed to find a sea route to Asia as the European nations quickly realized the wealth in monopolizing the highly profitable Asian trade. Portugal was the first nation to begin actively seeking a sea route by going around the southern tip of Africa. The experience in exploring the coast line of Africa in the early 1400s gave the Portuguese access to the lucrative African trade which had been dominated by the Muslims.

Background

Christopher Columbus was born in Genoa, Italy in 1451. He began his seagoing career at the age of 14 when he enrolled in Prince Henry's school of Navigation in Sagres, Portugal. He served on various ships as a messenger, common sailor, and privateer. This experience led him to the idea that with enough supplies and water on board, it was possible to reach Asia and the East by sailing west. For the Europeans sailing long voyages in open water was not practiced, as ships tended to hug coast lines.

Initiating

The project charter was clear. Columbus's principal objective was to reach the east by going west, or the "Enterprise of the Indies." Christopher Columbus planned his exploration for many years but had to wait for the right time to find the support (people), the funding, and the required advances in marine technology.

Planning

Columbus's plan to sail west was not original and he never claimed that it was. Columbus drew knowledge from the Greek and Roman geographers who theorized that there was only one body of water that connected Europe and Asia. Only the distance was disputed based on the size of the earth through descriptions contained in works including the 2ⁿᵈ Century manuscript Geography by Ptolemy; Imago Mundi (Image of the World) by Pierre d'Ailly, published in the early 1480s; and The Travels of Marco Polo, written in 1298.

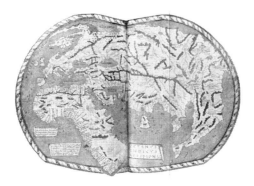

Figure 9.3: The map of the known world dates back to 1492.[4]

Several facts strengthened Columbus's faith in the feasibility of completing the project. The Azores and the Cape de Verde islands were the most westerly lands off the European continent at a distance of 1500 kilometers (800 miles). Westerly winds had washed up onto these shores curiously carved pieces of wood, seeds of unknown species, and the bodies of two men of strange color and facial features.

Columbus founded his theory on two mistaken propositions, that Asia stretched further eastward, and that Japan lay about 2,400 kilometers (about 1,500 miles) east of the Asian mainland. Columbus also underestimated the circumference of the earth. Columbus calculated that the Canary Islands lay only about 4,440 kilometers (about 2,760 miles) from Japan; the actual distance is about 19,000 kilometers (about 12,000 miles). Similar errors were made by other learned men of the time, including the Florentine geographer Paulo de Pozzo Toscanelli. No one else in Europe suspected that two vast continents lay in the way of a westward passage to Asia. No ship that was available in the 15th Century could carry enough food and fresh water on such a long journey.

Columbus's plan was based on his association with the Genoese community in Portugal and his expeditions to Africa. Both furthered his knowledge of Atlantic waters, and his trips to Africa brought him close to the Canary Islands, giving him knowledge of the Canaries currents. Also, while in ports in England, Ireland, Iceland, and other northern regions, Columbus may have heard stories of lands to the west of Iceland. Although the histories of the Vikings, who settled Iceland and Greenland in the 9th and 10th centuries, never became part of the knowledge base of Medieval Europeans. It is believed that stories of their encounters with the unknown islands in the northern Atlantic were widespread. He also read the accounts of Marco Polo, and the Geography of Ptolemy, the Greek geographer. Columbus's genius was his remarkable ability to gather information from around the Mediterranean and the Atlantic and combine his own experiences with the ancient theories from these books.

The technological challenges in planning the voyage of Columbus were numerous. The trip would not have been successful. Without developments such as the astrolabe and caravel, as well as advances in cartography. Columbus now had to wait for the opportunity to take advantage of this technology.

Finance and Support

The expense of such a voyage was substantial primarily in the funding of a small fleet of ships, the crew, provisions and equipment. The voyage needed a major sponsor and funding, and typically this was only available from the ruling monarchies and wealthy royal families. These projects demanded a return and required finding a commodity that could provide a significant payback for the high project cost. Columbus married a Portuguese noblewoman and was in Portugal for seven years. He waited for the right political timing and in 1485 he presented his plans to John II, the King of Portugal, a strong supporter of geographical explorations. Columbus proposed to sail out into the Atlantic, and search for a western route to the Orient, and then return. The King would equip three ships and grant one year's time. He also requested the title "Great Admiral of the Ocean", and governor of all the lands he discovered. In addition, he wanted one-tenth of all revenues of the discovered lands. The King submitted the proposal to his experts, the geographers of his court who pronounced it was a visionary scheme but, rejected it believing the route of 3,860 kilometers (2,400 miles) was far too short.

The King secretly dispatched a vessel to test the theory and a ship with Columbus's charts, sailed westward from Cape de Verde islands for a few days. Seeing nothing but a wide ocean the captain returned ridiculing the idea. Columbus learned of this and was disheartened by the deceit of the King. He decided he would not trust the King any further and so moved to Spain with his young son with the intent to present his plan to King Ferdinand V and Queen Isabella I. Spain lagged the Portuguese exploration of the Atlantic and open hostilities existed between the two countries.

At the time Spain was embroiled in a costly war to dislodge the Moors from Spain so at that point there was not much appetite for a costly project. With no luck he tried to generate interest with the rulers of other countries, but with little success. He had tried to get help from a number of sources like his old home-town of Genoa, Portugal, the Republic of Venice, some of the richest and most powerful of the nobles of Spain, and the King of England (whom he got his brother, Bartholomew Columbus, to go and see). The King of France was still left and was looking like his last option.

Columbus did generate interest with certain individuals. One of the leading sea captains of the City of Palos was Martin Alonso Pinzon. He became so interested in the project that he offered to lend Columbus money, enough to make an appeal to the Spanish monarchy. If Columbus was successful then Captain Pinzon would go into partnership with Columbus and help him prepare and sail.

After seven long years of continual lobbying and of negotiations, the opportunity opened up in 1492. Granada had just fallen and the Moors were expelled from Spain. Luis de Santangel, the court treasurer, convinced Columbus that he should make one last attempt to convince the King and Queen of Spain. Santangel realized that after the war Spain would be overflowing with hidalgos, young nobles who expected payment in land for military service. With not enough land to go around Columbus could provide an alternate solution.

Santangel intervened on the behalf of Columbus and set up the meeting with the King and Queen of Spain where Queen Isabella was finally won over. This was a time of national imperialism and economic competition between nation states seeking wealth from trade routes and colonies. Spain having undergone an expensive war was desperate for an edge over its European rivals. The King remained indifferent and pleaded the lack of funds. The Queen in her earnestness exclaimed, "I pledge my jewels to raise the money." But her sacrifice was not required. Santangel advanced most of the money, from the ecclesiastical revenues under his control. The remainder was made up of a syndicate of seven noble Genovese bankers, resident in Seville, linked to Américo Vespucci and funds belonging to Lorenzo di Pier Francesco de Medici and the friends of Columbus. The total cost of the project was three thousand crowns. Two vessels were all that Columbus asked for, and the pay for the crews.

Columbus had succeeded at last. It had taken him 17 years to bring the project to the point of execution.[5] The contract was signed on the seventeenth of April, 1492 between Columbus and the Spanish monarchy who granted Columbus the rank of Admiral of the Ocean Sea (Atlantic Ocean), and Viceroy and Governor of all the new lands discovered. Columbus had the option of buying one-eighth interest in the venture and get back one-eighth of the profits. The terms were unusually generous, but as his own son later wrote, the monarchs did not really expect him to return. The sovereigns also issued an order authorizing Columbus to press men into service, and pardon any prisoners from jails, giving them their freedom, if they went on the voyage.

Figure 9.4: First Voyage, Departure for the New World, August 3, 1492. LC-USZC4-1727.[6]

Executing (Journey)

Columbus called together the leaders of Palos so he could read the royal order to supply him with two of their ships. The leaders agreed to provide the ships, but refused to provide him a crew. Columbus had great difficulty in recruiting a crew as the experienced seamen of Palos objected to his plan and held back, unwilling to risk their lives in what seemed a crazy project. Nobody was willing to go. In the end the King's officers went to work and begged, threatened, and even seized some sailors forcing them to go. Even with an advanced payment they could not recruit a full crew for the two ships.

The friends of Columbus, Juan Perez and the friars of La Rábida, helped by actively persuading men to embark. Then Captain Pinzón who had promised Columbus help, tried his hand, and as an experienced and respected sea captain from Palos, he made the most progress with big promises to the sailors. He worked for weeks until at last he was able to do what even the royal order could not do, persuade enough local sailors to join the expedition. He even added a third ship to the Admiral's fleet. A crew of ninety sailors was assembled to man the three vessels. Only four convicts accepted the offer of a pardon.

Captain Pinzón even persuaded his two brothers of to join the voyage and were given officer positions. The names of the three vessels were the Capitana (changed to the Santa Maria), the Pinta and the Nina (Baby). The larger Santa Maria was selected as the flag-ship of the fleet, under the commander of the expedition, the now promoted Admiral Don Christopher Columbus.

The Santa Maria was a small carrack around 30 meters (100 feet) long and weighing about 100 tons, and with a crew of 40. The Pinta and Niña were caravels of around 20 meters (70 feet), weighing about 50 tons and crews of 25 on each. Most of the crew were Spanish and from Palos made up of largely experienced seamen and about 16 government officials. But there were no priests, soldiers, or settlers as this was a voyage of exploration and discovery, and not settlement. Columbus was so sure he was going to find a rich and civilized country, such as India, that he took with him a lawyer to complete all the forms, proclamations, papers for the kings and princes he expected to visit. He brought a secretary and historian to write out the story. He also brought a linguist who could speak almost a dozen languages, a jeweler and silversmith for trading, a doctor and surgeon, cooks and pilots.

On the 3rd of August, 1492, they took mass, and then set sailed from the Bar of Saltes, making for the Canary Islands. Life aboard the ships was not comfortable, there were no crew's quarters and no mess halls. Only the captains and pilots had very small cabins. At night the crew slept wherever they could find a vacant spot, tying themselves down to prevent being tossed into the sea. Prayers, songs, stories, chores, eating, and waiting filled the sailors' days, and stargazing under a new, unknown sky filled their restless nights.

The ships carried enough water and provisions for several months, at a time when two weeks at sea was a long voyage. Supplies on board included pigs and hens (to be killed aboard), water, vinegar, wine, olive oil, molasses, honey, raisins, rice, cheese, garlic, almonds, sea biscuits (hardtack), dry legumes such as chickpeas, lentils, beans, salted and barreled sardines, anchovies, dry salt cod and pickled or salted meats (beef and pork), salted flour.

Equipment included navigational instruments, nautical almanacs, charts, compasses, magnets, hourglasses, and rulers. Also trade items were taken like glass beads, brass rings, knitted caps, gold, silver, pearls, and spices.

Columbus was aware of the circular wind patterns across the Atlantic, a closely held fact. The trade winds, both the easterly, a brisk wind from the Canaries, and the prevailing westerly to the coastlines of Western Europe, blew from the middle latitudes of the North Atlantic, curving southward towards the Iberian Peninsula.

Navigation in the 15ᵗʰ Century was far from an exact science, although several navigational tools and aids were available the most important were compasses, astrolabes, hourglasses, maps, and charts. Although celestial navigation (finding direction by the positions of stars) was the favored method while sailing under familiar skies, a technique known as dead reckoning was more dependable on voyages in unknown seas.

Figure 9.5: The three Ships of Columbus had to stay in close contact in the vast Atlantic.[7]

In dead reckoning the position of the ship was determined by starting from its last known location. Then, by calculating the ship's direction, how fast it was going, and how much time had passed, the pilot came up with a new position. Pilots could calculate the distance traveled in an hour or a day by dropping a floating object in the water at the front of the ship and timing how long it took to get to the back of the ship. Knowing how long the ship was, the pilot could calculate how fast the ship was moving and, thus, how far they had traveled.

Columbus preferred dead reckoning over celestial navigation which he was never comfortable with. Above all, he learned to interpret the signs of nature, such as the behavior of birds, the smell of the air, the color of the sky, the condition of the sea, the pressure he felt in his joints, the appearance of floating debris, and more. Successful navigators survived by "reading" nature in this way. Columbus was expert at this and could even predict hurricanes accurately.

When the sailors saw the last trace of land fade from their sight, many were wracked by superstitious fears. To their dismay, the compass no longer pointed directly north, and they believed that the very laws of nature were changing. The trade-wind took them steadily westward but this worried them further as how they could ever return against it. Eventually, signs of land appeared, birds, and floating green plants were seen. The clouds near the horizon assumed the look of land, but they disappeared with only the broad ocean in front of them. The sailors, lost heart, and started to insist they turn back home.

Columbus, with wonderful tact and patience, explained all these appearances. But the more he argued, the louder the murmurs of discontent became as the crew secretly plotted to throw him overboard.

Columbus had a hard time keeping his crew in order and from mutinying. He had to win the crew over, day by day.[8] He employed several communication strategies. First, he used soft words and distracted them from the present danger. Second, he focused on their large hopes and aspirations, of riches and fame. Finally, he reminded them of what their princes would do if they attempted to come against him or didn't obey his orders. The most important part of his communication strategy was that although he knew their feelings he did not waver. He was absolutely resolute to the cause of completing the voyage which instilled confidence in the crew.

After 29 days of sailing without sight of any land, on October 7, 1492, the crew spotted shore birds flying west and then changed direction as if making for landfall. Soon, signs of land silenced the crews' murmurs of discontent. An artificially carved staff, and a branch of thorns and berries were seen floating by. From the Pinta they let out a joyful cry, "Land!"

In the morning a shore line lay before them. Columbus, with a small detachment of officers and men, landed bearing banners on Friday, October 12, 1492. Dressed in a splendid military suit of scarlet embroidered with gold, fit for such a historic occasion, he threw himself upon his knees and kissed the earth. With tears of joy he thanked God and then formally planted the cross, thus taking possession of the country in the name of Ferdinand and Isabella.

The islanders stood along the shore in numerous lines looking on with innocent amazement. Columbus found the land to be an island, which he named St. Salvador believing the island to be off the eastern coast of India. He therefore called the dark-hued natives, Indians. Careful inquiries were also made concerning the rich products of the East, such as spices, precious stones, and especially gold. The islander had a few golden ornaments which they readily bartered for hawks' bells. The islanders also introduced Columbus to "tobacco," herbs wrapped in a dry leaf. When questioned about it they said that it took away fatigue. In the quest for oriental treasures he split up his ships and searched and discovered other islands (Cuba, and Haiti).

The situation changed on Christmas morning in 1492 when the Santa Maria ran aground on the northern coast of Hispaniola whilst under the watch of a junior deck hand who failed to alert the sleeping crew. Badly damaged she was abandoned and some of her crew returned in the remaining 2 ships. Columbus was received by the native leader Guacanagari, who gave him permission to leave 40 of his men behind. Columbus founded the settlement of La Navidad using the salvaged materials from the Santa Maria.

Controlling and Monitoring

On January 15, 1493 Columbus, urged by his crew, relinquished the search for oriental treasures, and turned his vessels homeward by way of the Azores. Columbus may not have found the oriental treasures yet there can be no doubt of the importance, financially and commercially speaking, of the discovery of tobacco. In the end it proved more productive to the Spanish Crown than all the gold mines of the Indies.

Figure 9.6: Columbus 16th Century engraving.[9]

Fig. 199.—Discovery of America, 12th of May, 1492.—Columbus erects the Cross and baptizes the Isle of Guanahani (now Cat Island, one of the Bahamas) by the Christian Name of St. Salvador.— From a Stamp engraved on Copper by Th. de Bry, in the Collection of "Grands Voyages," in folio, 1590.

Figure 9.7: Columbus triumphs in 1492.[10]

On his return to Spain the reception was flattering in the extreme, where the whole nation took a holiday, and he was hailed with the ringing of bells. The King and Queen of Spain were enamored by their new and unexpected acquisition. As Columbus described the beautiful land he had discovered, the brilliant birds, the tropical forests, the climate, and the natives waiting to be converted to the Christianity.

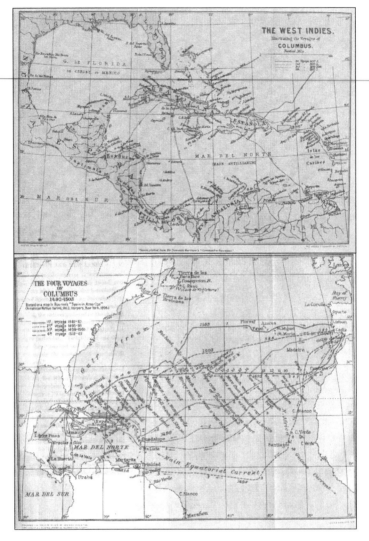

Figure 9.8: Columbus's four voyages of exploration.[11]

Closing

Columbus explored the world in search of new trade routes for
their goods and contacts with India and China, and a shortened
journey to traditional overland routes. His exploration across
the Atlantic Ocean had a profound impact. It led to general
European awareness of the American continents in the Western
Hemisphere and directly opened it up to European colonization,

large-scale exchanges of plants, animals[12], cultures, and ideas between the two worlds. On the darker side it also led to the deaths of millions of indigenous peoples from war, forced labor, and disease such as diphtheria, measles, smallpox, and malaria which they had not been exposed to.

After several attempts to establish a settlement on the island of Hispaniola, Columbus initiated the process of Spanish colonization which foreshadowed general European colonization of the "New World."

Key Players

Christopher Columbus (1451-1506)

Columbus completed two further voyages of exploration. Columbus left Cádiz, Spain, on September 24, 1493 to find new territories, with 17 ships carrying supplies, and about 1,200 men to colonize the region. He learned that the settlers from the settlement of La Navidad had mistreated the natives, who retaliated by killing all 40 of them. He spent a year and half in an ill fated effort to start another colony returning back for more men and supplies.

On May 30, 1498, Columbus left with six ships for his third trip, and arrived on Hispaniola on August 19ᵗʰ to find that many of the Spanish "settlers" were discontented and thought misled by his promises of bountiful riches. In reality many were adventurers only after the gold. Columbus repeatedly had to deal with these rebellious settlers, some even tried to kill him, so he had some of the ringleaders hanged.

In October 1499, Columbus sent two ships back to Spain. A number of returning settlers and sailors to Spain lobbied against Columbus at the Spanish court, accusing him and his brothers of gross mismanagement. The royal patronage started to waver. Columbus had asked the Court of Spain to appoint a royal commissioner to help him govern. The Court appointed Francisco de Bobadilla and he was given total control as governor. His authority went far beyond what Columbus had requested, and it disregarded the grant of governor-generalship to Columbus over all the lands discovered.

The new governor sent Columbus home in chains which shocked the Spanish nation at the indignity to such a man. He was arrested for a period and accused of governing tyrannically. "Even those who loved him [Columbus] had to admit the atrocities that had taken place."

Columbus was put into jail for six weeks before being released by King Ferdinand who summoned the Columbus brothers to hear their case. Satisfied by this he restored their freedom and wealth, and agreed to fund Columbus's fourth voyage. However, his position as governor was denied.

On May 11, 1502 Columbus left with four ships, for his fourth trip. He was denied port on Hispaniola by the new governor and proceeded to explore Central America and the Bay Islands off the coast of Honduras. Following storms 1503 he ended up beached in Jamaica. He sent for help but was ostracized by the new governor. Eventually, he was rescued and returned to Spain in ignominy in 1504.

In 1506 Columbus died a grieved and disappointed old man. At his request, his chains were buried with him, a touching memorial of Spanish ingratitude.

"It is now seventeen years since I came to serve these princes with the Enterprise of the Indies. They made me pass eight of them in discussion, and at the end rejected it as a thing of jest. Nevertheless I persisted therein... Over there I have placed under their sovereignty more land than there is in Africa and Europe, and more than 1,700 islands... In seven years I, by the divine will, made that conquest. At a time when I was entitled to expect rewards and retirement, I was incontinently arrested and sent home loaded with chains... The accusation was brought out of malice on the basis of charges made by civilians who had revolted and wished to take possession on the land.... I beg your graces, with the zeal of faithful Christians in whom their Highnesses have confidence, to read all my papers, and to consider how I, who came from so far to serve these princes... now at the end of my days have been despoiled of my honor and my property without cause, wherein is neither justice nor mercy."

Renaissance Engineers

This period saw the emergence of Renaissance Engineers or Pseudo Project Managers through men like Leonardo da Vinci, Francesco di Giorgio, Fillipo Brunelleschi, and Mariano di Iacopo, (called Taccola).

Leonardo da Vinci (1452 –1519)

He was the Renaissance man, a truly multi talented and skilled scientist, mathematician, engineer, inventor, anatomist, painter, sculptor, architect, botanist, musician, poet and writer. Leonardo the engineer developed concepts ahead of their time a helicopter, a tank, a calculator, and the double hull, and outlining a rudimentary theory of plate tectonics. As a scientist, he greatly advanced the fields of anatomy, civil engineering, optics, and hydrodynamics.

Chapter Wrap-up

Conclusions

Columbus was a man of patience and commitment. He had to overcome numerous obstacles just to launch the project. The most notable was the difficulty in getting the necessary backing for his project. He had to bide his time and wait for the right opportunity, a total of 17 years. He had to be very persistent, and resourceful, and to change allegiances to find the right sponsor/backer. He also had to wait for the right technology to come along, in the form of the caravel, and the astrolabe which he mastered. From a project perspective Columbus had to very much initiate the project, and drive it forward. The biggest challenges were in managing the crew who were very fearful of the voyage. They were motivated by riches and glory. The voyage proved the viability of Trans-Atlantic travel, and the trade winds, it took 29 days to reach the West Indies, and 224 days to complete a round-trip. It was inspirational and a catalyst to other European explorers, and spawned the voyages by Cabot, and Magellan. The lasting legacy from Columbus's voyage was overcoming the fears of crossing the Atlantic, and establishing viable routes using the trade winds.

Key Lessons

PMBoK® Guide Knowledge Areas:

- Integration Management

  - For Columbus's voyage the business case seemed very lucrative with potential land for resettlement, possible riches and spices. The go ahead decision from the sponsors was held back because of the perceived risks. When they finally agreed the monarchs still did not really expect him to return.

  - The project charter had to be approved by the monarchs who not only provided the financial backing, but permission in claiming new lands in their name through a Royal Charter.

- Scope Management

  - The principal requirements were to complete the return journey safely without loss of life, and return profit by bringing back goods.

  - The scope in terms of number of ships, men, provisions, and equipment was dictated by:

    - The likely daily progress which was constrained by the weather and conditions of the Atlantic.

    - The risks of the unknown mitigated by a fleet of three ships.

    - The availability of ships and finding a willing and committed crew.

    - The volume of provisions (food and water) that could be carried without spoiling for the 2000 mile journey.

  - Columbus was aware of the circular trade-winds and knew this clockwise wind pattern would reduce the project scope by weeks or months by not having to beat against the winds. This significantly reduced the amount of food and water to be carried.

  - Although the scope seemed straightforward in sailing for 3,000 kilometers (2,000 miles) and then returning, the inaccuracies in the estimates could have been catastrophic had the Americas not been the same distance away as what he thought Japan to be.

○ Columbus could verify the scope by closely measuring distances and keeping a daily log.

○ The scope of the project changed dramatically when one of the ships was lost. The return voyage was made much more difficult and dangerous. Around 40 men were forced into making a settlement which they were not prepared for at the outset of the project. This ultimately failed dismally when the settlement was wiped out.

• Time Management

○ The project was time dependent as the ships could only carry several months of adequate supplies, of food and water. Therefore, time overrode both project cost and scope. Columbus had to accurately measure distance so he could calculate the time and the ability to survive the return leg on the remaining supplies. If necessary he would have to turn back before the objective was reached.

○ Columbus estimated about 20 days of sailing which turned out to be 29 days.

○ The ships were maneuverable, and with flexible rigging traveled a respectable 160 kilometers (100 miles) per day under favorable conditions. They could sail in shallow water, a major advantage for exploration.

• Cost Management

○ Columbus's voyage required a significant budget but at great risk to the investors, and he did secure a number of private investors. Columbus struggled for a long time to get royal patronage, in the name of finding new lands and riches. It was the only way to get enough ships, sailors and secure resources, and cooperation from various organizations.

- ○ The Guttenberg's moveable typeface printing made books widely available in Europe by the 15ᵗʰ Century. This provided subject expertise at the finger tips of the project manager. It was now a lot easier and less expensive to have a book over an expert.

- Quality Management

  - ○ In the planning phase the project areas where quality was paramount included:

    - The accuracy of charts and maps.

    - The seaworthiness of ships, for example, the hull against leak, or the rigging against storms.

    - The securing of provisions (food and water) to last the journey. For example, olive oil and olives were stored in earthenware jugs. The rest of the food was stored in wooden casks like meat preserved in brine, or dry goods. All were stored in the hold, the driest section of which was reserved for casks carrying dry provisions.

  - ○ Daily inspections were made by a cooper (barrel maker), responsible for inspecting and keeping the casks tight.

  - ○ In executing the project, one of the most critical areas was navigation and before the invention of accurate clocks it was very difficult to find longitude at sea. This was one of the most significant problems continually faced by the project. The crude techniques of celestial navigation could readily put the ships off course. The pilots had to constantly re-check their findings using dead reckoning.

  - ○ The ability to accurately measure became life critical for the ocean explorers of the 15ᵗʰ Century, as counting the knots was the only practical way to measure distance and therefore longitude. Columbus figured it was a 3,200 kilometers (2,000 miles) journey to Japan from Europe and used this technique accurately in getting across the Atlantic.

- Human Resources Management

- ○ Columbus had great difficulty in acquiring a crew, as there was a genuine fear for their lives and reluctance to go. A few decades earlier it took the Portuguese 16 attempts to cross the equator south because sighting's of the North Star disappeared which petrified the seamen.

- ○ The crew was made up of mainly local boys from Andalusia, and nearly all experienced seamen. On the first voyage, the crew was reasonably well paid as follows:

  - Masters and pilots, 2,000 maravedis per month; able seamen, 1,000 maravedis per month; ordinary seamen and ship's boys, 666 maravedis per month. Total payroll was 250,180 per month. At the time gold was worth 3,000 maravedis per ounce.

  - Only four convicts took up the Spanish Sovereigns offered of amnesty if they signed up for the voyage.

- ○ The crew was reasonably well looked after with adequate provisions for the project. The ships were general purpose cargo vessels, uncomfortable, and not designed for exploration. The crew slept in cramped and damp conditions on deck or below on the ballast pile where cargo, the main anchor, and heavy armaments were stowed.

- ○ Columbus carefully planned the trip to East Asia right down to determining the right kind crew and skills that he needed on board based on his objectives and the scope of what he was doing.

- ○ Developing the acquired men into an effective team proved a challenge because of their fears and expectations of the unknown.

- Communications Management

  - ○ The project had a wide variety of stakeholders from the monarchy to Santangel (court treasurer), the leaders of Palos, personal backers, Captain Pinzón, and the crew itself.

- ○ Columbus had to sell the idea and get a broad interest from key stakeholders, and have the project adopted by a crew. He had to promise glory and paint a vision.

- ○ Columbus relied on Captain Pinzón to help with the adoption plan. As an experienced and well respected sea captain from Palos, he had influence and sway with local sailors. He worked for weeks to convince them that the voyage was feasible and could be completed safely. He was able to make inroads with big promises. Eventually, he was able to persuade enough local sailors to join the expedition. He even added a third ship which proved to be critical in the end.

- ○ Columbus had a hard time in keeping his crew in any order and from mutinying. He had to contrive how to win the crew over, day to day, and employed several communication strategies. First, he used soft words and distracted them from the present danger. He showed empathy to the crew to gain their respect. Second, he focused on the large hopes and aspirations in front of them, of riches and fame. Finally, he reminded them of what their princes would do if they attempted something against him or not obey his orders. The most important element of his communication strategy was that although he knew of their feelings of feel he did not waver from his objectives.

- ○ Columbus was consistent in his communication about reaching his objective. In fact, he was absolutely resolute to the cause that he would complete the voyage. He did not waver, and built up the crew's confidence as the voyage progressed.

- ○ The three ships had to stay in constant communication as any separation increased the risk substantially.

- ○ On the westward return Columbus kept two sets of distance figures in the log. One set was to allay the fears of the crew that they had sailed too far from Spain.

- Risk Management

  - Columbus faced the real risk of not finding a sponsor; it took him 17 years to find one. All of the potential sponsors, monarchs around Europe (Portugal, England, and France) felt the risk was too high and they would lose their investments.

  - Not having an accurate maps or knowing the distance to the Indies (based on an inaccurate estimate of the Earth's diameter) were major risks. Columbus accepted these risks because he was aware of the trade winds (Northeast and the Westerlies for a return journey). He was confident of steady winds in both directions if he traveled at the right time of year and at the right latitudes.

  - A fleet of three ships reduced the overall risk of failure. When a ship was lost, Columbus was still able to complete the journey. Had a further ship been lost the project would have failed.

  - Columbus accepted the risk of making an unexpected settlement after he lost one of the ships ran when it aground. Although they were able to create a settlement from the salvage it failed to survive its first year.

  - Columbus kept accepting increasing amounts of risk because he didn't want the voyage to fail.

- Procurement Management

  - Columbus had to procure enough ships to make the voyage feasible. The ships had to be well supplied and provisioned for at least two months.

  - Columbus entered into a contract with the Spanish Royals with very high personal demands including titles (Governor, Admiral, and a Knighthood) and 10% of the profits. At the time a single spice ship was worth a fortune. The Spanish Royals entered the contract because they did not think Columbus would return.

      ○ Columbus commissioned a "Book of Privileges" spelling out the exact agreements he had made. It also incorporated the Papal Bull *inter caetera* in which Pope Alexander VI extended Spain's rights to the New World.

Educators

- Discuss all the risks in Columbus's project and his approach in mitigating these?

- Discuss in the context of today's projects the importance of finding the right sponsor?

- Discuss Columbus's approach to managing his stakeholders and crew.

CHAPTER 10

The 16ᵗʰ Century

The period was dominated by the age of settlements and colonies spurred by exploration driven by the European powers. From a project management perspective European Colonization brings in vast wealth that is reinvested into further exploration. Large projects are sponsored by the crown and state more than through the church. The book industry expands and puts a vast array of subjects into the hands of a project manager New technologies improve the manufacture of glass and brass materials, and new equipment is manufactured from these materials.

In this Historical Period

- Brief synopsis of trends/changes (natural disasters, wars, technology, epidemics).

    a. Widespread printing of books leads to a communications revolution.

    b. European race to acquire colonies (Spain, Portugal, England, France, Holland) colonization of Africa, Asia, and Americas.

- Impact of changes

    a. Destruction of native cultures in South and Central America.

    b. Dissemination of knowledge through books (do-it-yourself) and Propagation of ideas like Protestantism when Luther publishes in 1517.

    c. Development of Mercantilism (International Trade).

- Major events

    a. 1514 Copernicus established a scientifically based heliocentric cosmology that displaced the Earth from the center of the universe.

    b. 1519-1522 Ferdinand Magellan the first navigator to sail around the world.

    c. Communications revolution helps the Protestant Reformation of the 16th and 17th centuries.

- New tools, techniques and breakthroughs

    a. New crops introduced from the new world.

    b. Paper money in Europe.

    c. Canon and firepower.

    d. Cartography.

    e. Pencil 1565.

    f. Miniaturization of clocks.

    g. New furnace for dirty coal spurs glass and brass making.

- Regions

    a. Europe and exploration of South America and Pacific Ocean.

- Significant projects

    a. Ferdinand Magellan's Voyage Round the World.

- Other Notable Projects

    a. Hardwick Hall (1591-1597).

    b. St Peter's, Rome.

    c. St Basil's Cathedral, Moscow.

- Key players

a. Ferdinand Magellan.

b. Michelangelo.

• Chapter Wrap-up

Trends and Changes

The world population in this period grew from 420 to 540 million. The nation-state continues to evolve in Europe.

Communications Revolution

The development of printing coincided with the availability of low cost paper. This combination was revolutionary and replaced the high cost of writing on animal skins or parchment. Printing with engraved copper plates also evolved and continued to be the standard until photographic techniques were developed. This Communications Revolution saw the widespread printing of books. The book industry expanded and made a vast array of subjects available, which spread knowledge and expertise. In Europe the Protestant Reformation of the 16th and 17th centuries was helped by the Communications Revolution.

European Colonization

For the next 300 years a race develops by European powers to acquire empires and become superpowers. It is a matter of time before Europeans make inroads into civilizations in Asia and Africa. This dominates the agendas of European Governments.

Impact of Changes

Destruction of Native Cultures

Indigenous cultures in South and Central America suffer destruction with wide spread colonization and this continues for the next few hundred years. The Spanish New World population in 1500 was about 50 million and by 1650 it had shrunk to 4 million despite Spanish immigration.[1]

Dissemination of Knowledge

New ways were developed to present, arrange and illustrate books. The transmission of technical information was more accurate as experts wrote books. Cheap and fast printing facilitated the spreading of ideas. For example, Luther's ideas spread like wildfire in 1517 (300,000 copies of his works were printed).

Development of Mercantilism

For the European powers the establishment of colonial trade (International) became an arm of governmental policy. Mercantilism, with its emphasis on foreign exports, was established and with it the acquisition and development of colonies as exclusive markets and sources of raw materials.

Major Events

Increase in Scientific Interest

The Renaissance sparked off renewed interest in science. Copernicus established a scientifically based heliocentric cosmology that displaced the Earth from the center of the universe. The Christian Church felt threatened and tries to suppress this.

Magellan Circumnavigates World

When Vasco da Gama and the Portuguese arrived in India in 1498. It became urgent for Spain to find a new commercial route to Asia. The Treaty of Tordesillas reserved for the Portuguese the routes that went around Africa. The Spanish Crown then decided to send out exploration voyages in order to find a way to Asia by travelling westwards.

New Tools Techniques and Breakthroughs

Agricultural

The introduction of new crops from the new world had a significant impact on the population in Europe, Africa and Asia.[2] Maize and potatoes are about twice as productive as wheat and barley.

Paper Money

Although available in China for over 500 years from the 9th through the 15th Century, paper money was not established in Europe until the 16th Century. Like in China, paper money facilitated economic growth, investments in exploration, and colonies.

Canon and Firepower

When gunpowder arrived in Europe from China, it evolved very quickly for its military uses. It was possible to attack from considerable distance. With this evolved ballistics a new science that used physics and geometry in calculating the trajectory of a projective. This led to the development of measurement devices, the quadrant[3] and the vernier which were also useful in surveying.

Cartography

The same devices quadrant and vernier were also employed in map making which became more accurate. With maps compass lines and other navigation aids were now included. New map projections were devised, and globes were constructed. New metal printing presses allowed for the mass production of accurate maps. Mercator issued a world atlas in 1585.

Miniaturization of Clocks

A significance development was in the form of a spring-driven mechanism to replace clocks powered by hanging weights. This made possible small clocks which could be taken from room to room, or in a carriage, eventually making possible the pocket watch.

It was not practical for the general public to consider minutes until the first mechanical clocks that displayed minutes appeared near the end of the 16th Century. Even though watches had appeared earlier they did not come into practical use for many centuries (17th/18th).

New Furnace

The development of a new underfed furnace for dirty coal kept glass free from soot and smuts. This spurred glass and brass making and the coal-mining industry increased its production. This led to lower cost glass used within building construction projects.

Risk

From the 16th Century onwards, the term risk further evolved and attained a beneficial meaning, for example in middle-high-German "Rysigo" was a technical term for business, with the meaning "to dare, to undertake a business and aspire for economic success".

Regions

The predominant region is Europe which is preoccupied with the exploration of North and South America, and the Pacific Ocean.

Significant projects

Ferdinand Magellan's Voyage Round the World (1519-1522)

Background

Ferdinand Magellan was born in Portugal in 1480. As a boy, he worked in the Queen's household and heard about all the discoveries that were being made in the new world and may have been present when Columbus visited the Portuguese court in 1493. In 1505, Magellan finally went to sea on a military expedition. Magellan came up with the idea of finding a new sea route to the Spice Islands but to realize this he needed to get support, financial backing, and patronage.

Initiating

The project objective was purely commercial, to find a Western Sea Route to the Spice Islands. The plan was to sail west around South America, cross the Pacific Ocean, reach the Spice Islands, and return to Spain thus circumnavigating the world. The Pope had divided the world into spheres of influence and he would stay in Spanish waters to the west. If he was unable to find the strait, he would sail around Africa. The business case was based on bringing back ship loads of spices that were worth more than gold by weight.

Magellan needed the full support from the Portuguese King. Unfortunately King Manuel would not fund this voyage because he saw no need for such a frivolous expenditure. He did not like Magellan's idea and rejected it.

Planning

Magellan's only other option was to approach the Court of Spain but in such a way where his Portuguese would not be held against him. He moved to Spain, married Diego Barbosa's daughter, renounced his nationality and became a Spanish citizen. Eventually, Magellan presented a plan to King Charles V that would give the ships of the Crown of Castile full access to the Spice Islands. He claimed that he knew of a secret strait through the new continent of South America.

Magellan, had a well painted globe in which the whole world was depicted, and on it he indicated the route he proposed to take, saying that the strait was left purposely blank so that no one should anticipate him.

Las Casas, the historian of the Indies, present in Valladolid when Magellan presented his plan to the King.[4]

The plan was to sail west, go through the strait, cross the Pacific Ocean, reach the Spice Islands, and then return to Spain.

King Charles listened to the proposal favorably and agreed with the plan which Magellan felt could not fail. Magellan was furnished by the King's orders with five ships, manned by two hundred and thirty-four men, and enough provisions for two years. The adventurers were granted a twentieth of the clear profit and the governorship of any islands they might discover to be vested to them and their heirs.

Executing (Journey)

Magellan's principal objective was to find the strait through the newly discovered and uncharted continent of South America to the Pacific an estimated 60 day sail. He could then just sail across the new sea all the way to the Spice Islands. This was a grossly under estimated three day sail.

On August 10, 1519 Ferdinand Magellan set sail from Seville with five ships under his command – *Trinidad, San Antonio, Concepción, Victoria, and Santiago.* The schedule had been delayed by 5 weeks, as the Spanish authorities had not trusted the Portuguese Magellan and held him back as they switched his Portuguese to a Spanish crew of 270 men.

Figure 10.1: Map of Ortelius: Magellan's ship Victoria.[5]

They sailed off the Cape de Verde Islands but got into a calm sea where they remained for seventy days without much progress. At last the wind picked up and they reached South America in early December, but since Brazil was Portuguese territory, Magellan had to avoid it. On December 13 he anchored south of the colony near present-day Rio de Janeiro and there the crew resupplied with an abundant supply of fruits, sugar-canes, and animals of various kinds. From there the journey was excruciatingly slow as he looked for a strait along the uncharted coast line. He had to explore every inlet, river, and bay. On January 10, 1520 they came across a huge bay and spent two weeks exploring the basin of the huge river Plate, 200 hundred miles across. Another explorer Juan Díaz de Solís, had previously sailed there in 1516 with 3 ships and a crew of 70 men. They had all died exploring the southern part of the new continent of South America. This was fresh in the minds of Magellan's crew.

As Magellan sailed farther south, the weather became colder with freezing nights and cool days which left the crew in serious danger of frostbite. The drop in temperatures saw a drop in the spirits of the crew.

Figure 10.2: The Strait of Magellan at the southern part of South America.[6]

On April 2nd, a mutiny broke out involving two of the five ship captains. It became obvious to the crew that Magellan had no idea where the strait was. He met the mutiny with force and quelled it quite brutally. The mutiny was unsuccessful because most of the crew remained loyal to Magellan. Reportedly those killed were drawn and quartered and impaled on the coast. Years later, their bones were found by Sir Francis Drake. The journey resumed and *The Santiago* was sent to scout the coast and was wrecked in a storm. The crew survived and made it safely to shore, where they were picked up and returned to Magellan.

On October 21st, the fleet reached Cape Virgenes and found deep inland, brine waters, concluding they had reached the passage. The four ships started through the 600 kilometers (373 miles) long passage that Magellan called the Estrecho (Canal) de Todos los Santos. On November 20th Captain Gómez deserted and returned with his ship to Spain.

On November 28th, the three remaining ships entered the South Pacific. These were the first Europeans to reach Tierra del Fuego. The journey had taken took 15 months much longer than the expected 2 months. The three ships waited at the entrance of the straits for the missing ship until the rendezvous time appointed had passed. With the severe cold weather Magellan steered towards the equator and a north-westerly course and with this morale improved.

Figure 10.3: A 1507 map showing the limits of the known world.[7]

Both the winds and the sea were calm, and as a result because of the calmness Magellan named it the Pacific Ocean.

On the 24ᵗʰ of January, 1521, two coral atolls were sighted but they did not land on these barren rocks. As they sailed further the crew finished all their provisions. The fresh water became so bad that it only could be swallowed with difficulty. Nothing was left to eat but pieces of skin and bits of feather. Scurvy set in where nineteen men died, and thirty were too weak to do duty. They anxiously searched for islands where they could obtain fresh provisions. Magellan continued to deal with the crew very harshly.

Figure 10.4: This 1581 map shows a more complete view of the continents.[8]

They continued to sail with a smooth and fair wind at a good rate, heading northwest they reached the equator on February 13ᵗʰ. After crossing the equator they arrived on the 6ᵗʰ of March at a cluster of three fertile inhabited islands the Marianas and Guam - 3 months and 20 days from the time of leaving the Straits.

On March 16ᵗʰ, Magellan reached the island of Homonhon in the Philippines with a crew of 173 left out of the original crew of 270 men. Here they resupplied. The needles of their compasses varied so much, and moved so irregularly, that they were often compelled to quicken them with a touch of the loadstone. Getting across the Pacific was an astonishing piece of navigation.

Magellan did not stop anywhere for long until he reached the central island of Cebu in April 1521. There he baptized the local chief, Humabon and 2,000 of his followers. The price of Humabon's conversion was aid in fighting an enemy chief, Lapu-Lapu of Mactan Island, which was, one mile away. Magellan was so confident of victory that he only took 60 men to Mactan.

Lapu-Lapu heard they were coming and assembled 1,500 warriors. Humabon brought 600 warriors to help, but Magellan told him to stay on the sidelines. His crew could do the job by themselves. The resulting battle was one-sided; the Spaniards never even got to Mactan's shore, and only 8 of the 60 men survived. Magellan was not among the survivors and was killed.

Magellan's death gave Humabon second thoughts about the alliance. He invited 24 officers to a banquet, plied them with palm wine and women, and then attacked them - killing all but 3. Now only 100 of the original 270 crewmen were left to complete the expedition. This was not enough to man all 3 ships, so they burned the one in worst shape, the Concepcion, and divided her crew and provisions between the other 2, the Trinidad and the Victoria.

It normally takes a week to sail from the Philippines to the Moluccas, but the crew had little idea where they were going, and they wandered aimlessly around Borneo and the Sulu Sea for 3 months. Finally they reached the Spice Islands and loaded a cargo of cloves, in fact, overloaded, for the Trinidad sprung a leak and could go no farther.

Controlling and Monitoring

Juan Sebastian del Cano (Elcano), the expedition's new commander, chose not to wait for repairs and took the *Victoria* alone, a wise move since *Trinidad* was captured by the Portuguese not long after that. On the return leg a further 65 men were lost to primarily scurvy. Ironically the precious cargo of cloves had vitamin C.

On September 6, 1522, 10 months and 11,000 miles later, *Victoria*, the one surviving ship, made it back to Spain with only 18 men out of the original 270 crew. Add to that the 17 men captured and later released by the Portuguese, and there were 35 survivors. The project took over 3 years and 235 lives were lost.

Closing

Magellan's expedition was the first to circumnavigate the globe and the first to navigate the strait in South America connecting the Atlantic and Pacific oceans. Magellan finally proved what all the great discoverers were really assuming, the roundness of the world.

The one ship that returned to Cadiz was loaded with spices from the Moluccas and these were sold by the Spanish Government for so large a sum of money that the king was remunerated for the whole cost of the expedition with a very large profit.

Other Notable Projects

Hardwick Hall Project (1591-1597)

This Tudor building project was unique for its substantial use of glass - "Hardwick Hall more glass than wall" the saying went, where the glass is nearly 50% of the façade. Robert Smythson the architect created a distinct landmark. The windows were exceptionally large and numerous for the 16ᵗʰ Century and were a powerful statement of wealth at a time when glass was a luxury. The massive windows stretched almost from floor to ceiling in every room on every floor. Glass in the Tudor period was an expensive material but its usage in the building of Hardwick Hall produced a spectacular effect on its vast array of windows. Some of the windows appear to illuminate just one room but they are cleverly constructed to light two rooms on two storeys. There are also some false windows which conceal chimneys. The use of so much glass made the interior of Hardwick Hall lighter and airier.

> "Clear glass was used by the Romans. Its production was rediscovered by the Venetians in the 15ᵗʰ Century and came over to England in the 16ᵗʰ Century".[9]

To make a pane of glass was a very time consuming and painstaking process. A blob of glass was blown into a cylinder shaped bubble which was placed on a cooling table and cut in half. A small piece of glass was thus produced. The small pieces of glass for the windows were joined together with lead.[10]

Figure 10.5: Hardwick Hall seemed to have more glass than wall. Glass made up 50% of the frontage.[11]

The building inspired a new architecture (glass in buildings) and changes in the manufacture of sheet glass in the 17th Century.

St Peter's Basilica Project (1506-1626)

The period witnessed an extraordinary series of projects that constructed palaces, villas, and churches. This completely transformed the city of Rome. It all started with St. Peter's Basilica Project when the ambitious Renaissance Pope Julius II tore down the most sacred shrine in Europe, the millennium old St Peter's Basilica built by the Emperor Constantine over the apostle's grave. Julius II's intent was to erect on the same site a stupendous mausoleum over the monument. The project began in 1506 and followed the design of Bramante. It followed a too

hasty execution which led to the collapse of two of the arches
under the dome. Over the next 40 years work progressed only a
little further as four architects and one pope were replaced.

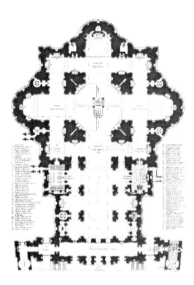

*Figure 10.6: Plan of St Peter's Basilica, Rome. The portion below the
line A, B, and the side chapels C, D, were added by Maderna. The
remainder represents Michael Angelo's plan.*[12]

In 1546, Michelangelo was appointed chief architect for the
reconstruction. Though aging and in poor health, he accepted
the appointment. He gave final form to the general design in
a simplified version of Bramante's plan with more massive
supports, and an unrivalled dome equal to that built by
Brunelleschi a century earlier. It measured 43 meters (140 feet)
in internal diameter, and with its two shells rises from a drum,
to a height of 123 meters (405 feet) to the top of the lantern. It
actually dwarfed the nearby Pantheon, smaller in diameter, but
stood higher, it was made almost entirely of heavy masonry. The
workforce placed three iron rings within the masonry to support
the dome.

In 1606 C. Maderna was employed by Paul V. to lengthen the
nave by two bays, and the atrium or portico were added (1629–
67), by Bernini.

St. Peter's was widely copied throughout Italy with churches
that have a Greek or Latin cross and a high dome rising from a
drum. In 1590 the dome of St. Peter's was finished, after nearly
a century of construction, and was the greatest achievement of
the Basilica.

*Figure 10.7: Interior of St Peter's Rome[13] the largest church in
existence. The central aisle is nearly 183 meters (600 feet) long, with
a stupendous paneled and gilded vault 25 meters (83 feet) in span.
The vast central area is capped by a majestic dome.*

Figure 10.8: St Peter's Basilica.

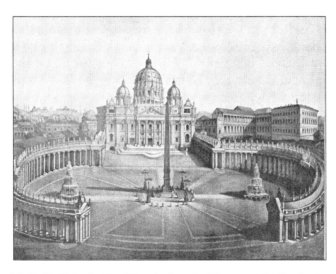

Figure 10.9: St Peter's Basilica is fronted by an elliptical piazza 240 meters across, enclosed by imposing colonnades (248 columns and 88 pilasters), and is the largest Catholic Cathedral in the world. There are over 140 life size saints crowning the cornice. The Vatican (the Pope's residence) is on the right.[14]

The project constructed the largest church in Europe, which was excessively massive. The project had all the hallmarks of a Gothic cathedral project in terms of longevity. It was an audacious project, lavish and opulent, and the cost was enormous. In a period of 176 years (1450-1626) the construction and improvement costs were over $63 million. It spanned 2 centuries, embroiled 27 popes, and the genius of the greatest artists of the time including Michelangelo, Bramante, Raphael, and Bernini. It provoked the Protestant Reformation, dividing the Christian world for all time. Gothic cathedrals reached up to heaven, but the basilica brought heaven to earth, and was the defining event of the high Renaissance.

St Basil's Cathedral Project (1555-1561)

The cathedral was built in the Red Square by the Ivan the Terrible (Tsar Ivan IV), to commemorate the capture of the Tartar stronghold of Kazan, in 1552. The design was based on strictly geometrical principles where the ground plan was in the form of an eight-pointed star, consisting of two superimposed squares, which represent the stability of faith, and the four corners of the earth. It was symbolic of the Christian Church, as a guiding light to mankind. It comprised of nine individual chapels, each had a unique onion dome, reflecting a victorious assault on the city of Kazan. Each onion domes is instantly recognizable and unique. The cathedral was built of stone and brick covered with stucco.

The cathedral was built by Postnik and Barma. Legend has it that Ivan had had the 2 architects blinded so that they could not recreate this masterpiece. However, the name Postnik, mentioned in the chronicles, was the author of many monuments after St. Basil's construction.

МОСКВА
Соборъ Василія Блаженнаго.

MOSCOU. № 79.
Cathédrale de St. Basile-
Blajennoy.

Figure 10.10: Old postcard - Cathedral of St Basil in Moscow.[15]

Key Players

Ferdinand Magellan (1480 – 1521)

He had the vision of finding a new sea route to the Spice Islands. He faced major challenges to realize this and had to bide his time to get the support, financial backing, and patronage. In austere circumstance the project was completed but with heavy loss of life including his.

Michelangelo (1475 – 1564)

Michelangelo di Lodovico Buonarroti Simoni was an Italian Renaissance painter, sculptor, architect, poet and engineer. He was a predominant architect with the St Peter's Basilica project in Rome.

317

Chapter Wrap-up

Conclusions

~~Like Columbus, Magellan needed to get project sponsorship from~~
the crown and state. Not only did he switch allegiances from
Portugal to Spain but, he also renounced his nationality and
become a Spanish citizen. Once he got the monarchs support
they downloaded all the project risks to him because they did
not trust him. They even swapped his Portuguese crew out for
Spaniards.

Magellan's voyage around the world fell into trouble partly
because he mishandled his communications. Magellan had
incomplete maps of the Americas, and an inaccurate estimate
of the circumference of the earth. He claimed he knew where
the passage to the Spice Islands was. He didn't and lied to his
sponsor (the King of Spain) and crew. Magellan's voyage around
the world was fraught with risk. During the voyage Magellan
didn't communicate constantly or effectively, which was vital
in heading off mutinies. Soon he had a full scale mutiny on his
hands which he overcame with a brute force.

Was Magellan a great Project Manager? The pro argument is
the voyage was probably undertaken well ahead of its time with
the levels of knowledge and technology that were available. At
this time the church was still claiming the world was flat. So, he
did reach the project objective, but at what cost? He misled the
stakeholders and crew on vital information, the location of strait,
and he used brute force to quell his crew which caused the loss
of life, resentment and desertion. He also took a huge gamble
and lost his own life and those of many of the crew. In reality
Magellan was constantly taking risks through the project.
He was poorly prepared not knowing the location of strait or
passageway, or distances across the Pacific.

The cathedral projects (of St. Basil and St. Peter)
delivered two structures that are world recognized symbols,
and landmarks in architecture. Unwittingly, the huge
costs associated with the St Peter's project may have likely
contributed to the coming of the Project Reformation as Julius
II bailed the project out by increasing taxes and authorizing the
sale of indulgences (remissions of temporal punishments due to

sin). Papal endeavors to create St. Peter's may well have played an indirect role in the Reformation movement.

Key Lessons

PMBoK® Guide Knowledge Areas:

- Integration Management

  - The voyages of exploration (for Columbus and Magellan) were significant projects that required integration of many disciplines. Both were pushing people and technologies to their limit.

  - The projects were driven by a financial payback that required substantial sponsorship and funding, only available through the ruling monarchies and wealthy royal families. The high cost demanded a high return and the emphasis was on finding a single commodity that could provide a payback. For example, Magellan's circumnavigation of the globe was a disaster in terms of lives lost but, the one ship that returned to Cadiz with a full cargo of spices remunerated the whole expedition cost and left a very large profit when sold by the Spanish Government. Both Magellan and Columbus had challenges getting the initial funding in place. They presented and pleaded their projects, and the returns these would bring to their royal sponsors. But a level of skepticism and lack of belief that the projects would succeed dogged them. When the final approval, was given the monarchs still did not expect them to return.

  - For Magellan the business case was very lucrative. The voyage really opened up a European scramble for more ambitious exploration, trade, and conquest.

  - Preparations for these voyages of exploration took more than a year.

  - Having royal monarchs as project sponsors increased the pressure on the project as Magellan was worried about failing them and the consequences of this.

- ○ The St. Peter's Basilica project is an example of a project that disregarded change control, as this project went on for two centuries with extensive changes and modification that drove up costs.

- Scope Management

  - ○ The principal requirements were to return to Spain with a spice load of cargo and proof of a Western Sea Route to the Spice Islands.

  - ○ Seeking a passage along the lengthy coast of South America (estimated at 3,900 kilometers (2,400 miles)) was very difficult and time consuming. Magellan lost two weeks just exploring the basin of the huge river Plate. He seriously misjudged the impact of this search, it radically changed the scope of the project, and he was unprepared for it in terms of number of ships, men, provisions, and equipment.

  - ○ Magellan set off with five ships but the scope of the project was shifted several times, quite dramatically, when ships deserted or were lost. The return leg to Spain was completed with only a single ship.

  - ○ Magellan had completely under-estimated the scope of the Pacific, estimating it only several hundred miles or three days to cross. This put the project into such serious trouble that it was almost unrecoverable.

- Time Management

  - ○ Because the scope changed the time estimates were grossly inaccurate. The journey just to the Pacific Ocean had taken took 15 months, much longer than the expected 2 months.

  - ○ Magellan's voyage was time dependent and dependent on adequate supplies of food and water. Therefore, time overrode both project cost and scope. The challenge for Magellan was the only flexibility he had was turning back before the objective was reached.

320

- With the St. Peter's Basilica project, time was secondary to scope and very much in line with other cathedral projects.

- Cost Management

  - To complete the project Magellan required a significant budget to pay for five ships with crew, supplies, and equipment.

  - Based on transatlantic travel Magellan had analogous estimates for the first part of the voyage across the South Atlantic.

  - Even though Magellan's ships hugged the South American coast line and could pull in shore to replenish provisions (food and water), he sapped the crew's morale. It took 15 months to complete this leg, and he had a full scale mutiny on his hands.

  - When Magellan cleared the straight he was anxious to complete the journey. He failed to stop and adequately replenish food and water, creating a bigger set of problems when crossing the vast Pacific Ocean.

  - So important was the St. Peter's Basilica project in Rome to the principal stakeholders that control of the project budget fell indirectly to the pope. Meeting the objectives and completing the structure far outweighed the project budget which ran out of control. The project sponsor believed he could fall back on the resources of the Catholic Church and in fact did.

- Human Resource Management

  - Magellan's crew of 251 men was double what Columbus had, and therefore he had greater challenges. He was very reliant on his captains to maintain morale and keep order.

  - The Columbus/Magellan voyages of exploration show a contrast in styles of management.

  - Acquiring and holding onto a project team was very difficult because of the perceived dangers. In Magellan's case one of the four ships deserted. He

took a more heavy handed approach with his crew
than Columbus. By not managing his crew properly,
and admitting not to what he knew, he quickly lost
their confidence and had a full scale mutiny that
almost ended in disaster.

- Prior to departure the project sponsors had ordered a
  swap out of the whole Portuguese crew for a Spanish
  one. The original crew would likely have been more
  sympathetic to Magellan and better tolerate his
  mistakes.

- Crossing the Pacific was done under extreme
  circumstances due to the shortage of provisions. The
  crew was forced to eat rats, sawdust and leather from
  the yardarms. After 3 months the three ships arrived
  at the island of Guam, in the Marianas, where they
  obtained fresh provisions.

- St. Peter's Basilica project went through 27 popes
  and a number of architects. This constant change
  destabilized the project and impacted the scope.

- Quality Management

  - When Magellan was planning the project, like
    Columbus, the areas were quality was paramount
    included:

    - The accuracy of charts and maps, even more so
      because of the greater distances.

    - The seaworthiness of ships especially the rigging
      against storms with the vast expanses of the
      Pacific.

    - The securing of provisions (food and water) to
      last across large stretches of the journey. The
      plan required landing on coast-lines to replenish
      and restock. The ship left with bread, beer,
      hard biscuits, fish and salted meat, all stored in
      wooden casks, and fresh livestock like pigs and
      chickens. Scurvy was going to be a major problem
      because of the lack of vitamin C.

- Daily inspections were made by a cooper (barrel maker) who was responsible for inspecting and keeping the casks tight.

- Magellan's voyage was faced with the problem of longitude that required measuring distance sailed over time. The precision in the measuring sextant was not accurate and was one of the essential devices that had to rapidly evolve and be re-engineered more precisely.

- With St. Peter's Basilica project the final deliverable was the creation of the premier church in the Catholic empire. For this the most exquisite levels of craftsmanship were required. Whatever was visible required the highest levels of quality.

- Communication Management

  - Both Columbus/Magellan had serious challenges with funding, and had to bide their time. They had to find and then carefully manage their fickle sponsors and stakeholders.

  - Magellan's communication management let him down. By misleading his stakeholders about the passage to the Spice Islands he lost the crew's confidence and trust.

  - Once underway Magellan did not have a way to communicate with his royal project sponsors where he could not share the problems encountered, like the increase in scope, and jointly review project alternatives.

  - During the voyage Magellan didn't communicate constantly or effectively, which was vital in heading off mutinies. When the crew realized that Magellan had lied he lost the confidence of the crew, and had a full scale mutiny on his hands which he overcame with brute force.

  - Magellan would likely have had a better relationship with the original Portuguese crew and communication would have been far easier.

- By drawing and quartering the mutineers Magellan was sending a very powerful message to the rest of his team relative to his tolerance for insubordination.
- St. Peter's Basilica project faced the communication challenges of an elongated building project (like the Gothic cathedrals). Concepts and ideas had to be shared along the time-line from architect to architect.

- Risk Management

  - Risks were downloaded to Magellan by the monarchy who didn't trust him and imposed constraints, such as the makeup of the crew (Spanish versus Portuguese).

  - Magellan's voyage was fraught with risk. The voyage required managing a vast array of risks predominantly financial, navigational, and environmental with the climate and weather, to running into the hostile groups (Portuguese or natives), to running out of provisions, leading to starvation and mutiny.

  - Magellan mitigated the risk by taking five ships. When a ship deserted he was still able to continue, although the overall project risks greatly increased.

  - By not having accurate maps of South America he was forced to explore every inlet, river, and bay along the coast in search of a passage to the Pacific. The risk of not finding the passage plagued him, and he failed to acknowledge the impact this was having on the crew.

  - Magellan, when compared to Columbus, was reckless in the execution of the project. He took huge risks with his own life and that of his crew's, without seeking advice. When he finally cleared the passage to the Pacific, he pushed ahead without taking adequate provisions on board which proved disastrous.

  - Not knowing the true distance across the Pacific Ocean was a massive risk he undertook. He grossly underestimated it as a three day sail based on an inaccurate estimate of the Earth's diameter.

- Magellan got too readily embroiled in a local conflict with natives without assessing the risks carefully enough.

- Through the project Magellan kept accepting risk because he didn't want to fail - that is return to Spain without finding the passage. Magellan was a risk seeker.

- With St. Peter's Basilica project the design was pushed to the limits resulting in the collapse of two arches.

- Procurement Management

  - When Royal consent was finally given to the project, Magellan entered into an agreement where he was nominated as captain of the expedition, and given five vessels with provisions for two years. He was nominated governor of any new lands found during the voyage and would receive one-twentieth of the profits.

  - Magellan, like Columbus, had to procure enough ships (five) to make the voyage feasible. The ships had to be well equipped, supplied and provisioned for.

  - With St. Peter's Basilica project the supply chain was massive considering the scale of the largest Catholic cathedral in the world. This included transportation of an Obelisk from Egypt.

Educators

- Discuss all the risks in Magellan's project and his approach in mitigating these?

- Discuss in the context of today's projects whether Magellan's voyage was a success or not?

- Was it undertaken ahead of its time considering the knowledge and technologies available to him?

- Discuss Magellan's approach to managing his stakeholders and crew.

- Discuss whether St. Peter's Basilica project set a precedent with project that ignored the costs. Was the project successful?

CHAPTER 11

The 17ᵗʰ Century and the Modern Age of Engineering

The period was dominated by the Europeans racing to colonize the world. Magellan's round the world voyage confirmed the size of the world and opened the door to the unknown world. New manufacturing techniques dropped the cost of a sheet glass and glass was widely incorporated into new building constructions. The birth of First Scientific Revolution saw the field of optics and atmospherics emerge, which spawned the first phase of Modern Engineering and its influence on project management. Large projects were now more likely to be sponsored by the crown and state, rather than the church that is undergoing dramatic reforms provoked by the Protestant Reformation and indirectly by St. Peter's Basilica project.

In this Historical Period
- Brief synopsis of trends/changes (natural disasters, wars, technology, epidemics).
  a. Divine right of kings leads to civil wars and monarchies start to lose their grip on power.
  b. Europeans race to acquire colonies and the colonization of North America, Africa.
- Impact of changes
  a. The development of democracy.

   b.   Destruction of North America native cultures.
- Major events
   a.   The first limited company, 1602.
   b.   The Great Fire of London, 1666.
   c.   Bank of England created, 1694.
- New tools, techniques and breakthroughs
   a.   The First Scientific Revolution.
   b.   Modern Engineering .
   c.   New Materials, manufacturing of large sheets of glass.
- Regions
   a.   Europe.
- Significant projects
   a.   Taj Mahal.
   b.   Palace of Versailles.
- Other Notable Projects
   a.   Blue Mosque in Constantinople.
- Key players
   a.   Sebastian De Vaubau – military engineer.
   b.   Gallelio.
- Chapter Wrap-up

Trends and Changes

The world population in this period grew from 540 to 600 million.

Divine Right of Kings

This idea evolved in Europe during the Middle Ages that kings were answerable only to God, the Absolutists. In the 17th Century national monarchs in England and France were asserting their authority in matters of both church and state. In England this led to a bloody civil war and is a precursor to revolutions across Europe.

European Colonization

The voyages of exploration opened the door to colonization and the exploitation of cheap colonial labor and slaves for the European powers. For the most part the Europeans used these as a workforce within the colony itself, and through the 16/17ᵗʰ centuries further inequities and exploitation grew with African slavery.

In the European race to acquire colonies and the colonization of North America, Africa, and Asia, Europeans set up settlements and forts. Through colonization international trade and the sophisticated procurement practices developed in selling goods on one continent and buying goods for the return leg.

Impact of Changes

The Development of Democracy

Both the French and English Monarchs refused to call parliaments (in England and France), which stirred up democratic forces and the dissolution of power.

Destruction of North America Native Cultures

As the Europeans settled and colonized North America, a repeat of history was inevitable. Native cultures were under threat similar to the Aztecs and Incas when the Spanish arrived a century earlier in South and Central America.

Major Events

The Great Fire of London 1666

The fire consumed about 2 miles² (5 kilometers²) of the city centre, leaving tens of thousands homeless. The fire was a catalyst for rebuilding the centre of London, and foundation of risk management. In the aftermath insurance companies protected their clients' property by forming private fire organizations who would only deal with buildings identified by fire insurance marks. This led to the birth of fire departments for cities without organized fire protection systems.

The First Limited Company 1602

The Dutch commercial expansion to the Far East, to get spices and set up trading posts, leads to the creation of the first limited company. So expensive were the ventures that merchants had to pool their resources together. In 1602 the Dutch East India Company was formed with a monopoly on trade to the Far East. The structure of the company was novel where citizens were invited to invest (unequally) through an issue of shares in the company's profits that were determined by supply and demand. This was the world's first true stock market. The company had to keep Spanish and English competitors at bay with 40 warships and an army of 10,000 soldiers.[1]

Bank of England 1694

The Bank was created to act as the English Government's banker. This was the first step in a national approach to the economy. The state bank did not directly promote industry but indirectly through hundreds of private banks.

New Tools Techniques and Breakthroughs

The First Scientific Revolution

This revolution developed a new understanding of the physical world through new instruments that could verify ideas by observation and experimentation, something the ancient cultures lacked.

> *"An inductive study of nature [is required] through experience and experiment."*
>
> —Bacon

Francis Bacon and Rene Descartes are seen as the founders of modern empiricism and rationalism. They both advocated that the ultimate purpose of theoretical science is to serve the practical needs of human beings. The century sees the new fields of optics and atmospherics emerge.

As the field of mathematics continued to evolve the beginning of the 17ᵗʰ Century saw the introduction of logarithms, the invention of the slide rule, and the development of a decimal system.

In 1604 Gallelio used mathematics to describe the law of falling objects at a rate of 32 feet/second per second. Every object accelerated at the same rate. Significantly, this was the first time that mathematics had been used to describe a law of nature. This was the emergence of the Scientific Revolution.

In 1610 Galileo looked at the skies using a telescope which he developed from a Dutch idea used for seeing objects at a great distance like incoming merchant ships on the horizon. When he turned the telescope to the heavens, and magnified the view by 1,000 times, he found a great numbers of stars which were not visible to the naked eye. He was able to distinguish the moons of Jupiter, Venus waxing and waning, spots on the Sun, and so proved Copernicus was right with his heliocentric theory (the sun at the center of the universe).

Galileo invented a simple thermometer and in 1643 a pupil of his Torricelli made a barometer to measure atmospheric pressure using a dish and an inverted tube of mercury.

The Reflecting Microscope was a powerful new microscope developed by A. van Leeuwenhoek in 1674. It was able to detect microorganisms as small as protozoa. Although microscopes were available in the mid-15ᵗʰ Century they were somewhat crude.

Isaac Newton combined experimentation with mathematics and scientific works to create a comprehensive set of physical laws around gravitation and astronomy. He also discovered that heat when applied to different elements gave off a characteristic color in the light spectrum, and each star had a spectrum of light. Newton composed Principia Mathematica during in 1686, probably one of the most important books ever written.

Modern Engineering

> *"The first phase of modern engineering emerged in the Scientific Revolution. Galileo's Two New Sciences, which sought systematic explanations and adopted a scientific approach to practical problems, is a landmark regarded by many engineer historians as the beginning of structural analysis, the mathematical representation and design of building structures."*
>
> —Source: History of engineering;
> http://www.creatingtechnology.org/history.htm#1

This marked the beginning of modern engineering with the formation of professional societies, the printing of treatise on engineering subjects in quantity, foundation of engineering schools, and specializations within the profession. Engineers also began taking advantage of the brilliant scientific discoveries of the time. The first handbooks on engineering appeared.

Steam engines appeared as a source of power (Thomas Savery), first harnessed in the 17th Century to operate mine pumps. They were somewhat limited until improvements were made by James Watt in the 1770s.

Daily Newspapers

In the last 500 years changes in communications were fast, broad, and significant. By 1650 the first daily newspapers were in circulation in Leipzig and this saw an explosive growth to become the main form of media for the next 275 years.

New Materials

Hardwick Hall inspired a new architecture and changes in the manufacture of sheet glass. The development of more sophisticated coal fuelled furnaces provided the required quantity of molten glass. New techniques like spreading and rolling out an even thickness on casting tables allowed for the manufacture of large sheets of glass. The Palace of Versailles incorporated this material on a vast scale not just in windows but mirrors, and as a result, started a building trend with this material.

Figure 11.1: Changes in glass making improved the quality and reduced the cost making it more viable for projects, 1706.²

Regions

Europe and Asia.

Significant Projects

Taj Mahal Project (1631-1648)

Background

By 1630 the architecture of the Moguls had reached the height of expression in the totality and balance of its qualities of construction, composition, detail, ornament, and settings.

Initiating

The project charter for the Taj Mahal was to build a mausoleum for the fifth great Mughal Emperor Shah Jehan. Taj means crown and Mahal means palace, hence the Crown Palace. He was one of the richest men in the world. Because it was to serve as his tomb after his death, the project had to be completed in his lifetime. It was located to the south of the walled city of Agra. The rationale for the project was no different to the Pharaoh

Khufu's pyramid of Giza. It was also a memorial to his wife, Mumtaz Mahal, who would have her mortal remains buried in this mausoleum.

The Taj Mahal incorporated and expanded upon many design traditions, particularly Persian and earlier Mughal architecture. Specific inspiration came from a number of successful Timurid and Mughal buildings.

Planning

The project demanded the varied talents, creativity, artistry, skills and abilities of many. A project core or creative team was formed of thirty seven men including designers and architects, who were all mentioned by name in the official Mughal histories.

- The chief architect (or plan drawer) was Ismail Afandi (a.k.a. Ismail Khan) who had worked for the great Ottomans in Turkey as a premier designer and builder of domes. He was also the architect of the Red Fort at Delhi.

- Mukrimat Khan and Mir Abdul Karim from Shiraz, Iran (Persia) were chief supervisors and administrators who handled the finances and management of daily production.

- Mohammed Hanif, Multan and Quandhar, master masons from Delhi, were supervisors of the masons.

- Qazim Khan, a goldsmith from Lahore who cast the gold finial that crowns the dome.

- Chiranji Lal, a lapidary from Delhi chosen as the chief mosaicist.

- Amanat Khan from Shiraz, the master calligrapher whose signature is inscribed on the Taj gateway.

Also included in this international team were sculptors from Bukhara, calligraphers from Syria and Persia, inlayers from southern India, stonecutters from Baluchistan, and a specialist in building turrets.

To this core was added a project workforce of twenty thousand recruited from across northern India and master craftsmen from Persia, France, Iran, Italy and Turkey to work on the project. Their names were recorded for posterity on scrolls.

The Mughal Emperor ordered for the best of artisans to create a unique design that could not be replicated. He played an active role in the project design and rejected hundreds of designs before he finally approved the blue print. He personally oversaw the artists as part of his daily routine. Most importantly once the project started the design was not changed.

> *"We know Shah Jahan was interested in architecture. We know he was interested in architectural decoration and design. Clearly, he was consulted. He was probably very interested in continually seeing the plans as they developed and commenting on them, and suggesting changes that might be made. The idea that he did any more than that, in terms of the design, is unrealistic. Clearly it's a building that was designed by professional architects who knew what they were doing, not by a prince and an amateur."*
>
> —Art historian Milo Beach[3]

The most significant challenge was the proximity to the river to the site and the high water table. This would affect the stability of the foundation.

As part of the planning the team also worked out the logistics for:

- The site preparation – drainage of the area and creation of a perfect square marble plinth, 55 meters (186 feet) on each of the four sides.

  - A 3 acre area was excavated, filled with earth and rubble, and leveled at 50 meters (140 feet) above the riverbank.

  - Wells were dug and filled with stone and rubble at the footings of the tomb because of the high water table.

  - Creation of a platform of red sandstone.

  - Perfect leveling of the square marble plinth, any slight inaccuracies would offset the Minarets.

- The production of brick

  - Vast quantities required for the inner core.

- The quarry operations
  - Red stone was brought from the neighboring towns.
  - A huge quantity of translucent white marble was brought from Makrana mines (Rajasthan) to Agra a distance of 350 kilometers (200 miles).[4]
- The transportation of the finished stone
  - The marble (2 ton blocks) was transported by pack animals were teams of twenty or thirty oxen were strained to pull the blocks on specially constructed wagons to Agra.
  - Over 1,000 elephants were used in the construction where a convoy of elephants hauled the marble blocks.
- The creation of accommodation for the project workforce
  - For a permanent workforce of up to 20,000.
- The construction of the ramps
  - To deliver the blocks to a precise position in the structure a 15 kilometers (9 miles) tamped-earth ramp was constructed to transport marble and materials from Agra to the construction site.
- The performance of the finished work
  - The marble blocks were raised into a desired position using an elaborate machine like post-and-beam pulley system.
- The removal of the ramps at the end of construction.

In the construction three types of materials (stones) were used:

1. Common stones: sang-i-Gwaliari (grey and yellow sandstone) sang-i-Surkh (red sandstone), sang-i-moosa (black slate) and sang-i-Rukhan (sang-i-marmar; white marble) were used in foundations, masonry and for giving finishing touch to the external surfaces.

2. Semi-precious stones like Aqiq (agate), Yemeni, Firoza (turquoise), Lajwad (Lapis- lazuli); moonga (coral), Sulaimani (onyx), Lahsunia (cat's eye), Yasheb (jade) and Pitunia (blood stone). These were mainly used for inlaying work.

3. Rare and scarce stones such as Tilai (goldstone), Zahar-mohra, Ajuba, Abri, Khathu, Nakhod and Maknatis (magnet stone) were used for bold inlay and mosaic work chiefly on floors, exterior dados and turrets.

The building was constructed with a brick core and a marble veneer, similar to the Roman project's like the Colosseum, and this lowered the costs. Workmen constructed a colossal brick scaffold that mirrored the tomb. Some of materials travelled a great distance and an extensive supply chain was created. For example, turquoise was brought from Upper Tibet, jasper from Punjab, jade and crystal from China, sapphire from Sri Lanka, Lapis lazuli from Afghanistan, Carnelian from Arabia. Other materials which were used included different kind of bricks, sweet limestone, tiles, and spouts to lead off water, reed glue, red and silver clay, and glass.[5] In all, 28 types of precious and semi-precious stones were inlaid into the white marble and used on the blooms of fuchsias, lily, honeysuckle and more.[6]

Executing

The construction of Taj Mahal began in 1631. Locating the site so close to the river was a major risk. The foundation was an engineering marvel where each massive peer, supporting the building, rested on deep rubble and a series of concrete holes, connected by arches. It was so effective that even today the building has not shifted. Wells were dug by the river so that any fluctuations in the ground water level were absorbed before reaching the foundation.

The double dome was constructed from concentric circles of bricks about 2 meters (6 feet) thick. It was 10 stories (50 meters or 80 feet) high and weighed 13,000 tons, and most remarkably was not supported by any pillars. It had an 18 meters (60 feet) diameter about half the width of Florence Duomo (42 meters).

The center and skeleton of the main building were made up of extra strong brick masonry in which massive white marble slabs were used on the headers and stretchers to give it a white marble outlook. This was held by cementing material made up of pieces of fossilized soil mixed with lime mortar.

The order of construction was:

- The Taj Mahal plinth and tomb.

- The Taj Gateway and garden.

- Remaining parts of the complex were built in stages.

The four Minarets at the four corners of the plinth were deliberately erected leaning outwards so in case of an earthquake they would fall away from the tomb in case of collapse.[7]

The Mughals believed that the precious and semiprecious stones affected the fortunes of different persons and places and this was calculated and strictly adhered to in the adornment. Marks were engraved on red stone slabs, pathways, stairs, plinths and pavements. These included Symbolic motifs (Swastika, Hexagon, Pentagon), Animated motifs (fish, bird), Geometrical motifs (triangle, Square, rectangle), and Floral motifs (leaves and petals of flowers).[8]

Water was drawn from the river by a series of purs, animal-powered rope and bucket system, into a large storage and distribution tanks, and then passed into three subsidiary tanks, from which it was piped to the complex.

Closing

The project was completed with the following main features:

- The Taj Mahal plinth and tomb, Pietra Dura, took 12 years to complete (1643). 4 Minarets, 43 meters (138 feet) each, were topped with 8 windowed cupolas. Immediately below the dome is the tomb of Mumtaz Mahal which is centrally lined with the main entrance, and that of Shahjahan, placed there by Aurangzeb to break up the symmetry.

- The massive red sandstone Taj Gateway, took an additional 10 years to complete although historical accounts list different completion dates. It stands 30 meters (100 feet) high, and is topped by small cupolas or chatris. It is Symbolic of the divide between the material and the spiritual worlds. The gateway is decorated in calligraphy with verses from the Holy Koran.

Figure 11.2: Taj gateway to garden court of the Taj Mahal.[9]

- The Charbagh or the Taj Garden, that enhances the over-all beauty laid out in the Persian Charbagh (four garden plan) style. The entire area of the Taj complex is 580 meters (1,902 feet) by 300 meters (984 feet) and the garden alone makes up an area of 300 by 300 meters (984 feet).

Remarkably the project was completed in 17 years, in the lifetime of the Emperor Shah Jehan. Much of the project effort went into the dome and the ornate finishing and adornments.

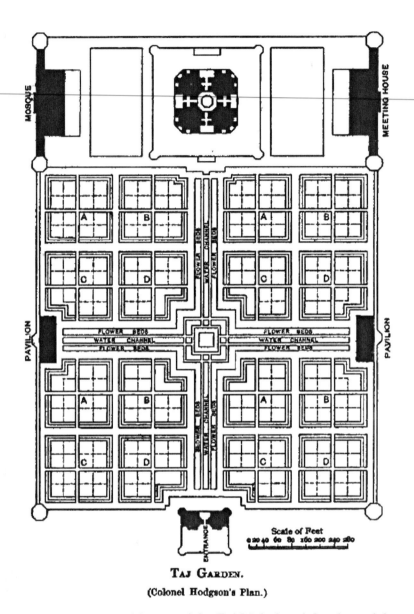

TAJ GARDEN.

(Colonel Hodgson's Plan.)

Figure 11.3: Plan and layout of the Taj Mahal and Gardens of the Great Mughals.[10]

Figure 11.4: The Taj Mahal, made of gleaming white marble and rested on an 8-sided platform of red sandstone measuring 40 meters (130 feet) long on each side.[11]

The project costs have been estimated at 32 million Rupees or hundreds of millions of dollars at today's currency rates. The output is the pearl of Indian architecture and a masterpiece of symmetry. It seems to be floating in the air from a distance. Today, most of the precious stones have been plundered.

Palace of Versailles Project (1661-1720)

Background

The struggle to unite the French nation had been going on since the fall of the Roman Empire. French Counts and Dukes were sometimes extremely powerful, and always determined to keep control of the land they possessed and their respective incomes. They often fought with the French King whenever he became a little too powerful. So, the King of France spent most of the year travelling through the provinces, meeting his subjects to make sure they were still faithful to him.

Initiating

King Louis XIV of France wanted a place where he could organize and completely control the Government of France by absolute personal rule. It was a straightforward project charter. As
a principal stakeholder he settled on Versailles, a royal hunting
lodge of his father because it was outside of Paris and also away
from the tumults and diseases of the crowded city. The project
had a clear political purpose of establishing the power of the
King of France over the French nobility.

In March 15th, 1661, Jean-Baptiste Colbert was named
Superintendent of Finances and was assigned to the project, as
well the construction of the Louvre, Fontainebleau, and other notable projects. Colbert's role in the project was certainly not easy
as he did not approve of it. He opposed the young king's obsti-
nately and expressed himself without reservation. But Louis XIV
had a dream and was not satisfied with Paris as a residence. So,
he told Colbert to make his dream come true, and Colbert had to
find some way to pay the cost.[12]

Figure 11.5: Frontispiece, Statue of Louis XIV, the project sponsor.[13]

Planning

Louis XIV was irritated by the splendors of the Chateau of Vaux-le-Vicomte built by his ill-fated minister, Nicholas Fouquet, for himself. Louis XIV was determined to surpass this chateau by one much more elaborate and to pale it into insignificance. Fouquet had employed the most renowned masters of this period namely, Louis Le Vau, the principal architect, and André Le Nôtre, the landscape architect and gardener, and Charles Lebrun, the painter and decorator.

On August 17, 1661, Fouquet gave an elaborate fête to celebrate the completion of the chateau, which Louis XIV attended. Within three weeks the host was a prisoner of the State, accused of embezzlement, and his estate was confiscated. Acting immediately to out-do the glories of Vaux-le-Vicomte, Louis XIV engaged Le Nôtre to plan the gardens and Le Vau to submit proposals for the enlargement and decoration of the Chateau at Versailles.

Executing

In September, 1663 Colbert reported with dismay that two years into the project he had spent millions of livres, and a good part was just for the construction of the gardens. With increasing frequency Louis XIV was in residence at Versailles and directly meddled in the project. He was dominant, directing minute details like the laying of floors, the hanging of draperies, the installation of art and decorative works. Builders and decorators took advantage and suggested one elaborate change after another, without any regard to the project cost, and despite the protests of Colbert to Louis XIV that they were exceeding all estimates.

Louis XIV also impatiently voiced disapprovals to the superintendents of any delays and so they strove feverishly to meet all his demands. Their eagerness to act cost the project a substantial sum, but when challenged by Colbert he only cared for the completion of his luxurious ideas. He was impatient to enjoy his creation and to invite his court there to celebrate its completion with both brilliant and costly fêtes. Every day a new requirement was tabled and it had to be incorporated into the project. There was no change control to the project scope. It was

Colbert's problem to finance the project. Effectively, the palace was built but not planned for over four phases. An example of the extravagance includeed:

> "An orangery of 1,250 imported orange trees was laid out, with a grand terrace, among the fruit and vegetable gardens. The original entrance court was greatly enlarged. Long wings terminated by pavilions bordered it with kitchens and quarters for the domestics, and stables with stalls for fifty-four horses. A long promenade, the Allée Royale, extended to a vast basin, the lake of Apollo."[14]

The first phase (1664-1668), Louis XIV gave orders for the building of small dwellings to be occupied by the favorites of his entourage. Also the highway to Paris "the Cours-la-Reine" was started. Already Versailles took on a more imposing aspect than ancient Fontainebleau, Louis XIVs other residence.

The scope of the project increased as workmen drained swamps and moved whole forests to construct reservoirs, ponds, fountains, grottoes, waterfalls, straying brooks in 250 acres of formal gardens. Thousands of troops were commandeered into the project workforce to divert the river to supply water to the pools and fountains. Cages were constructed for "the most splendid palace of animals in the world."

A great number of works of art were ordered for the adornment of the walks and gardens. Many mythological statues and busts were made in Rome and it was a thriving period for sculptors of France and adjacent countries. Colbert detailed expenditures of million livres of the nation's money for these. Many were removed after a short period as Louis XIV constantly demanded the work of the latest and newest artists, and all the newest novelties.

Figure 11.6: "View of Versailles from the Avenue de Paris", by Pierre Patel.[15] The immense project cleared 37,000 acres of land for vast formal gardens with geometrically arranged avenues, tree-lined terraces, woods, and canals, walkways, with over 1,400 fountains and 400 pieces of sculpture.

By 1668 the project approached completion but in 1669 the project changed direction as Le Vau, the architect, proposed new embellishments. Louis XIV had just turned thirty years of age, and was eager to achieve still greater improvements to mark the increasing prosperity of his reign. Again with no change control half-finished buildings were demolished and rebuilt. Immense structures arose, and once again artists flocked to Versailles. The palace and park were elaborately decorated.

The project scope further increased when a town was created on the vast acres purchased by Louis XIV. Houses were built for Colbert, now superintendent of the royal buildings, and for the officers of the Chancellery. Louis XIV bought the village of "Old Versailles" and made liberal grants of land to individuals who agreed to build houses there. Opposite the chateau nobles of the Court built their own mansions.

The second phase (1669-1672) saw a new structure provide lodgings for Louis XIV and members of his family. The main floor was given over entirely to two royal apartments for the King and Queen. It also saw the construction of the terrace, which was later enclosed to become the Hall of Mirrors.

Figure 11.7: The lavish Hall of Mirrors extends for 75 meters (240 feet) along a terrace overlooking the gardens. The marble walls are embellished with bronze-gilt trophies.[16]

Le Vau died in 1670 and the project was continued by the architect's pupils at great cost. Louis XIV maintained a lively interest in the project whether at home or abroad and eagerly read plans and listened to reports.

A third phase (1678-1684) involved over 36,000 laborers and craftsmen, and 6,000 horses. The workforce used immense hydraulic machines driven by the river. The north and south wings of the palace were built. The Hall of Mirrors was decorated with exquisite silver furnishings and reported to be one of the greatest rooms in Europe.[17] The project exploited the use of the new plate glass along the main building frontage and the inner elevation of mirrors. Louis XIV finally got possession of his new palace in 1682. "The State," exclaimed the Sun King, "it is I!"

Controlling and Monitoring

The fourth phase (1699-1710) concentrated almost exclusively on the construction of the royal chapel. Its decoration was completed after Louis XIV's death in 1715.

Fifteen thousand people resided in the many apartments within the palace which included the heads of state and military officials. Each member of the royal family had their own household as well.

Figure 11.8: The complete building had a floor space of 51,210 m², with 2,153 windows, 1,252 chimneys, 700 rooms, and 67 staircases. The project left France with a debt of 4000 million livres until 1774.[18]

In addition to being one of the world's most beautiful buildings it was also one of the most expensive and Louis XIV's disregard for costs placed such an economic burden on the people of France that it eventually festered into the French revolution.[19]

Closing

King Louis XIV said in his memoirs that Versailles was meant to be a place where everybody, not just people from the King's inner circle, could have access to the King. He needed a countervailing power to the court of mighty princes if he were to establish his absolute authority, and for this he needed to rely on the people. Not the masses but, educated people, intellectuals, scientists, and (mainly) writers.

This huge project, the redesigning of nature, the importation of huge quantities of water illustrates the Absolutist domination of nature. Just as Apollo the sun god controlled the light in the sky, his earthly incarnation King Louis XIV attempted to control nature. The project left a huge debt and was probably the most expensive project of all time. The French Government was bankrupt to the amount of 4,000 million livres (1 billion pounds or $6 billion) in 1774. The project was one of the principal causes along with costly wars, and the spending of the royals on luxuries. At the time a master carpenter earned yearly 200 livres, whereas a Paris parish priest earned 10,000 livres.

(Photo., French National Tourist Collection.)
The stately Palace of Versailles, near Paris.

Figure 11.9: The project exploited the use of the new plate glass along the main building frontage and the inner elevation of mirrors. This was widely copied across Europe.[20]

In 1788 the French Government was declared officially bankrupt. Louis XVI was forced to call for a meeting of the Estates-General which was a representative body of the government that had not met in 175 years. They met in Versailles at the Jeu de Paume, a tennis court, which became the backdrop for the French Revolution.

On the morning of October 6, 1789 a mob of angry Parisians, mostly women, marched to the Palace of Versailles demanding bread. They stormed the Palace, ran up the Queen's Staircase and broke into the Guard's Room, then into the ante-chamber. Marie Antoinette ran from her bedchamber into her private apartments towards the King's suite to find her husband and son. In an effort to quell public discontent Louis XVI moved his court to Paris.

The Palace of Versailles project is famous as a symbol of the system of absolute monarchy and its abuse. It also influenced other projects like the Winter Palace, in St. Petersburg, Russia.

Other Notable Projects

Blue Mosque Project (1609-1616)

This edifice in Constantinople was notable because of its overall scope and huge encompassing central dome, which is 23.5 meters

(77 feet) in diameter and 43 meters (141 feet) high at its central point. It was patterned on the Hagia Sophia (blue because the tiles adorning the walls of its interior). The project ran between 1609 and 1616.

Canal Du Midi Project (1666-1680)

This canal was significant as it was precursor to the canal revolution of the industrial eras. The unification of France under Louis XIV revived the vision for a canal joining the Atlantic to the Mediterranean. This canal would connect the Garonne River to the Étang de Thau on the Mediterranean. It would reduce a 3,000 kilometers (1,900 miles) sea journey around a hostile Spain.

The project was initiated in 1665 by Pierre-Paul Riquet, a rich tax-farmer in the Languedoc region, who persuaded Jean-Baptiste Colbert, the finance minister of Louis XIV to sponsor it. Riquet paid for one third of the project the rest by Louis XIV with an expected cost of 3 million livres.

The project's biggest challenge was getting enough water to the highest sections. This was solved by building the largest dam ever at Saint Ferréol on the Laudot River 700 meters (2,300 feet) long, 30 meters (100 feet) above the riverbed and 120 meters (390 feet) thick at its base. This was only the second major dam to be built in Europe.

A project workforce at its peak had 12,000 men (including over a thousand women) labored on the largest project in a century in Europe to create a 240 kilometers (150 miles) long canal. The project was nearly canceled after 10 years, when the canal had to thread through the mountains near Béziers. Riquet used explosives to build a tunnel. The project overran and cost over 15 million livres, bankrupting Riquet, who died just months before the Canal was opened. The canal created a commercial route and influenced the canals of first Industrial Revolution.

Key Players

Sébastien Le Prestre de Vauban (1633-1707)

A French military engineer whose innovative work on citadels revolutionized the art of siege tactics and defensive fortification. He devised outer defensive fortification, so the enemy could not start a siege, and effective offensive siege-craft. He designed fortifications for numerous French towns and outposts and his treatises on fortification and siege-craft were studied for more than 100 years. In 1703 he was made a marshal of France.

Galileo (1564-1642)

Galileo Galilei showed a remarkably modern appreciation for the relationship between mathematics, theoretical physics, and experimental physics. He made some original contributions to the science of motion through an innovative combination of experiment and mathematics.

Chapter Wrap-up

Conclusions

In this era there was a general view about architects:

> *"An architect was, in a sense, a kind of functionary. Architects and painters never achieved the kind of acclaim that placed them within the ranks of the nobility, for example. They were recognized, but they were never given an enormous amount of importance."*

> —Art historian Milo Beach

There are several parallels between the two significant projects of this age. Both were massive construction projects unparalleled in their expense and the transformation of the immediate environment around them, not with just monumental gardens but whole villages. There was an obsession over the control of everything. With the Taj Mahal project, the sponsor stood back as this project was completed in 17 years. The dome was half the diameter of the Pantheon and the Duomo.

In contrast the Palace of Versailles project dragged on for over 50 years. But it was truly significant in that it precipitated a disaster with the French revolution. The project was executed but not planned for, over the 4 phases. By the third phase, there were over 36,000 laborers and craftsmen, and 6,000 horses, an astounding workforce for its time. The project was completely out of control because of its principal stakeholder, the egomaniac Louis XIV, who just kept directly meddling in the project, expanding its scope, and disregarding the budget. The flagrant disregard for governance by Louis XIV reflected the supreme power he held. Colbert, his finance minister, was given accountability for the project but he gave up. They were both accountable to no one partly because the Estates-General, a representative body of the government, had not met in 175 years. The project left a huge debt and was probably the most expensive project of all time. Versailles can be blamed for most of the economic problems the country endured in the years leading up to the French Revolution. Yet the Palace of Versailles was still widely copied around Europe and a chateau building competition began between the European monarchs. For example, in the small courts of Germany, as ambitious as budgets permitted, the following projects were initiated: Schwetzingen near Heidelberg; the New Palace (Neues Palais) and Charlottenburg in Berlin; Herrenhausen in Hanover; the Residenz, Würzburg; Schönbrunn in Vienna; Esterhaz in Hungary. Also notably but on a similar scale, by the Czar (Russia) with the White Palace. In Italy, the Caserta near Naples, (by Luigi Vanvitelli, from 1752 onwards), and Stupenigi outside Turin, and the "Polish Versailles" in Wilanow.

Key Lessons

PMBoK® Guide Knowledge Areas:

- Integration Management

  - The project charter stated the Taj Mahal had to be built as a tomb in the life time of the Mughal Emperor. He was likely one of the richest men in the world hence the business justification was in place.

  - His presence and taking a hands-on-approach may have prevented excessive changes, and guaranteed the completion of the project in a relatively short time frame.

- ◦ Louis XIV became embroiled in the Palace of Versailles project on a daily basis and the governance of the Treasurer Colbert failed to check an avalanche of weekly changes to the project scope. Colbert had a vested interest in controlling changes to keep the project within a reasonable budget. Even though this level of governance was normally enough to keep a project like this under tight control. The project workforce took advantage of the situation.

- Scope Management
  - ◦ The direct involvement of the Mughal Emperor and the desire for the project to finish in his life-time better controlled the scope of the project. Once the design was done it did not change much.
  - ◦ The scope in terms of materials, and equipment was dictated by:
    - The height, breadth, and width of the structure.
    - The quarrying and transporting of marble and materials to site.
    - The complexity of the dome structure and the four minarets.
    - The proximity of the site to the river and the high water table affected the foundation, and required major stabilization.
  - ◦ The work breakdown structure laid out the project with the following:
    - Site preparation, drainage of the area and creation of a perfect square marble plinth.
    - Construction of ramps, quarrying operations delivery and transportation of marble and red stone to site, and the precise positioning of these within the structure. The production of brick for the inner core.

- Construction of the structure with ornate pieces on the exterior.
- Removal and ramp demolition, and site cleanup.

  - The Palace of Versailles project is a good example of a project were scope continuously changed and was not effectively managed crippling the budget and time-line. Part of the problem was there was no business case and it was driven by the vanity of Louis XIV. The schedule and costs were impacted dramatically but not much could be done as the project was under the control of the principal stakeholder Louis XIV and his changing whims.

  - Many of the contractors took advantage of the lack of change control and suggested extensions knowing that they would be approved.

  - A very similarly project was St. Peter's Basilica in Rome, lavish to the extreme, with no expenses spared.

- Time Management
  - The high profile and very prestigious Taj Mahal project was a showcase building built for the Mughal Emperor and had to be completed in his lifetime.

  - With the project the extensive supply chain had to work in lockstep with the project construction schedule. For example, the delivery of marble materials was critical to the schedule.

  - Even though Louis XIV wanted to finish the Palace of Versailles project in his life time the project continued into the next reign.

- Cost Management
  - The Taj Mahal was built from a concrete and rubble core, a key factor in reducing cost, and was then overlaid with a marble veneer, a very similar approach to Roman projects like the Colosseum.

  - Overseers of the project also managed the project budget which was a significant part of the management of the project.

- • Mir Abdul Karim handled the finances and is named in the official Mughal histories.
  - ○ The cost of extracting and transporting the marble and red stone was the most expensive major-activity in the project.
  - ○ Control of the Palace of Versailles project budget was assigned to Colbert (also Chief Treasurer), but was then usurped by King Louis XIV, who Colbert could not control.
  - ○ There was a complete disregard for the budget, for example, by La Vau's (the architect) replacement, his assistant, Francoise d'Orbay.
  - ○ Paying the large workforce of 36,000 was a logistical challenge.
- • Quality management
  - ○ The Taj Mahal project required incredible precision in creating the four Minarets which had to lean outwards in case of an earthquake.
  - ○ The foundation was built on a very high water table and required an elaborate system of draining. In both this and the Palace of Versailles projects the principal stakeholder was on hand to inspect the finished deliverable, and compare to the best of other architectures.
  - ○ Both projects were showcases for empires and for this the most exquisite levels of craftsmanship were required. Whatever was visible required the highest levels of quality. Overall the quality of construction was very high.
- • Human Resource Management
  - ○ With the Taj Mahal the project with its complexities demanded the acquisition of an international team of thirty seven men of varied talents, including designers and architects. This helped drive the project to completion.

- ○ The Palace of Versailles project incorporated many of the guilds (masons, glaziers, plasterers, gilders, painters, gardeners, fountainers) creating a skills shortage for other building projects. The logistics in housing and transporting a large workforce of 36,000 required much planning and organization. Skilled workers were brought in from distant places, for example, Venetian glass makers to make the mirrors.

- ○ The role of the architects was diminished to functionaries even though they were renowned masters.

- ○ In the Canal Du Midi project Riquet needed a reliable project workforce, so he implemented a social security system that paid above the going rate, provided affordable housing, and paid sick and vacation days. He also brought women into the project.

- Communication Management

  - ○ This was critical in the Taj Mahal project with the large international creative team across the continent. Language could be an obstacle but was not.

  - ○ Louis XIV's hands-on role in the project disrupted the hierarchy of communications flow, so that it went through him. There was also a major communications breakdown between him and Colbert who practically gave up in the later project stages.

- Risk Management

  - ○ With the Taj Mahal the architects were aware of the possibility of an earthquake. The four Minarets were deliberately erected leaning outwards so that in an earthquake they would fall away from the tomb.

  - ○ There was much risk in locating the site so close to river, as the ground was waterlogged and the foundation had to be protected. The risk was mitigated by boring a series of wells.

  - ○ Financial risks undertaken with the Palace of Versailles project over a long period led to massive cost overruns. But because the project spending continued to be out of control over a long period it led to the instability and eventual breakdown of the overall nation.

- ○ Insurance becomes more widespread and sophisticated in this period as ships were trading around the world. Lloyds of London was started in a coffee shop in 1687, where ship captains gathered to share information on past and upcoming voyages, routes, weather, and hazards.

- Procurement Management

  - ○ The land for the Taj Mahal was forcibly purchased from Raja Jai Singh in 1630.[21]

  - ○ With the Taj Mahal the logistical challenge was with the diversity of so many different materials brought from far and wide. For it was not just the core building material, like marble which required elephants for its transportation, and red stone from quarries more than 200 miles away but, the twenty eight types of precious and semi-precious stones that were inlaid into the white marble.

  - ○ The Palace of Versailles project incorporated a massive procurement process with a broad array of hundreds of suppliers. The cost of managing this was very high.

Educators

- Discuss in the context of today's projects the importance of the project manager controlling the project, and keeping interfering stakeholders at bay?

- The Palace of Versailles project was disastrous in terms of overruns of budget, time, and scope. Yet, it was revered architecturally and widely copied. What important lessons can be learned from this for today?

- Was there any way that Colbert could have kept Louis XIV from interfering? What else could have Colbert done?

- How did the acquisition of an international team of thirty seven men help with the execution and completion of the Taj Mahal project?

CHAPTER 12

The 18ᵗʰ Century

The 18ᵗʰ Century witnessed colossal changes in the Western World with the Industrial Revolutions. The field of Engineering became a more formalized profession with the French and British who began to take a more empirically oriented approach towards project management. In society there were major shifts of power away from the church and the crown, to the state. There was also a shift in wealth from the very rich (land owning) towards the upper middle and middle classes (merchants and commercial). Large projects were now more likely to be sponsored by state and commercial interests rather than the crown.

First Industrial Revolution 1700-1840 (Phase 1)

The Industrial Revolution began to pick up pace in 1750. The revolution started with the evolution of steam power and the development of the steam engine, which took many decades. This was a solution looking for a problem and it was applicable to many industries. It began in the UK with the introduction of steam power (fueled primarily by coal) and powered machinery (mainly in textile manufacturing). Factories were not new but the introduction of the steam engine and its interaction to existing machinery (looms, presses, stamps) modified the way workers interacted with it. Over ensuing decades major technological, socioeconomic and cultural changes were set in motion as a labor economy was partially replaced by machinery.

In parallel, with the Industrial Revolution the production costs of certain building materials like concrete, iron, and glass began to drop and so these gradually replaced more traditional materials. This happened first with iron and its use in bridges (Darby 1775, page 371) and then in railways circa 1825.

In this Historical Period

- Brief synopsis of trends/changes (natural disasters, wars, technology, epidemics).

  a. Ideas of political revolution threaten monarchies.

  b. Europeans continue to fight, protect, and exploit their colonies.

  c. Industrial revolution drives changes:

     i. In manufacturing.

     ii. In transportation, canals and railroads.

- Impact of changes

  a. Advances in transportation increases movement of goods and people.

  b. Rapid growth in cities.

- Major events

  a. American and French Revolutions, 1776-1789.

  b. Latin America 1810–1820.

  c. In 1812 Moscow burned for five days; 30,800 houses were destroyed causing $150 million worth of damage.

  d. Finance, the Gold Standard was introduced in 1816.

  e. First steam rail line open, 1825.

- New tools, techniques and breakthroughs

  a. Engineering (Civil and Mechanical) formulized through societies in the UK and France.

  b. In the West the development of new processes for the manufacturing of materials like concrete and iron.

    c. Steam engine.

    d. Machines for automating processes.

    e. Growth in management profession and techniques.

- Regions

    a. UK – Industrial revolution.

    b. Europe – Colonization, and overseas expansion.

    c. USA – Economic growth.

- Significant projects

    a. The Iron Bridge (River Severn) by Abraham Darby III, 1781.

    b. First Railway, Stockton Darlington Railway.

- Other Notable Projects

    a. Menai Suspension Bridge.

    b. St Paul's cathedral.

- Key players

    a. Thomas Savery, Thomas Newcomen, James Watt, and Richard Trevithick.

    b. Joseph Whitworth.

    c. George Stephenson, UK.

    d. Abraham Darby, UK.

- Chapter Wrap-up

Trends and Changes

The world population in this historical period grew from 600 million to 1.1 billion. A population explosion in China's pushed the population to nearly 400 million by 1800, caused by new crops like corn, sweet potatoes, better strains of rice, and improved transport of food along China's canals and waterways.

Ideas of Revolution

The English civil war created the first modern democracy, limiting the powers of a restored constitutional monarchy. It set in motion the notion that change was possible, which saw widespread revolutions through Europe.

Europeans and Colonies

The Europeans continued a race to acquire colonies around the world. They fought, protected, and exploited their colonies.

Industrial Revolution Drives Changes in Manufacturing

In the UK, factories were no longer dependent on the availability of water power, and with steam there were no restrictions on their location. Factories were located closer to natural resources, materials, and new industrial centers developed.

Industrial Revolution Drives Changes in Transportation

The new industrial centers increased production and there was an immediate requirement to improve transportation so trade could expand. This was enabled by improving roads and the introduction of canals which were built quickly and inexpensively. New "Iron Bridge" technology was used to raise canals above gorges, and rivers. Raw materials and finished goods were moved more rapidly and cheaply than before. The first railroads appeared as an evolution of the wagon ways.

Changing View of Human Resources

As the Industrial Revolution spread a more disciplined approach to business and management was required based on scientific research and principles. In 1776, Adam Smith wrote about the economic advantages of the division of labor in his work: *The Wealth of Nations*. Smith (1776) proposed that work could be made more efficient through specialization and suggested that work should be broken down into simple tasks. This would have the advantages of the development of skills, saving time, and the possibility of using specialized tools. Smith's suggestions led to many changes in manufacturing processes.

In 1832, Charles Babbage examined and expanded in his work the division of labor: *On the Economy of Machinery and Manufacturers.* He determined, as an advantage to the division of labor, that the amount of skill needed to undertake a specialized task was only the skill necessary to complete the task. Babbage analyzed and documented the manufacture of a pin and broke the process down into seven elements to illustrate his point. This study became important to employers in that they only had to pay for the amount of skill required to complete a task.[1]

In the early 19ᵗʰ Century, the working class began to question and defy the power of the aristocracy, and to form workplace combinations and trade organizations to provide a collective voice for their rights and improving workers' living standards. Governments for the most part fought this using legislation[2] which banned everything from meetings to combinations.

Impact of Changes

Advances in Transportation

The increased movement of goods and people led to an accelerated growth of new industrial centers, and cities.

Growth in Cities

New factories in industrial centers attracted massive migration into cities like Manchester, UK, and this led to a population boom. The population of England and Wales, which had remained a static 6 million from 1700-1740, rose dramatically after 1740.[3] The population of England more than doubled from 8.3 million in 1801 to almost 17 million by 1851.

Major Events

American and French Revolutions, 1776 and 1796

The American and later French revolutions set in motion cataclysmic change. The creation of the modern state led to nationalist governments and the ideas of republicanism spread around the world. These states produced a new kind of national territoriality within the world economy.

Latin American 1810–1820

In short time the Europeans lost imperial control of the Americas, both in the North and South, as the wars of independence flared. There was a de-colonization of countries away from the Spanish and Portuguese Empires.

Finance, Gold Standard, 1816

Gold was officially made the standard of value in England in 1816. At this time, guidelines were made to allow for a non-inflationary production of standard bank notes which represented a certain amount of gold. Bank notes had been used in England and Europe for several hundred years before this time, but their worth had never been tied directly to gold. In the U.S., the Gold Standard Act was officially enacted in 1900, which helped lead to the establishment of a central bank.[4]

First Steam Rail Line Open 1825

See First Railway Project (page 378).

New Tools Techniques and Breakthroughs

Engineering

The field of engineering became a formalized profession in France and the UK, but along different paths. The French developed university engineering education under the sponsorship of the government. The French, more rationalistic, spearheaded civil engineering with an emphasis on mathematics. The British, were more empirically oriented, and with the Industrial Revolution pioneered mechanical engineering. This saw the emergence of autonomous professional societies, with less government sponsorship than the French, but the information was shared more quickly and readily through organized meetings and journal publications. Technical training started to shift from apprenticeships to a university education.

Civil Engineering

Civil engineering is the oldest of the main disciplines of engineering. The first engineering school, the National School of Bridges and Highways in France, was opened in 1747. John Smeaton was the first person to actually call himself a "Civil Engineer" in 1768, and identified a new profession that was distinct from that of the military engineers.

These civil engineers built all types of structures, designed water-supply and sewer systems, designed railroads and highways, and planned cities. In 1828 the world's first engineering society came into being, the Institution of Civil Engineers in England.

Mechanical Engineering

Mechanical engineering was the second branch of engineering to emerge in the last part of the 1700s. The invention of the steam engine was the starting point for the Industrial Revolution. All types of machinery were being developed now and so a new kind of engineer, one dealing with tools and machines, was born. Mechanical engineers received formal recognition in 1847 with the founding of the Institution of Mechanical Engineers in England.

Precision Engineering

The use of high temperature kilns produced much higher quality metals. With these engineers produced very precise components for equipment and machines. For example, a very accurate sextant for navigation, or more close fitting and accurate pistons and cylinders for steam engines.

New Materials

Iron Production

Through the Industrial Revolution a number of processes were developed for making cast iron including a blast furnace that used coke instead of charcoal. Cast iron was an early form of iron with a crystalline structure and a carbon content of 3%-4% which made it hard and brittle. This characteristic was incorporated into the design of rail lines, and was used with great success in the later part of the 18th Century.

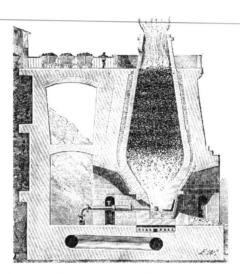

Figure 12.1: Coke Blast Furnace used for making cast iron in large quantities.[5]

Concrete 1756

British engineer John Smeaton rediscovered that limestone was the best mortar for underwater construction. The limestone had a high proportion of clay or hydraulic lime mortar.

Steam Engine

The Newcomben steam engine, mainly used in mines, was the first practical engine. James Watt improved it and in the 1780s this enabled the rapid development of efficient semi-automated factories in places where waterpower was not available. Organizations achieved more ambitious projects as engineers harnessed steam power.

Machines for Automating Processes

The introduction of machines for automating processes started in the textiles industry with inventions like the Flying Shuttle (1733), the Spinning Jenny (1767), and the Spinning Mule (1779).

Regions

The UK was in the midst of first Industrial Revolution which was soon to spread to the U.S. and Europe. Europe at the time was preoccupied with colonization, and overseas expansion. The U.S. continued to grow commercially and economically through agricultural exports (cotton, tobacco). In the modern looking map of the world below, all the continents were discovered and explored except Antarctica.

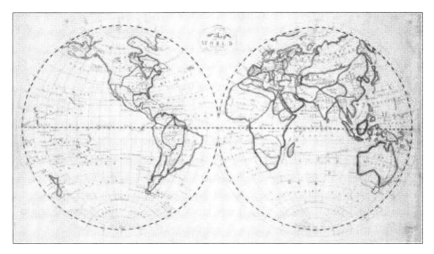

Figure 12.2: Map of 1811 showing all the continents with only Antarctica missing.[6]

Significant Projects

In the era of the Industrial Revolution the two projects selected below were very pioneering in how they pushed the limits of new technologies and showed ingenuity in overcoming project challenges.

Iron Bridge Project (1775-1781)

Background

The world's first Iron Bridge (River Severn) was built downstream from Coalbrookdale in Shropshire, a centre for iron production. The Severn River was the second busiest river in

Europe, and required a bridge to cross a gorge span of 30 meters (100 feet) and 18 meters (60 feet) above the river. There were six ferry crossings in the Gorge which moved people, equipment, and materials for use in iron production. Industry was always at the mercy of the river which sometimes ran too shallow in the summer, and too high and fast in the winter. The nearest bridge was two miles away at Buildwas. The ferry crossing was difficult and dangerous, especially in the winter. So, there was a very pressing business requirement for the project. In 1775 the first meeting was held to discuss a potential project. The following year in March 1776 the act for the project received Royal Assent.[7]

Initiating

The gorge was rich in coal, iron, and limestone and as a result Shropshire had more iron factories within a two-mile radius than any other city in the world. With the availability of this new material it seemed only logical to build the new bridge with cast iron. It would be considerably lighter than a stone bridge with a much larger arch span. There was a vested local interest in the project as it would help promote the use of iron, and exemplify a new application for it, solving a long-standing problem of crossing the river. To date cast iron had been primarily used for wheels and tracks in simple wagon-ways. The business justification was based on the use local materials, and the promotion of local products and industry.

At the beginning of the 18th Century, the industry was in near crisis as there was a shortage of wood, the main material used for making the charcoal that fired the blast furnaces. Iron producers were reduced to importing pig iron from abroad to keep the industry going. In 1709 Abraham Darby III, an iron master, discovered that coke could be used instead of charcoal for the smelting of pig iron, lowering the production cost of cast iron. He developed a new blast furnace at Coalbrookdale on the River Severn.

Cast iron is weak in tension but strong in compression and was the right type of material for building cast iron arched bridges. To date the Coalbrookdale iron works had produced the cylinders for the first steam engine used in mines in 1722, and iron wheels for the horse drawn wagon-ways in 1729, and iron tramway track in 1767.[8]

366

Figure 12.3: Inside of a smelting house (near Broseley) used for the production of cast iron.[9]

Planning

In 1775 Thomas Pritchard designed the Iron Bridge. Darby a local iron master was commissioned to cast and build the bridge. The bridge comprised of more than 800 castings of 12 basic types. The earliest estimate of costs for the project was £3,200, as drawn up by Abraham Darby and Thomas Farnolls Pritchard. When Darby defined estimates for the project he agreed to fund any overspend. In 1777 shares were issued to raise the required investments. Darby estimated 300 tons of iron would be required (at £7 a ton). The bridge was to be a toll bridge and provide an income to the shareholders.

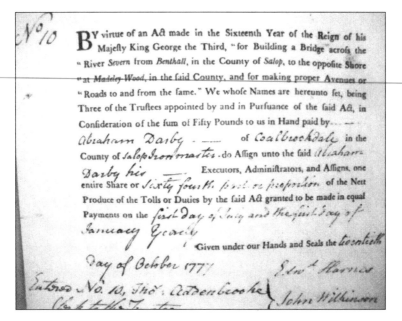

Figure 12.4: Iron Bridge Share Certificate. The Rev. Edward Harries contributed £525 in a share subscription of 1777 which raised £3,150.[10]

Executing

In 1777 the old furnace at Coalbrookedale was enlarged so as to increase iron production. The project team was unfamiliar with connecting large cast-iron pieces together so they resorted to the typical woodworking methods of the era. The joints cast would all be familiar to a carpenter like the mortise and tenons, and dovetails and wedges. This was common through the Industrial Revolution where an existing best practice was used with a new material or technology but usually in a less than efficient way. For example, iron ships were built on traditional lines (keel and ribs) right up to the Second World War.

There were 482 main castings and with the smaller deck facings and hand railings up to 1,736 individual pieces. This would have taken three months of continuous production in one of Darby's furnaces. All the large castings were made individually as they were all slightly different. This included the five large arch ribs (in 3 sections: lower, middle, outer) each cast in two halves.

As the castings were made work started on the foundations. Stone footings were built using local sandstone, and topped by iron base plates.

With this new technology less time was spent in erection and assembling the parts because of the pre-cast joints. The assembly required a pair of 21 meter (70 feet) wooden derrick poles which were stood in the river bed and acted as cranes. The castings were brought to the site by boat, from the foundry 500 meters (0.3 mile) downstream. It would have been extremely difficult to transport large loads over land using horses.

The assembly started with the five large lower arch ribs. Each arch rib was lifted from a barge into a base plate and rested against an inner vertical on either bank. The top of the arches were raised to the correct height until the two halves lined up and joined by a scarf joint and secured. This was a very dangerous job for the workforce balancing on a slender timber brace, and it required steady nerves. Ropes stopped the castings from tipping over. According to a newspaper report, the first arch spanned the River Severn on July 2, 1779.

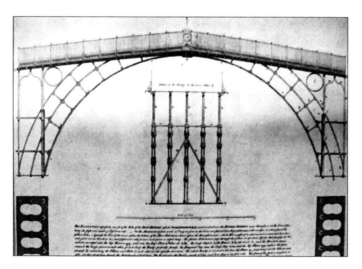

Figure 12.5: The design of the Iron Bridge incorporated three types of arch ribs (lower, middle, outer - seen on the left and right). The center of the image shows the five lower arch ribs connected (the width of the bridge) which rose from the base plates on either side of the bridge.[11]

Figure 12.6: The post card of the cast Iron Bridge shows more clearly the five arch ribs (the width of the bridge) rising from the base plate and the complexity of the interconnections between the iron castings.[12]

Two more arches were completed in the same way. Temporary timber braces made the structure rigid. These were later replaced by iron castings. All five arches were then braced by diagonal and horizontal castings near the base plates, which straddled the uprights. With all five lower rib arches in place, the ironwork was free-standing and strong enough to be used as a scaffold for lifting lighter castings. The rest of the middle frame was built next, starting with the middle ribs, followed, then by the outer ribs. The abutments were then built up to their final height.

The deck bearers were brought in from the now completed abutment on the north bank, cast in a temporary furnace next to the bridge. Made to measure each pair of straight deck bearers was linked at the centre by the Crown Bearer, which gripped and tightened the Crown Joint. All the joints on the bridge were then packed with iron blocks and wedges, which were sealed in with lead.

Deck plates were levered into place, starting with the centre one. They were located along the deck bearers by cast iron wedges and were topped by a road surface of clay and blast furnace slag.

Figure 12.7: Cast Iron Bridge over the River Severn near Coalbrookdale clearly showing the foundations and abutments.[13]

Controlling and Monitoring

There were no injuries during the production process, highly dangerous work, that took three months.

Abraham Darby bore the additional cost of 78 tons of iron (at £3,000 equivalent to £750,000 or $4.5 million today) plus the masonry abutments, and the assembly work. He overran the project budget and was in debt for the rest of his short life. The bridge was over-designed, and subsequent bridges used far less cast iron. The final cost of building the Bridge was £6,000.

Figure 12.8: A team of bow haulers on the towpath pulling a Severn trow (barge) upstream. View of the Iron Bridge over the River Severn, near Coalbrookdale.[14]

Closing

This was the first Iron Bridge of its kind. Completed in 1779 the construction had no precedent and therefore was based on techniques used in carpentry. It took six years to complete the entire project, mainly because of the delays in funding, but only two years were spent on the construction.

The Bridge was opened to traffic on January 1, 1781. Darby promoted the Bridge by commissioning paintings and engravings, but he had lost a lot of money on the project, which had cost nearly double the original estimates.

TABLE of TOLLS.

For every time they pass over this **BRIDGE**.

	s	d
For every Coach,Landau,Hearse,Chaise,Chair,or such like Carriages drawn by Six Horses,Mares,Geldings,or Mules._____	s	d
Ditto_____ by Four Ditto_____	2	0
Ditto_____ by Two Ditto_____	1	6
Ditto_____ by One Ditto_____	1	0
	0	6
For every Horse,Mule, Ass,pair of Oxen,Drawing or Harness'd to draw any Waggon,Cart,or such like carriage,for each Horse&c	0	3
For a Horse,Mule,or Ass, laden or unladen and not drawing	0	0
For a Horse,Mule,or Ass carrying double	0	2
For an Ox,Cow, or neat cattle	0	1
For a Calf,Pig,Sheep,or lamb	0	0
For every Horse,Mule,Ass, or carriage going on the roads and not over the Bridge, half the said tolls.		
For every Foot passenger, going over the Bridge	0	0

N.B. This Bridge being private property, every Officer or Soldier, whether on duty or not, is liable to pay toll for passing over, as well as any baggage waggon, mail-coach or the Royal Family.

Figure 12.9: A Table of Tolls for the Iron Bridge.[15]

This was the first bridge in the world to use cast-iron structurally and it rapidly became the focus of visitors from all over the world. It was an icon of what was to be called the Industrial Revolution. The Sunderland Bridge project, completed in 1796, continued to push bridge building projects forward. The iron arch was estimated to be 15 times lighter than an equivalent arch in stone, and its span of 73 meters (236 feet) was far in excess of any single-arched stone bridge that was existence.

First Railway Project

Background

The First Railway project was not unique but it was the first successful one from numerous attempts within a 20 year time frame. This technology was so far reaching and had such a significant impact on the forthcoming years of the century it deserves a mention as a significant project in its own right.

Tracked roads were used at least 2000 years ago, in quarries in Greece, Malta, and the Roman Empire, where animals pulled loads along cut stone tracks. The early forms of railway evolved in the mining industry in the 15ᵗʰ Century where wagons were pushed by miners along wooden rails or wagon ways.

Figure 12.10: An example of a medieval mining wagon the "Hund" running on a wooden tract.[16]

Gradually horse-drawn wagons were introduced and developed in Germany in the 1550s spreading across Europe in the late 16th and 17th centuries. On level ground one horse could pull 10.5 tons of coal for 38 kilometers (24 miles) using rails.[17]

Figure 12.11: An example of a wagonway with wooden rails and sleepers in County Durham, 1750.[18]

By the early 18ᵗʰ Century iron began replacing the wooden tracks and wheels. These systems were known as tramways (see figure 12.12). In 1802, William Jessop opened the Surrey Iron Railway, in south London. It was arguably the world's first public railway, although it was horse-drawn.

Figure 12.12: An example of a horse drawn tramway with iron rails and wheels.[19]

In the middle of the 18ᵗʰ Century stationary steam engines were generally available in mining for wagon-ways that had steep uphill sections, and these would employ cable power for the inclined sections. British troops in Lewiston, New York used a cable wagon-way to move supplies to base before the American Revolutionary War.

In 1812 a number of engineers were simultaneously building locomotives in the UK:

1. William Hedley built two locomotives in 1814-15 which ran on Losh iron plate rails. (William Losh - Walker Ironworks). Initially these were four wheelers, but this caused extensive damage to the track and they were rebuilt as eight wheelers.

2. George Stephenson designed his first locomotive in 1814. It was a travelling engine designed for hauling coal on the Killingworth wagon-way. This locomotive could pull 30 tons of coal up a hill at 4 mph (6.4 kilometers/hour), and was the first successful flanged-wheel adhesion locomotive (see Figure 12.9d).

Cast iron rails were still in their infancy and exhibited excessive brittleness. Together with William Losh, Stephenson improved the design of cast iron rails to reduce breakage.

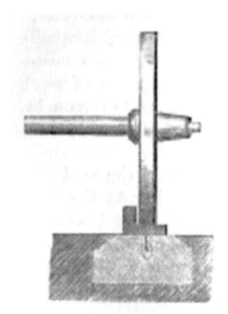

Flange rail.

Figure 12.13: In the late 18ᵗʰ Century, English civil engineer William Jessop designed edge rails to be used with flanged wheels.[20]

Initiating

The district west of Darlington, in Durham, was one of the richest mineral fields in the North of England with vast stores of coal. However, the lack of good roads to market made the cost of transport of coal in carts, or horses very expensive, and almost the closed mine. So, there was a strong business justification for a transportation system.

Initially, in the days of canal building James Brindley (Bridgewater Canal) was consulted about a canal. This was superseded, in 1813, when John Rennie completed another survey and published a report where he recommended a canal and priced the project at £95,600 for the section between Darlington and Stockton. This scheme was also scrapped due to the lack of financial support. In 1818, a Welsh engineer George Overton published a report favoring a railway (horse drawn) scheme at a cost of £124,000. There was strong and concerted opposition from landowners. From the supporters of the latter, a project sponsor emerged, Edward Pease, a successful local wool merchant, with very good connections. He wanted to create cheaper, more efficient, and more reliable transport for merchants and workers between the mines of South Durham and the North Sea at the mouth of the River Tees. He had woolen mills and other interests that would benefit from this. Pease was a Quaker, who was known for his business enterprise, and was interested in a creating a network of railways in the UK.

> *"[Pease was] a thoughtful and sagacious man, ready in resources, possessed of indomitable energy and perseverance, he was eminently qualified to undertake what appeared to many the hopeless enterprise of obtaining an Act for a railway through such an unpromising district. One who knew him in 1818 said, "he was a man who could see a hundred years ahead." [21]*

In 1821, the project charter was finally set when a parliamentary bill was finally passed to build 40 kilometers (25 miles) of track. It would connect the collieries near Bishop Auckland to the River Tees at Stockton, passing through Darlington.

> *"In getting up a company for surveying and forming a railway, Mr. Pease encountered great difficulties. The people of the neighbourhood spoke of it as a ridiculous undertaking, and predicted that it would be ruinous to all concerned. Even those most interested in the opening of new markets for their coal, were indifferent, if not actually hostile."*

Pease assembled a team for planning and executing the project, and with some difficulty. Pease also had to find the funding for what was seen as a risky project.

"The Stockton merchants and ship owners, whom it was calculated so greatly to benefit, gave the project no support; and not twenty shares were subscribed for in the whole town. Mr. Pease nevertheless persevered; and he induced many of his friends and relations to subscribe the capital required."

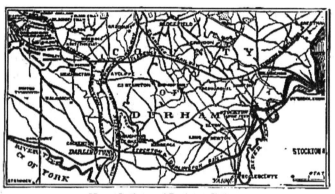

Map of Stockton and Darlington Railway.

Figure 12.14: Map of the Darlington to Stockton-on-Tees railway a distance of 40 kilometers (25 miles) of track.[22]

George Stephenson was a self-taught engineer with the desire to increase his steam railway experience, and a wish to share some larger practical field knowledge he had already gained. Attracted by the project he approached Edward Pease who was much impressed by him and offered him the position of Chief Engineer in the project responsible for its execution on a basic annual salary of £300.

Planning

As railways evolved they started to encroach on a transportation niche held by canals. A horse drawn barge could pull up to 40 tons. The railways could dramatically raise the speed, increase the load, and had a cost advantage. Project director Edward Pease met with George Stephenson to make design decisions on:

1. The comparative merits of cast and wrought iron rails.

2. The gauge of the railway.

3. How the system would work either through the use horse or engine power.

These decisions affected the formation and working of all future railways. They decided on wrought-iron rails for two-thirds of the railway, the remainder in cast-iron. This was because they could be produced in much longer lengths and less likely to crack under heavy weights. This was also a financial consideration.

Stephenson proposed a 4 feet 8½ in (1.435 meters) gauge for the line, and this was adopted as a standard throughout the world. It was the same gauge of wheels for common vehicles of the country, carts and wagons.

Initially, this was going to be a horse drawn tramway with two cable-worked inclines at the western end with stationary engines. In 1821, the plans changed when the two men surveyed the line to see if anymore improvements could be made to Overton's original report. This was one of the influences that Stephenson brought to the project. In comparison to horse drawn traffic steam locomotives were expensive, but after tests they proved to be viable and economic as they pulled more wagons and pulled them faster.

> *"I was so satisfied that a horse upon an iron road would draw ten tons for one ton on a common road, that I felt sure that before long the railway would become the king's highway."*
>
> —Edward Pease

Stephenson's first estimate for the project, in accordance with the instructions of the project directors, was £6,200 for stationary engines, not including any locomotives.

Executing

The construction began in 1821. The line was relatively level through Darlington to Stockton. In 1822 the decision was made to use steam locomotives. It now became Stephenson's priority to concentrate his efforts on the building of high-precision steam locomotives for use on the line. The project needed a team of laborers for this and contracts for the project were highly sought after.

"Navvies (Laborers) were paid up to three shillings a day, which was quite handsome for the time. Their board and lodging was only nine shillings a week – and that included one shirt washed – so many were able to send money back to their families in Teesdale (a lot of lead miners had been attracted to the district to do the digging).

Young boys received 8d a day for drilling two holes in 24 stone blocks which would become sleepers – the holes were for nails which tied the rail on top of the stone to wood beneath it.

Most of the navvies were local, but Stephenson brought in specialist explosives men from his native Northumberland.

George Stephenson worked them hard. He ordered that they left their lodgings in Darlington in the dark so that they could be ready on the ground to begin their work the second the day's first rays of sunshine illuminated the gloom."

—Source:http://www.darlingtonandstocktontimes.co.uk/history/
railway/sadr/

A quarry at Brusselton won the contract to deliver the blocks of stone for the rails, with 8,000 blocks required by March 1st and another 8,000 every 2 months until 64,000 were delivered. These came from a quarry close by where they could be conveyed as needed. In May, a ship arrived at Stockton from Portsea, Hampshire, carrying 9,200 oak blocks.

By the summer of 1823, 22 of the 26 miles of rail had been laid. However, the two rivers Skerne and Gaunless had to be crossed with the first railway bridge in the world. The bridge had to be designed to carry the weight of fully laden train (100 tons). For the Gaunless River Stephenson designed an Iron Bridge which was built in a local factory and erected in 1824. It was truly innovative as it combined both arch and suspension principles of construction, with both cast and wrought iron, with different properties, compression versus tension and bending. The load carried by the bridge was shared where the outward thrust of the arch was counteracted by the inward pull of the elements.

As part of the project Pease and Stephenson jointly established a company in Newcastle to manufacture locomotives using skilled mechanics. This was a large investment and risk in an immature technology that was essential to the project.

On November 7, 1823, the project committee handed Robert Stephenson and Company its first order: four stationary engines (two 30 horsepower, two 15 horsepower) for use on the Brusselton and Etherley inclines.

On September 16, 1824, the project committee ordered the first 2 "travelling" locomotives at £500 each, known as No 1 Locomotion and No 2 Hope. For Pease, the principle was simple: he had provided the money so that the Stephenson's would deliver.

Figure 12.15: On September 16, 1824, Stockton and Darlington Railway project ordered No 1 Locomotion and No 2 Hope at £500 each.[23]

Closing

On the 27th of September 1825 the railway opened, dignitaries were carried on the opening journey in the first purpose-built carriage, the first time passenger traffic had been run on a steam locomotive railway. Up to 40,000 people had traveled to see the event. A locomotive driven by Stephenson, *Locomotion No 1*, pulled an 80-ton load of coal and flour 15 kilometers (9 miles) in two hours, achieving a speed of 24 miles per hour (39 kilometers per hour) on one stretch. The horse-powered opponents tried to derail the locomotives by offering £100 to buy these as scrap but there were no takers.

Figure 12.16: On September 27th, 1825, the Stockton and Darlington Railway project was operational.[24]

"The engine started off with a train of 38 carriages in all and such was its velocity, that in some parts the speed was frequently 12 miles an hour! By the time it reached Stockton there were about 600 persons in the train or hanging on to the waggons, which must have gone at a safe and steady pace of from four to six miles an hour from Darlington."

As the railway went operational Stockton and Darlington Railway owned the tracks but did not operate trains. In fact, it was run in the traditional manner of the wagon-ways where anyone could operate steam or horse-drawn trains by paying a fee. This was similar to the operation of canals, where canal companies were often forbidden from operating any barges. However, this proved chaotic as fights broke out when rival operators came into conflict over right-of-way on the tracks. Mixing faster steam-drawn and slower horse-drawn traffic slowed the operation down and a collision was inevitable that could have serious consequences. As steam technology became more reliable, horse-drawn traffic was abandoned and new operating methods were developed with timetables and other forms of central organization.

The project directors never contemplated sending more than 10,000 tons of coal a year to Stockton, and looked for their profits almost exclusively from the land sale. The railway proved highly successful and in the course of a few years the annual shipment of coal, was more than 500,000 tons. Instead of being a subordinate branch of traffic, it grew into the main branch traffic and the land sale was merely subsidiary.

Figure 12.17: "The Rocket" Locomotive, 1825. (Smiles: Life of George Stephenson.)[25]

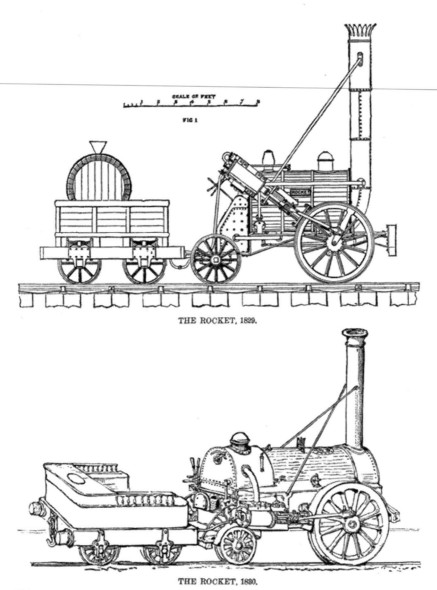

THE ROCKET, 1829.

THE ROCKET, 1830.

This sketch of the Rocket I made at Liverpool on the 12th of September, 1830, the day before the opening of the Liverpool and Manchester Railway, while it remained stationary after some experimental trips in which George Stephenson acted as engine driver and his son Robert as stoker.　　　　JAMES NASMYTH.

Figure 12.18: "The Rocket" Locomotive, 1830 represented in two and three dimensions.[26]

The railway the world's first permanent steam locomotive pulled public railway. It provided a great public benefit to the local inhabitants of the district and opened up entirely new markets for coal. The high profits derived from the traffic yielded large dividends to the project sponsors who had invested their capital. This encouraged other railway entrepreneurs and stimulated similar enterprises in other districts notably the commercial men of Liverpool and Manchester, where the next line was built.

Other Notable Projects

In 1783 the first successful hot-air balloon flight in France. Although very limited in controlling height and direction, this started a race to deliver a viable air transportation system.

Menai Suspension Bridge Project

The island of Anglesey off the coast of Wales had no fixed connection to the mainland and all movements were by ferry or by foot at low tide.

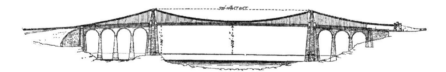

Figure 12.19: Menai Suspension Bridge had a span of 180 meters (570 feet) and a height of 32 meters (100 feet) under the main span.[27]

The bridge was designed by Thomas Telford, and it was one of the first modern suspension bridges in the world. The span was unattainable at that time with any other system, and allowed for tall sailing ships up to 30 meters (100 feet) of clear space under the main span.

In 1819 Telford began construction using concrete and iron. The stonework was finished in 1824 which was followed by the monumental task of raising the sixteen massive chains that would hold up a 189 meter (579 feet) length of road surface between the two towers, the central span. Tunnels were driven into solid rock on either shore to anchor the chains which were drawn up to the top of towers and left to hang down to the water level. The central section of chain, weighing 23.5 tons, was then loaded onto a raft, maneuvered into position and connected to the dangling chain. The project workforce of 150 used block and tackle to draw the chain up to the top of the tower to complete the span. The remaining 15 chains were raised in a similar manner over the next ten weeks. Rods were then hung from the chains and bolted to iron bars that were used as the base for the wooden road surface. The bridge was opened on January 30, 1826 to great fanfare. Its completion, along with other improvements to the road by Telford, reduced travel time from London significantly.

Figure 12.20: Menai Suspension Bridge painting.[28]

The bridge suffered damage in a storm but remained in good condition and is one of the most graceful of bridges ever built.

Bridgewater Canal Project

In 1759 a 40 mile (65 kilometers) canal was built in the UK. It is recognized as the first of the Industrial Revolution and it started a boom in canal building in the UK. Within one hundred years there was a vast network of 1,000 miles of canals that covered the whole of the UK. With this came a Transportation Revolution.

Key Players

1. Thomas Savery, Thomas Newcomen, James Watt, and Richard Trevithick, and their steam engines.

2. Joseph Whitworth's development of screw-cutting and other machine tools, standardized screws, nuts and bolts. This accelerated the pace of development of machinery for the mass production of industrial goods.

3. The railways, George Stephenson, Isambard Kingdom Brunel (1806–1859) and others.

The Darby Family (1750-1800)

The first of a succession of iron manufacturers, Abraham Darby was the son of a Quaker farmer and served his apprenticeship with a maker of malt-kilns. He began a business in Bristol around 1700 with three Quaker partners and set up a works of malt-mill making but added brass and iron founding.

He acquired premises at Coalbrookdale, in Shropshire, close to supplies of low-sulphur coal and in 1709 he produced a marketable iron in a coke-fired furnace. He demonstrated the superiority of coke in cost and efficiency by building much larger furnaces than were possible using charcoal. The quality of Darby's iron made it possible for him to manufacture thin castings that could compete successfully with brass in such applications as the manufacture of pots and pans.

The steam engine (Thomas Newcomen) in 1712 created a new market opportunity for iron. By 1758 Darby, succeeded by his eldest son Abraham (1711-63), cast at Coalbrookdale more than 100 Newcomen cylinders. In 1779 Darby's grandson,

Abraham Darby III (1750-91), completed the world's first Iron Bridge.

George Stephenson (1781-1848)

Stephenson was one of the most renowned British engineers of this age. He was determined to create an overall railway solution that worked. As an engineer by training he took the best ideas, improved upon them, and made them work. He refused to give in, and continuously made improvements. For example, he improved Trevick's engine, and also worked on the replacement of wooden rails with malleable cast iron rails.

Stephenson is best known as the Chief Engineer for the following railways: Manchester & Leeds, Birmingham & Derby, Normanton & York and Sheffield & Rotherham. His designs revolutionized transportation that saw the birth of railroads.

Stephenson was at the forefront of the Industrial Revolution, and when the Stockton & Darlington project was completed in 1825, he was thrown into the public eye. Large crowds saw George Stephenson at the controls of the Locomotion as it pulled 36 wagons of goods, and people. So successful was the line in reducing the cost of transporting coal he was appointed engineer and provider of locomotives for the Bolton & Leigh railway. As the Chief Engineer of the Liverpool & Manchester railway he also solved a large number of serious engineering problems like crossing the unstable peat bog of Chat Moss, a nine-arched viaduct across the Sankey Valley and a two-mile long rock cutting at Olive Mount.

The Stockton & Darlington projects as Pease so presciently said made Stephenson's fortune, when he died 27 years later, he left £140,000 or $840,000.

Chapter Wrap-up

Conclusions

The projects from this period are significant in having to deal with all the challenges that new technologies and materials can present. This is best exemplified by the Iron Bridge project where Abraham Darby assumed far too much of the risk by agreeing to pay for any overruns. More time was spent at the forge in preparing the new materials, and then delivering these to site. Although not much different to the ancient practice of working with stone which was quarried, marked and transported, and made ready for assembly. Another major challenge was the initial assembly of the very large and heavy arches and ribs into a framework. The rest of the assembly was simplified by the interlocking nature of the pieces. The project broke Darby financially but it was widely copied making iron the material of choice for bridges. The bridge was a breakthrough as the iron arch was 15 times lighter than a stone equivalent, and its span of 73 meters (236 feet) was far greater than any single-arched stone bridge.

The Stockton and Darlington Railway project faced so many challenges that it was a surprise the project ended up being so successful. These challenges included local resistance to the project, lack of business support, choice of several new technologies, and most of the risks sitting with the project directors. One of them Edward Pease induced many of his friends and relations to subscribe the capital required for the project, further increasing the risk to him personally. Decisions had to be made relative to the technology options available namely, powering the railway (horse, cable, and locomotive), cast versus wrought iron rails, and the gauge of the railway. The testing in the pilot proved that steam was viable and economic, as locomotives pulled more wagons, and faster.

The railway became hugely profitable, up to 50 times greater than the original projections. After its proven success, the 2nd railway project (Manchester to Liverpool) was deemed so important that the Prime Minister opened it. From this point onward most projects started to take a more rigorous empirical approach and were now estimated and measured using scientific techniques. The project also set a precedent for future projects in how opposition to it was handled and empowered by the railway act.

The Menai Suspension Bridge project highlighted how materials and technology had evolved during the Industrial Revolution in a relatively short time. The project delivered a structure that was a quantum leap forward in span at 180 meters (570 feet). At the time bridges were generally limited to 46 meters (150 feet) span. It would have been impossible to reach this span with earlier materials, and is a good example of how the Industrial Revolution was a catalyst for innovative projects and designs.

All three projects benefited from the evolution and advances in both civil and mechanical engineering which allowed for the introduction of new materials in a controlled manner. The characteristics of the materials could be explored scientifically using mathematics and how they would behave in the real world. Practical thinking became scientific in addition to intuitive, as engineers developed mathematical analysis and controlled experiments.

Key Lessons

PMBoK® Guide Knowledge Areas:

- Integration Management

    ○ For the Iron Bridge project, the business justification was based principally on the promotion of local products and industry through the use of local materials.

    ○ The First Railway project required the integration of many project knowledge areas. It also required the disciplines and skills in engineering including surveying, road building (laying out the path for the track), bridge construction, construction of iron rails, and locomotive building.

    ○ The business justification was vastly under estimated (by a factor of 50) of how profitable the steam venture could be.

    ○ The First Railway project was planned to be flexible to changes that were likely to occur. There was no accurate business model to base the project on, so the payback projections were very inaccurate. The nearest point of reference was the canal and its operation, but this proved in execution to be completely inadequate, as it operated at a much lower speed and was not used to ferrying passengers. In the later stages of the project a new business model emerged.

- ○ Pease shared Stephenson's vision for the impact of the project and expressed it as:
  - *"You will see the day the railways will supersede almost all other methods of conveyance in the country – when mail coaches will become the great highway for the king and all his subjects. The time is coming when it will be cheaper for a working man to travel on a railway than to walk on foot."*
- ○ The First Railway and Iron Bridge projects were significant in pushing materials and technologies to their limit.
- Scope Management
  - ○ The Iron Bridge project scope was dictated by:
    - The height based on passing boat traffic.
    - The load bearing capacity based on the expected road traffic.
    - The availability of materials and the transportation of cast iron pieces to site.
  - ○ The project demonstrated the difficulties in estimating the overall scope accurately with the introduction of new technologies and materials.
  - ○ The bridge was over-designed, and subsequent bridges were designed with much less iron as lessons were passed on.
  - ○ The First Railway project scope was dictated by a number of deliverables that had to be carefully considered. These included surveying, digging cuttings, road building (laying out the path for the track), bridge construction, construction of iron rails (casting), and locomotive building.
  - ○ The scope of First Railway project changed as technologies were proven and became viable, i.e., from horse drawn to locomotive. When Stephenson invited Pease to ride on the locomotive footplate he was impressed by the experience "in truly visionary terms."

- ○ With the First Railway project, there was no accurate business model to base the project on, so the projections were very inaccurate. The nearest point of reference for transportation was the canal.

- ○ For both projects the scope was decreased by having the foundries close by (500 meters (1,650 feet) for the Iron Bridge).

- Time Management

  - ○ The Iron Bridge project required a carefully coordinated erection sequence which was hazardous as the bridge was very unstable at the outset.

  - ○ The production and delivery of cast iron had to be carefully scheduled with the erection sequence to stay within timelines.

  - ○ The First Railway project was under significant time constraints, as there was tremendous competition and business pressure to have the first working railway because it would lead to new business opportunities.

  - ○ The project required the careful coordination and scheduling of many engineering discipline activities like surveying, construction of roads, bridge, iron rails, and locomotives.

- Cost Management

  - ○ In the Iron Bridge project the use of local materials like coke, iron, and sandstone kept the costs down.

  - ○ Abraham Darby III grossly under-estimated the amount of iron and was obliged by the contract to pay for this. The bridge was over-designed, and subsequent bridges used far less cast iron.

  - ○ For the First Railway project Pease and Stephenson had to raise the finances which could only be supported by a share issue. They were also then hit by the extra costs of having to jointly establish a company in Newcastle to manufacture locomotives.

This immature technology was a large extra investment and carried a lot of risk but, it was essential to the project.

- ○ Stephenson employed a practice where the spoil, from cuttings and embankments, was used to build up neighboring embankments. It was transported by tipping wagons instead of carting away by road, a more expensive proposition.

- ○ Stephenson employed cheaper materials were possible for example wrought over cast iron rails, and stone over oak blocks (for sleepers).

- Quality Management

  - ○ For the Iron Bridge project the quality standards for the casting joints had to be high to ensure a good fit, as there was far less flexibility than working with timber.

  - ○ Similarly, for the First Railway project the quality standards for the casting of the rails and locomotives had to be high, and meet certain tolerances.

  - ○ Quality for both projects was very much controlled in the foundry, and as a result was a lot easier to do than on site. This was one of the significant changes for projects using the new materials of the Industrial Revolution.

- Human Resource Management

  - ○ In both the Iron Bridge and First Railway projects it was essential to have the right project team, workforce, and adequate skill sets. Locomotive production required a workforce of skilled mechanics, and engineers.

  - ○ The technologies were bleeding edge with a steep learning curve. The workforce had to adapt and get to grips with it in a short time frame. Much improvisation was required as the technology was trialed and went through a pilot.

  - ○ The success of the First Railway project resulted from the ability of Pease (project sponsor) to recognize the character and genius of Stephenson.[29] He was the chief engineer and architect, and given the latitude and space to bring his ideas to fruition.

- Pease was careful himself not to be elected to the management of the project, although he had immense influence on the project.

- Communication Management

  - Promoting the First Railway project locally was necessary to build up project support both with businesses and with the local population. Because this was a first there was no reference point to the advantages of the technology in economic terms.

  - The opening of the First Railway was done with much fanfare in terms of it as an event and it attracted up to 40,000 people. Dignitaries were invited, and for the first time the public could travel in purpose-built carriages pulled by a steam locomotive, at great speed for its time. This greatly helped promote the railway and get its acceptance.

  - The project set a precedent in that it thwarted the fierce opposition to it namely that of the landowners. Parliament was conscious of the need to support the railway's progress and the railway act allowed the project to go where it pleased to survey a line. The project was empowered to acquire sufficient land by a form of compulsory purchase.

  - Edward Pease, head of the board, managed all the communications with the board. He would regularly update them on all significant progress in the project. Promoting the First Railway project locally was necessary to build up project support both with businesses and with people.

  - Darby promoted the Iron Bridge by commissioning paintings and engravings. With time the bridge became an icon and was a popular attraction.

- Risk Management

- Darby assumed too much of the financial risk by agreeing to pay for overruns which had cost nearly double the original estimate.

- Another risk was the use of large cranes (a pair of 21 meters (70 feet) high), where wooden derrick poles were stood in the river bed. These were for lifting the pre-cast joints into place from boats brought to the site.

- With the First Railway project most of the financial risks were with the project directors. Pease induced many of his friends and relatives to subscribe the capital required, and share the risk.

- Pease and Stephenson had to assume all the financial risks in their personal investments in establishing a company to manufacture locomotives.

- The risks with using steam technology were high as the technology was still immature and unproven but essential to the project.

- The risk of not selecting the right technology and making a costly mistake was high, for example horse versus steam.

- The risk of not getting support as there was local opposition to the project who thought it was unfeasible, and wrote the project off. Business people were not supportive even though it had a very high potential payback.

- The risk of selecting the incorrect business model, as choosing between goods and cargo versus people proved difficult.

- With the Industrial Revolution new materials and technologies had a significant impact on projects of the era in terms of increased risks. Steam engines on ships were a high risk and caused great casualties. For example, when high-pressure steam engines appeared 2,563 people were killed, with a similar number injured, in 233 steamboat accidents between the years 1816 and 1848. In 1852, Congress created the Joint Regulatory Agency of the Federal Government.

Eventually, uniform codes and regulations for the construction and operation of high pressure boilers were adopted.[30]

- Procurement Management

  ○ For the Iron Bridge project the contract was written so the risk of under-estimates was carried by the seller. Abraham Darby III was in debt for the rest of his life and bore the additional cost of £3,000 himself, a small fortune worth about £750,000 today.

  ○ For the First Railway project specialization in specific materials, machines, and products required extensive procurement management.

  ○ In both projects there was reliance on procuring materials locally. At the time iron industries were located close to coal mines so both were close to hand.

Educators

- Discuss in the context of today's projects why these projects were successful considering the major challenges that they initially faced.

- Discuss the importance of running a pilot when working with new technologies and materials. How would this have helped the First Railway project?

- In developing a business justification for the First Railway project was there a business model other than the canal to base the project on?

- Did Darby, Pease and Stephenson assume too much of the financial investments and risks?

CHAPTER 13

The 19ᵗʰ Century

The 19ᵗʰ Century witnessed even greater changes than in the previous century. The Industrial Revolution continued to expand and evolve changing the economies and societies of nations, and through globalization spread to other parts of the world. With this came the evolution of management principles in the business world that became the backbone of modern project management today. In society, the shifts in power continued towards the state, and wealth was driven by a rapidly growing industrial class, superseding the land owning classes. There is a growth in mega projects supported by the state but driven by commercial interests. The crown took a less relevant role. Projects became more ambitious and larger in scale as confidence and experience in new technologies was mastered.

First Industrial Revolution 1840-1890 (Phase 2)

The development of all-metal machine tools in the first two decades of the 19ᵗʰ Century spurred the second phase of the first Industrial Revolution. These machine tools could cut metal on an industrial scale with precision and accuracy. This in turn vastly evolved steam engines where high-pressure steam could be used in a great many new ways. For example, cotton, railways, road transport, farming, and even shipping could exploit steam power to the fullest extent possible.

The machine tools enabled the manufacture of more production machinery and manufacturing activity rose in scale, becoming more mechanized and spreading to numerous industries like food, textiles, lumber, furniture, paper, printing, etc.

In parallel, the Industrial Revolution brought the production costs down of materials like concrete, iron, and glass. These all had a significant impact on projects of the era and started to replace more traditional materials. This happened in different industries and construction projects. For example, the era saw the use of iron in bridges (Darby 1775, page 387) and in shipping (SS Great Britain in 1843), and iron and glass and concrete in buildings (Crystal Palace 1851, page 418). There was a learning curve as the project workforce had to learn how to work with these new materials, and evolve new practices. Without the adequate skill sets, like skilled mechanics, these projects would not have been possible.

Heavy machinery and equipment also changed with the likes of steam powered locomotives, tractors, cranes and shovels that facilitated the building of railroads, tunnels (under the River Thames 1825-1843) and canals, including the Suez and Panama Canals. This shift happened very quickly across many different industries.

The Industrial Revolution impacted many other industries including established ones like agriculture with the introduction of steam powered machinery, that vastly improved agricultural methods and created surpluses to feed the growing municipal populations. It also lowered the cost of transportation and started to dominate and influence work life in different ways. As a result, labor had to react to these changes and the technical advances that were occurring across all industries. These advances created a need for improved work methods, increased productivity, and higher levels of quality.

In this Historical Period

- Brief synopsis of trends/changes (natural disasters, wars, technology, epidemics)

  a. Industrial revolution drives changes.

    i. In transportation on land (railroads) and sea (iron ships).

       ii. In increasing the automation of factories.

       iii. In communications.

- Impact of changes

    a. Globalization and commercial trade.

    b. Increased urbanization.

    c. Major emigration to the West.

- Major events

    a. Revolutions of 1848.

    b. U.S. Civil War, 1861-1866.

    c. The Great Chicago Fire sparks a building boom, 1871.

- New tools, techniques and breakthroughs

    a. The Second Scientific Revolution.

    b. Communications, the Morse telegraph 1840, and the first telephone 1876.

    c. New materials.

    d. Industrial revolution drives new technologies.

    e. Growth in the management profession and techniques.

- Regions

    a. UK, Europe, U.S.

- Significant projects

    a. Transatlantic Cable 1857-1866.

    b. Transcontinental Railroad 1863-1869.

- Other Notable Projects

    a. Crystal Palace 1850-51, the first modern 20ᵗʰ Century building.

    b. London City Sewer Project – 1853-1860.

    c. Suez Canal 1869.

    d. Eiffel Tower 1889.

    e. First mechanical computers, 1840-1890.

- Key players

    a. Isambard Kingdom Brunel.

    b. Theodore Judah.

    c. Cyrus Field.

    d. Hermann Hollerith.

- Chapter Wrap-up

Trends and Changes

The world population in this period grew from 1.1 to 1.6 billion.

Industrial Revolution Drives Changes in Transportation

The development of steam-powered ships, and railways on
a massive scale significantly improved transportation links
across land and sea, and lowered their costs. This freed up the
movement of materials, goods, and people, and led to a global
economy.

Industrial Revolution Sees Automation of Factories

The rapid development of steam engines and their application to
factory machinery previously powered by water caused changes
on a previously unimaginable scale.

Industrial Revolution Drives Changes in Communication

The application of steam power in the printing industry caused
a massive expansion of newspaper and popular book publishing.
The Industrial Revolution drove a communications revolution
with the Morse telegraph (1840) and then Telephone (1876),
both essential tools for project communications management. In
1870[1] the invention of the typewriter further changed written
communication as it was rapidly adapted by the business world.

Impact of Changes

Globalization

Major social change was spurred with the rapid mobility of people, merchandise and information brought about by railroad, ships, and telegraph.

Impact on the Work Force

Factories were not new but the introduction of the steam power, and the engine and its interaction to existing machinery (looms, presses, stamps) modified the way workers interacted with it as the rate of production stepped up, and it caused a change in the relationship between worker and employer.

The late 19th Century also saw the industrial welfare movement a voluntary effort by employers to improve the conditions in their factories and extended into the employees life outside of work. The employer provided assistance to employees to purchase a home, medical care, or for education.

Increased Urbanization

A rapidly expanding workforce of thousands needed to be taken care of in terms of housing, health, welfare, and education. All this brought in new institutions, establishments, and organizations.

Major Emigration to the West

The movement of goods and people increased, through advances in transport, at greater speeds. The industrialization of the Americas saw a vast increase in trade and required a growing labor force.

Major Events

The Great Chicago Fire 1871

The fire destroyed most of the city, an area of about four miles (7 kilometers) long and 3/4 mile (1 kilometer) wide, encompassing more than 2,124 acres (8 kilometers²). More than 17,500 buildings were destroyed initiating a massive building boom.

New Tools Techniques and Breakthroughs

The Second Scientific Revolution

The sciences became more mathematical and theoretical, and data collection was more prevalent. Separate sciences began to merge and collaborate, for example, biology, electricity, mechanics, magnetism, and chemistry all contributed to the discovery of electromagnetic induction and field theory.

New Developments

As both industrialization and science progressed, real collaboration became increasingly common in the 19th Century. Faraday's and Maxwell's work on energy fields led directly to new developments, as did Maxwell and Hertz's work on electromagnetic waves.

Electrical Generators and Motors 1832

The development of electrical generators and motors was also a logical outcome of Faraday and Maxwell's work. The first dynamo was built by Hippolyte Pixii in 1832. The first commutator-type direct-current electric motor was built by the William Sturgeon in 1832.

Communications - The telegraph 1839-42

Wheatstone and Morse's work on the telegraph followed shortly on Faraday's demonstration of magnetic inductions. The first commercial electrical telegraph was constructed by Sir William Fothergill Cooke for the Great Western Railway in the U.K. in 1839. This was followed by Morse in 1842.

Elevators 1860

The electrical motor was first used by Otis and was significant as it allowed buildings to rise in height.

Communications - Radio 1897

Maxwell and Hertz's work on electromagnetic waves or radio waves (1888) directly inspired Marconi to develop the radio.

New Materials

Manufacturing of glass and steel took large steps forward as building materials in their own right.

Patent Plate Glass 1838

The new process of glass production and the annealing of glass provided a simple and economical method for making glass sheets of any size or thickness. The invention of the patent plate glass (1838 by J.T. Chance) allowed the building of the Crystal Palace.[2]

Bessmer Steel 1856

Steel took over from iron when Henry Bessmer proved that pig iron could be decarbonized, thus producing an immensely strong material.

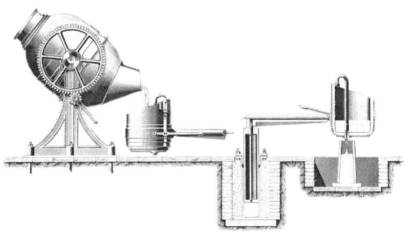

FIG. 11. EARLY FORM OF BESSEMER CONVERTING PLANT AT SHEFFIELD

Figure 13.1: Bessemer Converter decarbonized iron to produce steel.[3]

Industrial Revolution Drives New Technologies

The momentous changes brought about by the Industrial Revolution and its repercussions required new thinking and solutions at a more macro level. This new industrialized world of the factory system with mass production of manufacturing and assembly required a system to supply large quantities of raw materials, resources, man power, equipment and organization. It also needed more sophisticated systems of storage, transportation, and distribution.

The Factory System

This was a system of manufacturing that grew in the 19th Century. Goods were made by workers gathered in a factory rather than handcrafted by craftsmen at home. The principals of the system lay in breaking the overall process into smaller activities and creating cells were tasks would be repeated. The first factories manufactured products like pulleys for sailing ship rigs, and firearms (muskets), and later textiles in mills.

First Assembly Line

Five hundred years before Henry Ford thought about the assembly line, the Italian harbor (Arsenal of Venice) was a classic example of an assembly line operation, though not acknowledged as such. Venice developed methods of mass-producing warships in the Arsenal, including the frame-first system to replace the Roman hull-first practice. The new system was much faster and required less wood. At the peak of its efficiency in the early 16th Century, the Arsenal employed some 16,000 people who apparently were able to produce nearly one ship each day, and could fit out, arm, and provision a newly-built galley with standardized parts on a production-line basis not seen again until the Industrial Revolution.

Transportation Networks

The Industrial Revolution required an advanced system for transportation and distribution. This was first brought about by the canal networks, in the 18th Century, and then the railway systems, in the 19th Century. As trade expanded globally, so did the development of ocean going steamship liners. From

404

all this evolved some of the largest projects ever sponsored by governments namely the transcontinental railroads of the U.S. (1869), Canada (1870), Russia (1917), and the super-liner and dreadnought races of the early 20ᵗʰ Century.

In the early 1860s the bicycle evolved from a simple push device to one with a mechanical crank drive for pedals on an enlarged front wheel.

Portable steam engines provided belting power on farms to run threshing machines, circular saws, etc. This Frick model steam engine operated regularly from 1877-1949.

Figure 13.2: Frick portable steam engine of 1877.⁴

Growth in Management Profession and Techniques

As business expanded so did the management of production which saw the introduction of techniques like standardization, quality-control, cost-accounting, interchangeability of parts, and work-planning.

Insurance and Risk

The emphasis was on enhancing security for businesses significantly during the Industrial Revolution of the 19ᵗʰ centuries. This saw the advent of limited liability that shielded company board directors from personal liability if the business failed.

Regions

The Industrial Revolution spreads from the UK to Europe and to the U.S. Through the 19th Century it spread around the world.

Significant Projects

Transatlantic Cable Project (1857-1866)

Background

As early as 1842 Professor Morse declared a submarine cable connection between America and Europe was possible. By the mid 1850s, telegraph cables stretched along most of the rail lines in the UK and U.S. This form of communication was land based and across short bodies of water only, like the English Channel in 1850. It did not cross large bodies of water because of the many challenges in implementation and the reliability of the cable. For example, manufacturing a strong enough cable that was resilient to water. If the cable snapped recovering it would be very difficult.

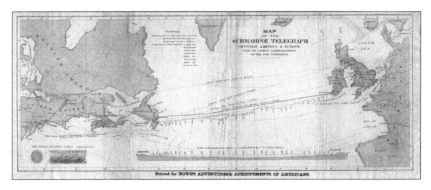

Figure 13.3: Planned route of submarine telegraph lines 1858.[5]

Initiating

The first transatlantic telegraph cable was attempted in 1857 by an American named Cyrus Field and an English engineer named Charles Bright. Field, a rich industrialist, had a vision of global communications. Supported by the British[6] and U.S.

Governments, funds were raised from both American and British sources by selling shares in the Atlantic Telegraph Company. Field himself supplied a quarter of the needed finance estimated at £88,000 ($500,000).

Planning

A special survey was made along the proposed route of the cable and revealed that the proposed route was possible, showing a maximum depth of 4,400 meters (3 miles). The technology was proven and land cables had existed for 30 years. But submerging a cable meant it would be inaccessible and therefore the quality had to be vastly superior. A cable was specially constructed and consisted of 7 copper wires covered with 3 coats of gutta-percha (a form of natural rubber used for wire insulation), and wound with tarred hemp. This was overlaid with a sheath of 18 strands, each of 7 iron wires, in a close spiral. It weighed nearly 550 kilograms/kilometer (1.1 tons per nautical mile), was relatively flexible and able to withstand a pull of several tons.

Executing

Prior to laying the Transatlantic Cable, a pilot project undertook the laying of a cable across the Cabot Strait a distance of 110 kilometers (69 miles). The cable was to connect Newfoundland to Cape Breton and then the rest of Canada to the United States. This step was to be much easier than crossing the Atlantic, but it proved extremely difficult. In the end the pilot cost almost the whole project budget and more funding was needed.

The first attempt across the Atlantic was on August 5, 1857 when the cable was started from the Southwest Coast of Ireland. Several cable-laying ships were used the converted warships HMS *Agamemnon* and USS *Niagara*. On the first day, the cable broke, but was grappled and repaired. A few days later it broke again over the 'telegraph plateau,' nearly 3,200 meters (2 miles) deep, and the operation was abandoned for the year.

The second attempt using a different approach began the following summer when the same two ships met in the middle of the Atlantic. The two cables were spliced together, and on June 26 1858 the cable was dropped and the ships headed out in opposite directions. The cable broke, after less than 5.5 kilometers

(3 miles), and again after 87 kilometers (54 miles), and then a third time when about 370 kilometers (200 miles) of cable had been run out of each vessel. The attempt was abandoned and the ships with the cable returned to Queenstown, Ireland.

The third attempt was on July 17, 1858 when the ships met again mid Atlantic. The cable ends were found and spliced on July 29th. The cable ran out easily this time. *Niagara* arrived in Trinity Bay, Newfoundland on August 4th and the next morning the shore end was landed. *Agamemnon* too had a successful run and on August 5th arrived at Valentia Island, and the shore end was landed and completed the link. The first official telegram was sent from Queen Victoria to the U.S. President, James Buchanan, on August 16th. The following month the cable was destroyed when excessive voltage was applied to it in an effort to achieve a faster telegraph operation.

Public and investor confidence was undermined in the project, and this delayed efforts to restore a connection. It was not until 1864 that Field succeeded in raising the necessary finance to try again.

The technology was improved in this period where Lord Kelvin recommended a new type of cable for the next transatlantic attempt, which included using lower voltages, thicker insulation, and more sensitive detection, and for this he developed a sensitive mirror galvanometer. The new cable core consisted of seven twisted strands of pure copper, coated with compound, then covered with four layers of gutta-percha, alternating with four thin layers of the compound cementing the whole. This core was covered with hemp saturated in a preservative solution, and then spirally wound with eighteen single wires of soft steel, each covered with fine strands of manila yarn steeped in preservative. The weight of the new cable was 980 kilogram/kilometers (2 tons) per nautical mile), or nearly twice the weight of the old. The new and improved long cables had been successfully laid in the Mediterranean and the Red Sea.

In 1865 a fourth attempt was undertaken with just one ship *Great Eastern* to lay the new cable. Her immense hull was large enough to carry all the cable required. She was fitted with three iron tanks for the 4,260 kilometers (2,300 miles) of cable required, and her decks furnished with the paying-out gear.

At noon on July 15, 1865, the ship left Ireland but the attempt failed on July 31ˢᵗ when, after 1,968 kilometers (1,062 miles), the cable snapped and the end was lost.

Undeterred Field got back to England he issued another prospectus, and formed the Anglo-American Telegraph Company to lay out a new cable and recover and fix the broken one. On July 13, 1866 *Great Eastern* started once more, is a fifth attempt. Despite weather problems on July 27ᵗʰ reached Trinity Bay. Friendly telegrams were again exchanged between Queen Victoria and the United States President. On August 9ᵗʰ *Great Eastern* put to sea again to find and grapple the lost cable of 1865, sixth attempt, some two-and-a-half miles down. After several weeks the lost cable was fished out with a grapnel and brought to the surface where it took 26 hours to get it safely on board. The recovered cable was then spliced to a fresh cable in her hold and paid out so there now were two working telegraph lines.

Controlling and Monitoring

The first message on the 1858 cable took over 17 hours to transmit or 2 minutes to transmit just one character. The 1866 cable could transmit eight words a minute over 50 times faster and the service generated £1,000 ($6,000) in its first day of operation. Much of the credit for this was owed to design improvements by Lord Kelvin. It was not until the 20ᵗʰ Century that message transmission speeds over transatlantic cables would reach even 120 words per minute.

Closing

The project was ahead of its time pushing materials and technology to their limits. Without Field's commitment and extraordinary leadership it would have taken longer (years to decades) to complete the project. The multiple attempts required and the trials and tribulations that Field experienced over the next 12 years would have been enough to shake anyone's faith, but Field persevered. Field put money into the project and became very rich where by 1880 he was worth $6 million.

In 1870 telegraph lines were connected from the UK to India (under the Eastern Telegraph Company). In October 1872

Australia was linked to the rest of the world. London became the world centre in telecommunications and 11 cables radiated out to form a "live" girdle around the world. By 1892, when Field died, there were a dozen cables across the Atlantic.[7]

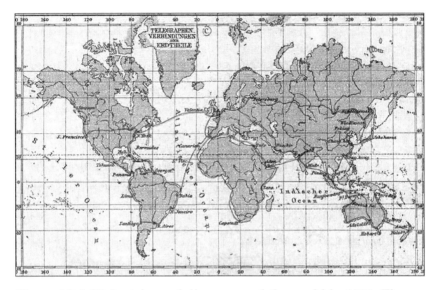

Figure 13.4: Major telegraph lines around the world in 1891. There were over a dozen cables that spanned the Atlantic.[8]

First Transcontinental Railroad Project (1863-1869)

Background

The U.S. Transcontinental Railroad, crossing the continent from coast-to-coast, had been discussed for at least 30 years in government circles. As early as 1832, many wanted to connect California to the rest of the Union. Many reputable engineers testified before legislative committees that the project was impractical. The main challenges related to the 800 miles of almost uninhabited country over the Sierra Nevada Mountains and across an alkali desert, most of it unexplored territory. On March 3, 1853 Congress gave $150,000 and authorization to the Secretary of War, Jefferson Davis, to order the Topographical's to explore the American West and "ascertain the most practical and economic route for a railroad from the Mississippi River to the Pacific Ocean."

410

Initiating

This was one of the first mega projects organized by the U.S. Government. The civil war had provided much experience in organizing major projects. President Abraham Lincoln initiated the project through the Pacific Railroad Act - an official project charter to build both a transcontinental railroad and telegraph line. The business case for the government was based on the cost savings for the government in military transportation with a net gain of $50 million over the cost in a seven year time frame.[9] These figures did not include the major economic impact of opening up the American West in terms of increased trade, or decreased postal Costs.

> *"...the necessity that exists for constructing lines of railroad and telegraphic communication between the Atlantic and Pacific coasts of this continent is no longer a question for argument; it is conceded by everyone."*
>
> —August 16, 1856, Mr. Denver of the House Select Committee on the Pacific Railroad and Telegraph

Planning

The project was breaking new ground in terms of the distances covered in crossing these mountain ranges and plains, and this required an enormous project budget, the likes not seen before. The government had to initiate the funding through both funds, government bonds, and land through grants (see Figure 13.5 below). It had to act as a catalyst for the project persuading financiers to join in. Many thought the project was "impossible." The bonds were to be repaid after project completion. They loaned $16,000 for each mile of track laid in the flat plains, $32,000 for each mile of track laid in the Great Basin, and $48,000 for each mile of track laid in the mountains. The land grants were evenly split between the railroad companies and government, so both would benefit financially when the land value rose after the project. Each company received 10 sections (6,400 acres) of public land grants, mineral rights excluded, on each side of the track for each mile of track built.

An initial capital of $1 million was raised for the surveys. The railroad was to be built by two competing companies, the Central Pacific Railroad (CPRR) and Union Pacific Railroad

(UPRR) from the east and west simultaneously. As the companies completed project milestones they were granted the land, a healthy competition opened up between them, to race to an undetermined meeting point.

Railroad	Character of Bond	Payable in	Amount Issued	Date of Issue		Est. Acres of Land Granted
CPRR	California State aid	Gold	$1,500,000	July 1, 1864		
"	United States subsidy	Currency	$25,885,120	Jan., 1865	— July, 1869	
"	First Mortgage (A, B, C, and D)	Gold	$6,378,000	July 1, 1865	— July 1, 1866	
"	First Mortgage (E, F, G, H, and I)	Gold	$19,505,000	Jan. 1, 1867	— Jan. 1, 1868	
"	**Central Pacific Bond Subtotal**		**$53,268,120**	July 1, 1864	— July, 1869	
"	**Central Pacific Land Grants**			July 1, 1862	& July 2, 1864	**7,997,600**
UPRR	First Mortgage	Gold	$27,237,000	Jan. 1, 1866	— July, 1869	
"	United States subsidy (second mortgage)	Gold	$27,236,512	Jan., 1866	— July, 1869	
"	Land-grant mortgage (first mortgage)	Currency	$10,400,000	Apr., 1867	— 1869	
"	**Union Pacific Bond Subtotal**		**$64,873,512**	July 1, 1864	— 1869	
"	**Union Pacific Land Grants**			July 1, 1862	& July 2, 1864	**12,000,000**
	Total Bonds		**$118,141,632**	July 1, 1864	— 1869	
	Total Land Grants			July 1, 1862	& July 2, 1864	**19,997,600**

Table 13.1: The project was driven by mortgages and bond issues. The railroad companies (Central and Union Pacific) were incented forward by land grants.

During the planning phase there were many obstacles and hardships. Initially the U.S. congress was deadlocked from 1845 until 1862 over whether to use a northern or southern route. This lasted until the departure of Southern Senators during the Civil War. The project scope was enormous and the challenges began with surveying the vast American West and locating a practical route across the Sierra Nevada Mountains of California. Other challenges lay in pulling together a disparate multi-trade workforce. In addition to track laying which employed 25% of the workforce, the project also required hundreds of blacksmiths, carpenters, engineers, masons, surveyors, teamsters, telegraphers, and cooks.

Theodore Judah (Chief Engineer) put together an assignment of tasks to the trades and their costs.[10]

#s	Job description	Wage
	Transit party for running Lines - Salary per month	
1	Principal Assistant	$250
1	Transit man	$150
2	Chainmen, each $50	$100
2	Axemen, each $40	$ 80
2	Flagmen, each $35	$ 70
1	Stakeman	$ 35
	Party for taking Trigonometrical Observations of points along the line	
1	Transit man	$100
2	Tape men, each $40	$ 80
	For sketching Topography.	
1	Topographer	$100
1	Draughtsman	$100
	For running Levels	
1	Leveller	$100
2	Rodmen, each $50	$100
1	Axeman	$ 35
	For running test Levels.	
1	Leveller	$ 80
1	Rodman	$ 45
1	Axeman	$ 35
	Miscellaneous	
2	Hunters, each $40	$ 80
2	Teamsters, each $50	$100
2	Wagons and Horses	–
1	Steward or Commissary	$ 75
1	Cook	$ 50

Table 13.2: The average cost of provisioning these men was $1.50 per day each or $3,150 per month for all.

1	Chief Engineer, 1 year, at $10,000	$10,000
1	Asst. Engineer, at St. Louis, at $3,000	$ 3,000
1	Asst. Engineer, at San Francisco, at $5,000	$ 5,000
1	Draughts man at each place, at $2,400	$ 4,800
1	Asst. Draughts man at each place at $1,500	$ 3,000
4	Parties' salaries, at $20,000 per year	$80,000
4	Parties' subsistence, at $16,250	$65,000
	Instruments	$ 5,000
	Stationery and office expenses	$ 3,000
20	Horses, at $100	$ 2,000
8	Wagons, at $200	$ 1,600
	Tents, camp equipage, &c.	$ 5,000
	Travelling expenses, incidental, &c	$13,000

Table 13.3: The total project workforce was between 20,000 to 25,000.

Executing

In 1863 the Central Pacific began laying track eastward from Sacramento, California. Two years later the Union Pacific started laying track westward from Omaha, Nebraska in July, 1865.

By 1864 Cost Management became the predominant project issue. The CPRR laid 20 miles of track before running out of money. Since the North and the South were deeply engaged in the Civil War, the price of equipment was becoming greatly inflated and some was becoming more difficult to acquire. During the building of the railroad, the price of one ton of rails went up from $55 to $115. The price of one keg of black powder went up from $2.50 to $15. A large locomotive once cost $10,000, but due to inflation, a small engine cost $14,000. The Federal Government paid in "greenbacks" that were not trusted and converted to only $.57 on the dollar. As a result, the Big Four,[11] the major financiers, developed the Contract and Finance Company, in which they pocketed $63 million, held $100 million of stock, and had power over nine million acres of land from federal grants. They did, however, play the leading role in organizing and permitting the building of the Central Pacific Railroad.[12] The financiers had risked their entire fortunes but rigged the process to make another, larger fortune with the project.

414

The project approach was to build first and improve later, as the railroad became operational. The emphasis on a rapid pace kept the critical path of activities on track. Hence, the use of wooden trestle bridges, which could be rapidly put together. From 1840, trusses, chiefly of timber but with wrought-iron tension-rods and cast-iron shoes, were adopted in America. In addition, new technologies were introduced like swivel wheel trucks which enabled trains to negotiate tight curves and for use in switchbacks to climb steep inclines.

The project had a substantial supply chain and there were many obstacles and hardships that faced it. The principal one was the transportation of materials and equipment from the East coast by ship (via Panama or around Cape Horn at the tip of South America a distance of 30,000 kilometers (18,000 miles)) for the CPRR. The materials for the project included the iron, the spikes, the tools to dig, the powder to blast, the locomotives, the cars, the machinery, everything had to make an expensive and hazardous eight months' voyage, before it could be landed in San Francisco, and had then to be reshipped 193 kilometers (120 miles) to Sacramento by water. The project depended upon the promptness with which all this material was shipped. The only thing plentiful in California was wood for the ties.

> *"The number of ties varied from 2,260 to 2,640 per mile, depending upon alignment and grade. ...The total completed length of the sheds and galleries was about 37 miles, the building of which consumed 20,000,000 meters (65,000,000 feet) of board measure lumber, and 900 tons of bolts, spikes, and other iron."[13]*

The project workforce was obtained from everywhere including some from California which was thinly populated, had very high wages, and could not really spare workers due to gold mining. The project workforce came from New York and the East Coast. As the project got underway so desperate was the labor shortage that the U.S. Government initiated a scheme with the Chinese Government where a Chinese workforce of 10,000 was procured for the project. The incentive for individual workers had to be high to travel over 5000 miles. They were paid $35 per month, in gold, and with work camps provided they could readily save $20 per month, a substantial amount to take home.

In the rugged Sierra Nevada Mountain range, they had to blast 15 tunnels out of solid rock and build bridges across ravines and rivers. The railroad was built by hand prior to mechanized construction equipment and without dynamite, only gunpowder was available. Also on the slopes the line passed through dense forests up to 46 meters (150 feet) in height. The cost of cutting trees, clearing brush, and grubbing out stumps was an item of considerable expense. Large numbers of pack animals were used to carry nearly all supplies and hay and grain over steep mountain trails to the construction camps. During the winter the Sierra Nevada blizzards, snow drifts, and avalanches, required the construction of 60 kilometers (37 miles) of snow sheds.

In crossing the desert eastward, towards the mountains, the arid conditions and lack of forests required the transportation of water for men and animals, on one occasion up to 40 miles. It was necessary to have the heavy work in Palisade Canyon done in advance of the main force, so 3,000 men with 400 horses and carts were sent there over the desert, a distance of 482 kilometers (300 miles), along with hay, grain, and all the supplies.

Other project problems included hostile Plains Indians, smallpox breakout, the UPRR financial scandal (Credit Mobilier), and corrupt politicians. Confidence in the project was low, for example, out west even in towns were the project was going to provide significant benefit, workers demanded their pay before they performed the day's work.

Figure 13.5: The ceremony for the driving of the golden spike at Promontory Summit, Utah on May 10, 1869.[14]

Controlling and Monitoring

Over time the daily construction rate increased from one to two, to five miles a day. The progress became so rapid it astonished engineers, capitalists, and governments around the world[15] who were closely watching its progress. The leading Central Pacific road crew set a record by laying 16 kilometers (10 miles) of track in a single day.

The overall construction time was three years, six months, and ten days. The project was completed eight years ahead of schedule, but was still running into cost control issues with the budget. The closing ceremony was delayed because of the kidnapping of Thomas Durant, the vice president and architect of the Union Pacific by the unpaid workforce who revolted and demanded their back pay. The New York Board of Directors wired the $500,000 necessary to pay the subcontractors and Durant was freed.

The eventual railroad was 2,858 kilometers (1,776 miles) long (690 miles built by CPRR and 1,086 miles built by UPRR) with the joining of the rails at Promontory Summit, Utah on May 10, 1869. The line went from Omaha, Nebraska (UPRR) to Sacramento, California (CPRR), connecting with other railroads from the east (Boston, and New York via Chicago, Illinois or St. Louis, Missouri) to span the continent from east to West Coast. About 200,000 net tons of iron total were used just for building the railroad from Omaha to Sacramento. In 1868 the trip from San Francisco to New York cost $150 for first class, and $70 for second class.

Closing

A commission appointed by an 1898 act of Congress determined the overall cost to be $58.8 million in 1899. In 1909 the last of the government debt was duly paid. The railroad established a transcontinental mechanized transportation network and was considered the greatest American technological feat of the 19th Century. It also paved the way for other projects like the Canadian (1885) and Trans-Siberian (1905) Transcontinental Railroad.

Other Notable Projects

Crystal Palace Project (1850-1851)

Background

The advances in sheet glass and iron saw John Cladius London write his book *Construction of Hothouses*. He paved the way for a competitive boom to build ever-larger glass houses used as conservatories and hothouses including the conservatory at Kew Gardens in England and Jardin d'Hiver in Paris.

Initiating

The project justification was based on the increasing competition on British manufacturers who needed a showcase to promote their products. The Great Exhibition of 1851 provided the venue but required a building to house over 14,000 exhibitors with 100,000 exhibits from around the world. It was a matter of prestige as in 1851 the U.K. was the leading industrial power in the world and the organizers wanted the Crystal Palace to symbolize this stature.

Planning

Planners had been looking for strength, durability, simplicity of construction and speed. In a very competitive climate the design attracted 233 entrants and was won by Joseph Paxton. He had experimented with glass and iron in the creation of large green-houses, like the Great Conservatory at Chatsworth (1836) and had seen their strength and durability. This was the first time that glass was considered a climatic membrane that controlled the encapsulated space and light, rather than windows held in a confined load bearing frame. His £79,000 design was well under the £100,000 ($15 million today) limit imposed. A most important requirement was that the structure could be moved after the exhibition to another location without scarring the landscape.

Figure 13.6: Shows the assembly of modular sections. A row of elms trees was incorporated into the structure.[16]

Executing

The Crystal Palace was built in 8.5 months in 1850-1851 by about 5,000 Navigators (laborers used in canal projects) with up to 2,000 on site at once). It was a glass, iron and concrete pavilion about one third of a mile long. Its short construction phase was due to the interchangeable prefabricated parts, a modular building of standardized factory made components. The glass panels were held by cast and wrought iron trusses, and cast iron columns. This acted like a stressed skin envelope that strengthened the whole structure.[17]

The central exhibition hall, near the 8 meter (27 feet) tall Crystal Fountain, enclosed a full-size living elm trees in the park. The great engineer Brunel praised it as it showed how far iron technology had progressed in the 70 years after the first Iron Bridge project. All the parts were prefabricated off-site, transported to Hyde Park and assembled from a kit of parts in seven months. Over 550 tons of wrought iron, 3500 tons of cast iron, and 400 tons of glass (900,000 feet) were used.

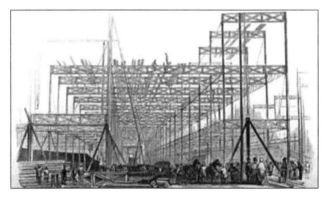

Figure 13.7: General view of the works in progress, base plates for the cast iron columns were set into concrete foundations.[18]

Controlling and Monitoring

The Palace's 92,000 meters^2 (990,000 feet^2) of exhibition space were of Gothic proportions and were used to display the latest technology from the Industrial Revolution. It also housed the World's first dinosaur exhibition. The Great Exhibition building was 564 meters (1,850 feet) long, with an interior height of 33 meters (108 feet). The exhibition drew 6 million visitors.

Figure 13.8: The front entrance of the Crystal Palace, Hyde Park, London that housed the Great Exhibition of 1851, the first World's Fair.[19]

Closing

It was the first modern building, in its use of materials, that heralded in the 20th Century.[20] It initiated the spirit of the new technological era, focused on lightness of materials, and flexibility of structures. The project was widely copied with national railway stations and enclosed shopping arcades in Milan, Paris and Hamburg.

Its designer, Joseph Paxton, along with other colleagues formed the Crystal Palace Company. As the company issued shares and raised £1.3 million ($8 million), it was able to buy the palace from its builders, the engineers of Fox and Henderson.[21] The building's future after the fair was uncertain because of its unprecedented scale.

London City Sewer Project (1853-1860)

Background

John Snow, a British physician and William Farr, a British epidemiologist, regarded as one of the founders of medical statistics, collected statistic on the Cholera epidemic of 1853.

Due to industrialization the population of London had exploded. In the early 19ᵗʰ Century the River Thames was practically an open sewer, with disastrous consequences for public health, including numerous cholera epidemics. In 1849, there was a major outbreak that nearly killed 15,000 people.

John Snow, a British physician and a leader in the adoption of anaesthesia and medical hygiene, studied the cholera outbreak and proposed that the transmission mechanism for cholera was by hands. At the time little was known about bacteria. At the same time, British epidemiologist William Farr supported the miasmic theory, which stated that cholera was spread through polluted air and not by polluted water.

In 1853 another epidemic killed over 10,000 people. William Farr collected the statistical evidence to support his miasmic theory. Analysis of his data showed people living close to the River Thames, at lower elevations, were more susceptible to cholera. The theory was the lower altitude air was more polluted than the regions at higher altitude, and he used this data as evidence for the miasmic theory. Farr collected details about water companies, their water quality, and mortality rates. His results proved that people who consumed water supplied by Southwark & Vauxhall and Lambeth water companies were more susceptible to cholera, because these companies used water from the Thames River as their source. Though William Farr did not agree with the theory derived by John Snow, he helped him with his statistical data.

Snow completed a much more elaborate investigation of the effect of the water-supply in Soho. The London epidemic of 1854 which suddenly killed 600 local inhabitants gave him an opportunity to test his theory. By charting the incidence of the disease, he showed that over 500 cases occurred within 10 days over a radius of some 229 meters (750 feet) centered on London's Broad Street. By talking to local residents he identified the outbreak's source as the public water pump on Broad Street. Snow found a leaking cesspool situated only three feet from the Broad Street well.

A work house in that area had its own private well, and there were only 5 deaths among its 535 inmates. A brewery on Broad Street likewise never used the water from the Broad

Street pump. Its workers drank beer and there were no cases among its 70 workers.

Through a spot map Snow illustrated how cholera cases were centered around the pump. He also used statistics to illustrate the connection between the quality of the source of water and cholera cases. He showed that companies delivering water from sewage-polluted sections of the Thames to homes increased the incidence of cholera. This pattern convinced the local council to disable the well pump by removing its handle. As soon as the pit was sealed off and the water filtered off the problem disappeared.

Figure 13.9: A portion of John Snow's map showing cholera cases in the area surrounding the Broad Street Water Pump.[22]

Initiating

Drastic action was needed and the tipping point followed the 'Great Stink' of 1858. The backed up River Thames caused thousands of upper class residents to flee the City. Parliament remained in session but the windows of the building were draped with curtains soaked in chloride of lime, to prevent the closure of Government.

Planning

Proposals to modernize the sewerage system had been put forward in 1856, but were shelved due to a lack of funds. However, after The Great Stink, Parliament realized the urgency of the problem and resolved to create a modern sewerage system. The government called in top engineer Joseph Bazalgette to create an underground complex of sewers.

Executing

Joseph Bazalgette, a civil engineer and Chief Engineer of the Metropolitan Board of Works, designed an extensive underground sewerage system that diverted waste to the Thames Estuary, downstream of the main centre of population.[23] Six main interceptory sewers, totaling almost 160 kilometers (100 miles) in length, were constructed, some incorporating London's lost rivers.

The intercepting sewers, constructed between 1859 and 1865, were fed by 720 kilometers (450 miles) of main sewers that, in turn, conveyed the contents of some 21,000 kilometers (13,000 miles) of smaller local sewers. Construction of the interceptor system required 318 million bricks, 670,000 meters^3 (880,000 yards^3) of concrete and mortar, and excavation of over 3.5 million tons of earth. Bazalgette and his team built 82 miles of intercepting sewers parallel to the River Thames, and 1,770 kilometers (1,100 miles) of street sewers at a cost of £4.2 million. Bazalgette used 318 million bricks to create the underground system and dug up more than 2.5 million meters^3 (2.8 million yards^3) of earth.

Controlling and Monitoring

Work began on this ambitious enterprise in 1859 and was complete by 1868, a major achievement for its time. Bazalgette drove himself to the limits in realizing his subterranean vision. The job was made harder by having to work alongside the developing underground and overground railway systems. Originally built to serve 2½ million people, the sewers were already serving 4 million by their completion.

Closing

The 19ᵗʰ Century London city sewer project (1860) wasn't undertaken until there was overwhelming statistical evidence that cholera was directly linked to the sewage polluting the Thames River, and that it was used for drinking water. The project was significant as it exemplified the use of empirical evidence in establishing a business case to support a project.

By 1866 most of London was connected to a sewer network devised by Bazalgette. He is generally recognized as having saved more lives than any other single Victorian public official. Today the extended system serves a population of eight million and is essential to the smooth-running of London.

In 1866, there was another major epidemic for which William Farr produced a monograph to explain the death rate which provided a comprehensive analysis of the epidemic, treating it as a complex social and medical phenomenon.

Edward Jenner's and Louis Pasteur's work on inoculation and their germ theory of disease was widely accepted and William Farr's statistics added credence. As a result large engineering projects (public health measures) were initiated to collect and treat sewage so as to eliminate the causative agent of cholera.

Snow's study was not just a major event in the history of public health, which is regarded as the founding event of the science of epidemiology. It was also significant to projects generally. The use of statistical (quantitative evidence in that it could be measured) influenced a business case for a major project. Why is this so significant? The business case was built on empirical evidence.

Suez Canal Project (1858-1869)

The first Suez Canal was built more than three thousand years earlier in the 13ᵗʰ Century BCE possibly Ramses II.[24] It ran from about the midpoint of the modern canal straight west, and connected to the Nile River. In about 550 BCE, the Persian King Darius I rebuilt the canal after conquering Egypt.

In 1854, Ferdinand de Lesseps created a company to construct a canal. The dimensions of it were 22 meters (75 feet) in bottom width, 58 meters (175 feet) in surface width, and a depth of 8 meters (26 feet).

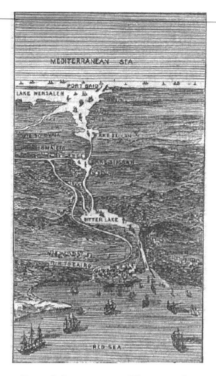

*Figure 13.10: Suez Canal from 1881. The canal cut through 3 lakes that made up almost **19** miles (**30** kilometers) of the total length which was **100** miles (**160** kilometers) in length, **75** of which were excavated.[25]*

Figure 13.11: Suez Canal section south of El Guisr, showing construction trains.[26]

The excavation took nearly 11 years. It was opened in November 1869 and cost about $100 million, or a million dollars a mile including harbors. This low cost was due to the fact that the cut was made through a stretch of level sand. The laborers employed were chiefly natives, estimated at 30,000, and mainly were drafted by Said Pasha, the Khedive of Egypt, a large stockholder in the enterprise. He practically forced his subjects to work on the project in much the same manner as Rameses of old.[27] The project was a major success. However, in a short time frame the British Government took control of the canal as it owned most of the stock.

Eiffel Tower Project (1887-1889)

The tower was built by engineer Gustave Eiffel as the entrance arch for the Exposition Universelle, a World's Fair and this drove the project's business justification, similar to the Crystal Palace project. Considering this was the first structure of this size and the rudimentary means available, this was a record speed.

On site Eiffel ingeniously anchored the tower in wet soil by sinking air tight iron chambers to the bed rock. He put hydraulic pistons in the cylindrical shoes for the four columns so they could be adjusted precisely. A project workforce of 300 construction workers joined together the 18,038 pieces of puddled iron (pure structural iron), using two and a half million rivets. The assembly was made up of 5 meters (16 feet) prefabricated sections prepared in Eiffel's factory on the outskirts of Paris. Two thirds of the total rivets were assembled in these prefabricated sections. Each of the 18,000 pieces were used to construct the Tower were specifically designed and calculated, traced out to an accuracy of a tenth of a millimeter. All the metal pieces of the tower were held together by rivets as workers assembled this gigantic erector set. The pieces were hauled up by steam cranes, which themselves scaled the Tower along runners intended for the Tower's elevators.

It took 5 months to build the foundations and 21 months to finish assembling the metal pieces of the Tower, and this was a record speed. The installation of public elevators with slanting tracks along various angles created many technical questions, as the project team had no real experience in running elevators to such heights and with such weight.

The tower was 300 meters (1000 feet) tall and was the world's tallest tower until 1930 when New York City's Chrysler Building 319 meters (1,047 feet) took over.

Figure 13.12: Partial construction of the Eiffel Tower which was to dominate the Paris skyline to this day.[28]

One of the most significant facts about the tower was it was mathematically designed to reduce wind resistance.

First Mechanical Computers (1840-1890)

The Babbage System Project (1840-not completed)

Both Charles Babbage and Herman Hollerith had remarkably strong business cases to get their mechanical computers completed. These were purpose-built systems that solved particular problems; hence, the high payback. Although both systems were soundly engineered, only Hollerith ever completed his and implemented it into government census bureaus around the world. Babbage's project ground to a halt as he was faced by a wide range of technical issues.

Hollerith System Project (1882-1887)

The U.S. census was faced with the challenge of trying to complete the 1880 census before the next census count started at the beginning of the next decade. As the population increased the race to finish was becoming very close. The 1880 Census took nine years to complete at a cost of $5.8 million dollars but their main concern was that the 1890 census would not be completed before the 1900 Census. Herman Hollerith joined the U.S. Census Bureau as a statistician in 1880. He was faced with the problem of analyzing the large amounts of data generated by the 1880 census. Hollerith was led to look for ways of manipulating data mechanically, suggested by a colleague Dr John Shaw Billings.

In 1882 Hollerith moved to the Massachusetts Institute of Technology to teach mechanical engineering. He investigated Billings's suggestion and examined some technologies like Jacquard loom and realized that the punched cards stored information very efficiently. Another idea was related to how a train ticket collector punched tickets, and the ease with which information could be punched onto cards. Hollerith began to experiment with paper tape, rather than cards, and pins that would complete an electrical contact when going through a hole. Through trial and error Hollerith realized that cards were better suited for the purpose.

Hollerith moved in 1884 to a post in the U.S. Patent Office in Washington, D.C. a brilliant move in that he could patent his own inventions. His first patent in 1884 (there were 30 patents from the United States during his career and many overseas patents) was for a system to convert the information on punched cards into electrical impulses which activated mechanical counters. Over time he improved the machines which read the cards through better electrical connections by placing mercury beneath. He also increased the amount of information handled on each card.

Hollerith's system was first tested on tabulating mortality statistics in Baltimore, New Jersey in 1887 and again in New York City. It also competed for the 1890 U.S. census contract with two other systems, and by showing that it could handle data more quickly it won convincingly.

Hollerith now turned to manufacturing the punches (Pratt and Whitney) and counting devices (Western Electric Company). Everything was in place by June 1890 and the first data from the census arrived in September of that year. The counting was completed by December 12, 1890 having taken about three months to process instead of the expected time of two years if counting had been done by hand. The total population of the United States in 1890 was found to be 62,622,250. The system also gathered new data such as the number of children born in a family and still alive and the number of people who spoke English.

The Hollerith system was clearly a great leap forward. It saved the U.S. $5 million by completing the analysis in a shorter time and with less manpower. The system was again used for the 1891 census in Canada, Norway and Austria and later for the 1911 UK census. Honors came to Hollerith from all sides for his outstanding invention.[29]

When Hollerith took on the project to automate the count he had to meet a very specific deadline. The project was well executed as he took a very pragmatic approach with a very specific end goal in sight.[30]

Key Players

Isambard Kingdom Brunel (1806-1859)

Brunel was the most renowned British engineer of this age. He is best known for the creation of the Great Western Railway, a series of famous steamships, including the first with a propeller, and numerous important bridges and tunnels. His designs revolutionized public transport and modern day engineering.

When Brunel completed the London to Bristol railway (1841) he demanded precision in a consistent and uniform gradient. It is still one of the flattest tracks in the UK. Soon nicknamed as 'Brunel's billiard table', it was completed at a total cost of £6,500,000.[31] On this track high speed records were set. Brunel showed that new technologies, like the railroad, brought in new requirements for precision and quality.

Brunel was very much at the forefront of the Industrial Revolution and his projects innovatively solved long-standing engineering problems like building of the first tunnel under a navigable river, the first propeller-driven ocean-going iron ship and the largest ship ever built. His steamship the *SS Great Eastern* was instrumental to Transatlantic Cable project. He was very much ahead of his time and general acceptance of his innovations often lagged far behind.

Joseph Paxton (1803 – 1865)

He started his career as a garden boy and at the age of 20 was the Head Gardener at Chatsworth, one of the finest landscaped gardens in the UK. There he designed a series of pre-fabricated and modular buildings for use as glass houses. These could be quickly produced and assembled into a variety of configurations. This became a test ground for him as he experimented with prefabricated glass and iron structures. As a pioneer of these techniques he made a bid for the Crystal Palace project of the Great Exhibition of 1851. The successful completion of this led to further work on public parks in Liverpool, Birkenhead, Glasgow, Halifax, and also significant country houses around Europe for the likes of the Rothschild's family.

Theodore Dehone Judah (1826–1863)

He was the Chief Engineer at the Central Pacific Railroad and surveyed the best route over the Sierra Nevada Mountains. He then lobbied in Washington, D.C. on behalf of his company, and helped with the passage of the 1862 Pacific Railroad Act. He also helped raise the finances through the Sacramento merchants. However, he was marginalized by these financiers after the act.

Cyrus Field (1819–1892)

He was a very successful businessman and financier who led the Atlantic Telegraph Company to successfully lay the first telegraph cable across the Atlantic Ocean in 1858. As a principal project stakeholder he persisted with the project and drove it to a successful outcome with a number of attempts.

Hermann Hollerith (1860-1929)

He took a very pragmatic approach to solving problems with a very specific end goal of meeting a deadline that would have been overrun using previous methods. Hollerith adopted existing off the shelf technologies where he could so he kept the development time down. The 1890 Census was completed in seven years at a cost of $11.5 million, but it was a more complete census and took two years less than the 1880 census. Hollerith was paid $750,000 for the rental of his machines.

Chapter Wrap-up

Conclusions

The second phase of the first Industrial Revolution, coupled with a scientific revolution, evolved new materials and technologies that opened up new project avenues in transportation (railroad, ships, canals, bridges) and construction (buildings). The projects were driven by engineers (Bright, Paxton, Judah, Eiffel, De Lesseps, Hollerith, and Brunel) and project directors (Field, Durant). They all had vision and exuded great confidence in taking these projects on and delivering successfully. For example, the Transatlantic Cable had a number of setbacks but Field persevered and was repeatedly able to raise the finance to complete the project. He also persevered with the technology buying a new cable and ships.

The latter part of the century was a testament to a line of projects in the use of new materials (iron, glass, and concrete), coupled with the use of technologies (steam, electrical, communications), and taking on projects of significant scope like the Transcontinental Railroad, and Transatlantic Cable project. The projects were based on scientific principals and empirical methods. These projects were very successful and this in turn inspired confidence in taking on the mega project of the 20th Century. The U.S., UK, and France were at the forefront of these projects. The rule book was rewritten by these projects.

Key Lessons

PMBoK® Guide Knowledge Areas:

- Integration Management - became more important in the 19th Century with the need to support projects with more technological complexity and a great number of dependencies. For example:

  ○ Laying the Transatlantic Cable was a daunting prospect as the sheer scope of the effort was mind-boggling. The business justification showed a tremendous payback on creating a new communication link through this new technology. Messages could be sent in real time, slowly at a transfer rate of eight words a minute, but at a price (initially it cost $100 to send 10 words). The project required a major investment that could only be achieved through a share issue in the Atlantic Telegraph Company, both in the U.S. and U.K.

  ○ The project required the integration of many disciplines. From the precision engineering of the submarine cable, to the expertise required in corrosion protection, or in electrical and conductive materials, to its manufacture in vast lengths. It required large vessels with payout gear that could carry 9,200 tons of cable. Other disciplines from an implementation view included communications methods, navigation, and underwater cable recovery. The technologies used were continuously evolved and improved. From projects like this, systems engineering evolved, driven by the requirement to manage numerous disciplines in a complex project.

○ The Transcontinental Railroad project required substantial project planning. One of the major challenges was coordinating the vast quantities of materials, supplies, equipment and men required. For example, 200,000 net tons of iron were used for building the rails from Omaha to Sacramento, and up to 2,640 ties were required per mile. As the project was initiated from the west, half of the U.S. iron had to be transported from the east to the West Coast by ship, around the Cape Horn of South America, a distance of 30,000 kilometers (18,000 miles), with up to thirty vessels at sea at any one time. The estimated cost was a daunting $60 million project, even though the business case showed a substantial return for the U.S. Government. The cost savings just in military transportation was $50 million over the cost in a seven year time frame and did not include the economic impact of opening up the American West in terms of trade, or decreased postal services.

○ The U.S. Government had to pull out all the stops to get this project initiated, but even then suffered the humiliation over the lack of confidence. Newspapers were constantly circulating stories, and the project workforce demanded pay in advance. Eventually these problems were overcome through substantial bond issues, land grants, and incentives to get the railroad companies competing to build track.

○ The project directed the two competing railroad companies to manage their own project budgets carefully as they raced to complete their tracks. As a result, the U.S. Government could focus on other activities like raising the finances for the project.

- The Crystal Palace project, had to be completed in a short time frame with the world exhibition pending, pushed the limits of technologies, with new materials, concrete, iron and large sheet glass, not used in construction before on such an unprecedented scale:

  - The project charter required the delivery of a very large building of strength and durability which could be constructed quickly with a readily assembled workforce at a fixed cost. It also specified it could be easily removed after the exhibition so Hyde Park could be restored to its previous state.

  - National prestige drove the business case for the project and this was to deliver a building that would exhibit the best of national output (manufactures and industrial goods) of all nations to the world. Half of the Great Exhibition would be devoted to Britain and its Empire and the other half to nations from around the world.

  - The repetitive, simple, and modular design reduced the need for change control. In addition, the desperately short time-frame put the pressure on to avoid any changes.

- The London Sewer project was significant because this was one of the earliest examples were the business case was based on empirical evidence.

- Both Charles Babbage and Herman Hollerith had remarkably strong business cases to get their mechanical computers completed. These were single purpose built systems that solved particular problems hence the high payback.

- Herman Hollerith succeeded in completing his mechanical computer project, in contrast to Charles Babbage, because he took a more pragmatic approach with a very specific end goal even though the engineering capabilities were similar for both projects. He was a more effective project manager.

- Scope Management

  ○ To define the scope and challenges of the Transatlantic Cable project the team ran a pilot that laid a cable across the Cabot Strait. It proved extremely difficult. By the time a working cable was installed almost all the project budget was spent.

  ○ The Transcontinental Railroad project required substantial project planning, and coordinating to manage the vast quantities of materials, supplies, equipment and men required. The logistics of this including the 30,000 kilometer (18,000 mile) supply chain were as difficult as the construction. The latter was affected by the difficulties of the terrain (mountain and desert) and the environment.

  ○ Defining the scope of the Crystal Palace project was simplified by the fact that the structure was modular and made up of standard parts. The project reviewed previous large constructions using a similar approach (greenhouses) and scaled these up to come up with a relatively accurate scope.

    - It required a space of Gothic proportions to house large exhibits, and encompass in the structure a row of fully grown elm trees.

    - Working with new materials like concrete, iron and large sheet glass, reduced the volume of required material.

    - Scope was controlled by awarding a fixed price contract that was delivered by a fixed and approved plan.

  ○ Empirical evidence helped define the scope of the London Sewer project in determining how extensive the underground complex of sewers was going to be.

- Time Management

  ○ The Transatlantic Cable project was a one shot affair, as any failures in rolling out the cable would stop the project. There were a number of stops and starts to the project when the cable snapped and the ships had to return to port. However, Field always kept the project going with the necessary investments.

○ The Transcontinental Railroad project approach was to build first and improve later as the railroad became operational. This approach established a revenue flow early and kept the project on schedule by focusing on the critical path activities. Quite remarkably the project was completed eight years ahead of schedule.

○ The project schedule was very much impacted by the environment and seasonal weather, especially through the mountains.

○ The Crystal Palace project reflected how interchangeable prefabricated parts made from cast iron and glass, could be easily manufactured, quickly delivered, and very readily assembled. This project had a relatively short schedule of 8.5 months. These materials dramatically reduced project scheduling when compared to traditional methods and materials.

  • An equivalent project, but using brick or stone, would have taken 3 to 5 times to complete. The site construction used pre-fabricated components, some cast less than 24 hours earlier, exemplifying how the schedule was crashed as the rate of assembly became closely aligned to the rate of production.

○ With some projects there was no flexibility with timing and deadlines like the U.S. census which had to be completed before the next census started, within a decade. Hollerith took on the project to build a system to automate the counting and had to meet this deadline.

○ Similarly William Farr and John Snow in 1860s raced against time to find a solution to the Cholera epidemic raging through London. Their conclusive evidence led to the necessary funding of the London Sewer project to build sewers diverting sewage from the River Thames. There was a push to compress the project schedule.

- Cost Management
  - By the 19th Century mega projects like the Transcontinental Railroad and Transatlantic Cable project required a mega budget, and could only be supported by a share issue, government bonds, and land grants. This was complicated by the fact that two governments, the U.S. and UK, were collaborating on the project and raising the necessary investments.
  - For the Transcontinental Railroad project a history of railway project experience was critical to the estimates. Although this was a considerably larger project, the teams were able to make parametric estimates in the planning phase.
    - The extensive surveys calculated the scope of the various routes and the degree of difficulty in constructing them. The work effort for the multi-trade workforce was estimated and put together in a simple project plan that showed the assignment of tasks to the trades, and their labor and provisioning costs.[32]
  - The costs escalated out of control as the Civil War raged. The project was plagued by inflated costs, and confidence in the project dropped to the point that the project workforce demanded their pay before they even performed the day's work.
  - Cost Management was tightly managed through the Crystal Palace project. An initial project budget of £100,000 ($15 million today) was set but none of the design tenders could meet this. The project was made financially feasible by two factors.
    - First, a plunge in material costs of concrete, iron and large sheet glass, through changes in production brought about by the Industrial Revolution.

- Second, the impact of pre-fabrication (a by-product of the factory system), and the ability to assemble standardized parts and modules, quickly and cheaply. The final project costs were £150,000 ($22 million today), the equivalent structure in brick or stone would have been 3 to 6 times the costs.

  ○ The London Sewer project controlled costs by:

  - Using concrete (Portland cement) on a massive scale (880,000 cubic yards of concrete and mortar).

  - Not treating the sewage but relying on emptying reservoirs when the high tide was going out to sea - a simple but effective solution.

- Quality Management

  ○ For mega projects like the Transcontinental Railroad and Transatlantic Cable Quality Management started in the factories with the manufacture of components. The high cost of shipment demanded a reduced out of the box failure rate.

  ○ One of the complexities of the Transatlantic Cable project was to ensure that the cable functioned properly before it was laid. Any repair would be extremely difficult. Therefore, the quality of the cable was paramount, and so was the operation of laying it as it could easily snap as happened in 1857. This required a recovery and repair operation the following year.

  ○ The U.S. Government mandated precision engineering through enforced standards for the Transcontinental Railroad project. For example, the track gauge had to be within a tolerance as had the incline (limited to grades of 116 feet per mile), and curves (as sharp as a 400 feet or 120 meters radius). These were inspected through the life of the project.

  ○ According to the railroad commissioner

  - *During construction of the Central Pacific... Railroads and their early subsidiaries, the Federal*

> *Government appointed commissioners to inspect the work prior to the awarding of subsidies. Operating under the auspices of the Secretary of the Interior, the commissioners reported on the progress of construction, condition and valuation of the line and appurtenances, facilities, and equipment in use, and made recommendations for improvements, route changes, etc. The investigations...were carried out in sections generally of forty mile increments.*[33]

- ○ Many projects from this era using the new materials of iron and glass, required very high tolerances. For example, the Eiffel Tower was specifically designed with accuracy to a tenth of a millimeter.

- ○ With the Crystal Palace project quality management was paramount as the design required a greater accuracy, and degree of tolerance than with other materials. The quality control started in the factory which was responsible for delivering uniform iron pieces, and modules. The cast iron columns were tested on site as part of the on-site quality control. The structure comprised of 24-foot bays or modules, repeated 77 times, simplified quality control. The approach was very similar to the repeating arches of the Roman Colosseum. Prefabrication of components improved the quality of the overall structure, as the components were typically molded.

- ○ The London Sewer project was triggered by John Snow's use of spatial analysis that resolved a ghastly public health problem. This was a precursor for the broader use of statistical analysis in investigating problems, and within projects.

- • Human Resource Management

  - ○ As the production costs of materials like concrete, iron, and glass dropped their usage increased across different industries and projects. The project workforce had to learn how to work with these new materials, and rapidly evolve new practices, that required a steep learning curve. Without the adequate skill sets, like skilled mechanics and engineers, these projects would not have been possible.

- In the Transcontinental Railroad project when the U.S. Government procured a substantial Chinese workforce from the Chinese Government the incentives and rewards for individual workers had to be high to travel the 16,000 kilometers (10,000 miles) for the project.

- The Crystal Palace project required a huge quantity of glass, 280,000 meters^2 (900,000 feet^2), and this posed a manufacturing problem for the glass manufacturers which they only solved by taking on additional French and Belgian glassblowers.

- The workforce comprised of 5,000 laborers with 2,000 on site, at once. Because of the repetition of the pre-fabricated modules unskilled workers in teams could quickly and accurately assemble these. There was a learning curve as the workforce had to learn how to work with the new iron and sheet glass materials, and evolve new practices. Paxton showed great skill in his ability to coordinate his crews of workers and was recognized for his compassion. The equivalent structure in brick or stone would have required a much larger and far greater skilled workforce.

- Communication Management

  - Mega projects like the Transcontinental Railroad and Transatlantic Cable required constant and proactive communications to keep the project in the public eye so investors and public funding could be found. Both projects were hampered by reputable engineers testifying up front that the projects were impracticable and unfeasible.

  - The Transcontinental Railroad was presented to the public as a project of national great importance as it connected California to the rest of the Union

  - The term "Crystal Palace" was coined to promote this new glass structure as part of the communication strategy to get public awareness up to the significance of the project. External communications management was important in gaining public support for the project and the exhibition. The Times newspaper had campaigned against the Exhibition committee's own proposal of a fixed brick structure, and heavily promoted the Crystal Palace.

- ○ With the London Sewer project the publication of statistics was a catalyst in getting broad based support for the project.

- • Risk Management

  - ○ The Transatlantic Cable project ran a high risk of the cable snapping as it was being laid. This risk was accepted by the project by incorporating a procedure to search for and grapple the ends of the cables.

  - ○ The financial risks were high as the project would involve procuring several ships, or one large enough ship, for several months. Manufacturing a strong enough cable that was resilient to water was very expensive. Recovering the snapped cable was a very expensive operation that could take weeks.

  - ○ The first Transcontinental Railroad project was funded with U.S. Government issued bonds and was estimated to have cost a staggering $60 million. The financial risks were enormous, which plagued the project throughout, to the point that confidence was so low in the project that the project workforce was demanding a day's pay in advance.

  - ○ The U.S. Government had to come up with creative incentives that pushed the two competing railroads into a race to build the most rail track to an undetermined meeting point. This in itself was contingency planning as the risks were spread around.

  - ○ As the companies hit construction milestones, they were granted sizable parcels of land along the track to race. The two were into a race to earn the most money and obtain the most land by 1865.

- ○ The companies at the same time had to manage the many risks. For example, the project schedule was affected by difficulties in getting equipment, materials, and supplies to the construction camps, and the overall shortage of laborers on the west side.

- ○ The transportation of materials and equipment to site, for the CPRR, was a risk. It came from the East coast by ship (via Panama or around Cape Horn a distance of 30,000 kilometers (18,000 miles). It readily delayed the schedule because of the six month journey required.

- ○ Other risks to the project included environmental for example building through the rugged Sierra Nevada Mountains in winter meant facing blizzards, snow drifts, and avalanches in extremely cold temperatures.

- ○ The construction risks were enormous. Over fifteen tunnels had to be blasted out of solid rock, and wooden trestle bridges were built across ravines and rivers (hundreds of feet high).

- ○ The project had to also cross a very inhospitable desert.

- ○ The U.S. Government was in a state of war with Native Americans and the project faced all sorts of dangers and hostilities because of this.

- ○ As the Industrial Revolution progressed, greater risks were taken with the size of new structures using materials such as iron, glass, and concrete. For example, the Crystal Palace project used these materials on an unprecedented scale. This public exhibition building was based on the concepts of a greenhouse. So, there were many risks that ranged from the ability to use the new materials effectively, to staying on schedule, to the safety aspects of the project (the overall size of the structure), and financial risk in the success of the project.

- ○ Paxton had a model structure completed where workers walked around it as part of a test. Army troops were recruited to march past it in unison. The model held well and this helped prove the structure was safe.

- ○ William Farr and John Snow in the 1860s used statistical analysis to find a solution to the raging Cholera epidemic through the quarters of London. Their conclusive empirical evidence led to the necessary funding of a project to build sewers. This led to a more empirically based approach to projects, and in turn a quantitative approach to risk analysis in projects.

- Procurement Management

  - ○ Procurement contracts were at the core of the Transcontinental Railroad project which was based on two competing companies receiving land grants.

  - ○ The U.S. Government also initiated a scheme with the Chinese government were a substantial Chinese workforce was procured for the project with up to 10,000 workers. The incentives and rewards for individual workers had to be high to induce them to travel over 5,000 miles for the project.

  - ○ Much thought had to be given to the transportation. At any point there were thirty ocean-going vessels moving the supply chain along. On land there was a high dependency on the use of pack animals to the most inaccessible places in the mountains and desert.

  - ○ The Transcontinental Railroad procurement processes were further complicated by the elongated delivery. For example, East coast steel mills had to ship materials (via Panama or around Cape Horn) a distance of 30,000 kilometers (18,000 miles). The only local material in plentiful supply in California was timber. This was used for the ties and the trestle bridges.

  - ○ For the Crystal Palace project the building committee put out to tender a design to which 245 submissions were made. These were all unsatisfactory, none meeting the tender criteria, so the committee proceeded with their own composite traditional design. None of the 19 bids submitted by contractors came in under the £100,000 budget ($15 million today). Paxton delivered a design, in just 13 days, modelled on his greenhouse at Chatsworth. Paxton published his plans in the Illustrated London News and had a very favourable public response. The design was accepted by the building committee.

○ The use of contractors in all the aforementioned mega projects required good planning and maturity of processes to execute.

○ The use of new materials, like iron and glass, were used in projects on an unprecedented scale, to the point where only share issues could fund these projects. For example, the Crystal Palace Company was formed and issued shares that raised £1.3 million. Similarly, the first Transcontinental Railroad project was funded with U.S. Government issued bonds as was the First Transatlantic Telegraph Cable project.

Educators

- Discuss in the context of today's projects the importance of using empirical evidence with the London Sewer project to create the business justification.

- Discuss Cyrus Field's multiple setbacks with the Transatlantic Cable project, his perseverance, and whether it was wise to continue to raise finances to complete the project?

- Discuss the U.S. Governments approach to the Transcontinental Railroad project particularly the funding through land grants, and the competition that opened up between the railroads, to race to an undetermined meeting point.

- Discuss the impact of the Industrial Revolution on project management.

CHAPTER 14

The 20ᵗʰ Century

T he 20ᵗʰ Century witnessed colossal changes across the
world, a period of wars and conflicts around the world,
with two major Industrial Revolutions. The century
started in the midst of the second Industrial Revolution and
the emergence of new technologies and disciplines namely,
chemical and electrical engineering. So many changes and
innovations required an even more structured approach to
business, management, and projects. In society the shifts of
power continued to the industrial classes and business world.
The state's role is to run mega projects of national interest. The
crown virtually has no role in these.

> *"The hundred years after 1900 were a time of unparalleled
> progress. In real terms, it has been estimated [that] average per
> capita global domestic product increased by little more than 50%
> between 1500 and 1870. Between 1870 and 1998, however, it
> increased by a factor of more than six and a half."*
>
> —Niall Ferguson, The War of the World

Second Industrial Revolution (1890-1940)

This second Industrial Revolution gradually grew to the
exploitation of fossil fuels as a source of energy and to include
the chemical industries, petroleum refining and distribution,
electrical industries, and, the automotive industries and internal
combustion engine. German Chemistry extracted nitrogen

from the atmosphere to make explosives, and fertilizers, that increased crop yields. The revolution sees a transition of technological leadership from the United Kingdom to the United States and Germany.

In this Historical Period

- Brief synopsis of trends/changes (natural disasters, wars, technology, epidemics).

    a. Mass adoption of the factory system in the West was a catalyst for growth.

    b. Greater confidence in deploying the technology of the 19th Century.

- Impact of changes

    a. Mass urbanization and growth around manufacturing centers like Manchester, Detroit, Massachusetts, the Ruhr, and Lille.

- Major events

    a. San Francisco Earthquake and Fire, 1906.

    b. World War I, 1914-1918.

    c. New Deal in the U.S.

- New tools, techniques and breakthroughs

    a. First mechanical computer, 1890.

    b. First automobile, 1890.

    c. Tesla coil transmits radio waves 50 miles, 1895.

    d. Typewriter, 1898.

    e. First manned powered flight, 1903.

    f. Plastic, 1907.

    g. First modern luxury transatlantic liner, 1912.

    h. Pierre du Pont introduces the principle of return on investment, 1915.

    i. The Television, 1927.

j. The first "word processor,"[1] 1936.

- Significant projects

  a. Panama Canal, 1880-1914.

  b. Empire State Building, 1929-1931.

  c. Hoover Dam, 1931-1935.

- Regions

- Other Notable Projects

  a. Titanic, 1907-1912.

  b. Race to the South Pole 1911-12.

  c. Golden Gate Bridge, 1933-1937.

  d. Public works projects in 1930s to address unemployment.

- Key players

  a. Frederick Winslow Taylor.

  b. Henry Gantt.

  c. Frank Crowe.

- Chapter Wrap-up

Trends and Changes

The world population in this period grew from 1.6 to 2.2 billion. The spread of mass production is best exemplified by the meteoric growth of automobiles. From the first horseless carriage in 1880 the total number of automobiles in the U.S. topped 4,000 in 1900, and by 1911 rose to 600,000, 1915 to 895,000, and by 1927 to 3.7million.

Impact of Changes

Mass urbanization saw a massive growth in industrial cities around manufacturing centers.

Major Events

Sinking of Titanic 1912

~~The least likely disaster of the new century the sinking of the~~
liner was headlined as a major event around the world. Through
Titanic's construction project the elevation of expectations that
this was "the greatest ship ever built" instilled a sense of supreme confidence. This led to compromises in the implementation stage and allowed for catastrophic mistakes to be made
like pushing the ship to its operational limits in a bid to beat
Olympic's best crossing time.

World War I (1914-1918)

For most politicians the world war was something that they
never envisioned happening:

> *"New economic factors clearly prove the inanity of
> aggressive wars. ...Because of the interlacing of nations,
> war becomes every day more difficult and improbable."*

> —Source: Lectures in 1910 by Viscount Esher, chairman of the
> UK's "War Commission" and senior advisor on foreign policy
> and the military; he believed that the Angel doctrine was as
> accepted in Germany as in the UK[2]

The consequences of industrialized nations fighting such
a war would lead to bankruptcy. The world war mobilized
continents with huge armies and resources into a global conflict
which proved to be a prolonged war of stalemate. It manifested
the industrialization of war and leveraged mass production in
weapons and ammunition, and mass transportation to mobilize
and mechanize vast armies. By 1918 the logistical operation
supplying the British Expeditionary Force was the largest the
world had ever seen.

New Deal (1933 and 1938)

The Wall Street crash (1929) and the onset of the Great
Depression drastically affected profits, voluntary welfare
programs, and millions of jobs. In reaction to this U.S. President
Franklin Roosevelt set up the New Deal where public works

projects emerged to keep unemployment down and provided assistance by creating jobs. Technocrats and planners in Washington were to plan and run mega projects. Projects included road building, the Hoover Dam, Golden Gate Bridge, and San Francisco-Oakland Bay Bridge. New social programs, including old-age pensions, labor standards, and minimum wages for some industries were developed

Other governments took a similar approach of state initiated public works mega projects. These included the autobahns in Germany, ocean liners (Queen Mary and Queen Elizabeth) in the UK, and the Maginot Line in France.

New Tools Techniques and Breakthroughs

Engineering (Chemical and Electrical)

Chemical Engineering

In the 1800s, industry started using more and more chemical processes in many areas such as metallurgy, food production and textiles. At the end of the 19ᵗʰ Century, the increased use of chemicals in the manufacturing industry eventually created a new industry whose main function was the production of chemicals. A "Chemical Engineer" was involved in the design and operation of these new chemical producing plants. It wasn't until the development of the petroleum industry that chemical engineering became recognized as a unique engineering discipline.

Electrical Engineering

In the 1800s research in electricity grew, from the original electric cell invented by Alessandro Volta in 1800 to the Gramme dynamo and electric motor that were invented in 1872. Electrical engineering developed in close collaboration with these developments in chemistry and physics. This led to the development of electrical power and organizations could achieve more ambitious projects by harnessed electricity. An electrical engineer was involved in the design and operation of these new electrical plants.

Other Engineering

By the end of this period many other forms of engineering were appearing. For example, marine engineers tamed the peril of ocean exploration. Aeronautic engineers turned the ancient dream of flight into a travel convenience for ordinary people. Control engineers accelerated the pace of automation. Industrial engineers designed and managed mass production and distribution systems. College engineering curricula were well established and graduate schools appeared.

The Second Industrial Revolution

The very late part of the 19th Century saw the second Industrial Revolution emerge (1890-1940) with a number of new emerging technologies. It was dominated by electricity and chemicals, and the internal combustion engine. It brought in telecommunication (telephones), electrical devices, the internal combustion engine, transportation by land (automobiles), sea (ocean going liners), and air. Epitomized by mass production of consumer goods and the mechanization of manufacture it served the needs of an increasing population.

Light bulb 1879

The electric light was invented simultaneously in 1879 by two people: Thomas Edison in the United States and Sir Joseph Wilson Swan in England. Swan created the first electric light bulb, but could not maintain a vacuum in the bulb. Edison's lamp consisted of a thin platinum filament housed in a glass vacuum bulb, and he created a small version for domestic use.

Industrial

In the 19th Century there was an effort to standardize work. The Gilbreths (Frank 1868-1924 and Lillian) worked with standardization and method studies. In one example, bricklayers were observed and it was determined that no two used the same technique or set of motions. With standardization of technique they were able to improve productivity by over 100%. With his wife they studied the work habits of various manufacturing and clerical jobs to determine how else it could be increased.

Their work was carried into factories to increase the efficiency of workers.

Frederick Winslow Taylor 1856-1915 was the first to scientifically analyze each part of an individual's work so that the most suitable person could undertake it, scientifically chosen, and trained to perform it in a certain way. This required good co-operation between management and workers, with a clear division of labor and responsibility between them.

> *"Taylor's impact has been so great because he developed a concept of work design, work measurement, production control and other functions that completely changed the nature of industry. Before scientific management, such departments as work study, personnel, maintenance and quality control did not exist."*
>
> —Source: (http://www.accel-team.com, 2004)

In 1911, the scientific management movement was started by Frederick Winslow Taylor with his book *The Principles of Scientific Management* which stated the principal objective of management should be to secure the maximum prosperity for the employer, coupled with the maximum prosperity for each employee. Taylor believed that management should use the techniques used by scientist to research, analyze, and test work skills to improve the efficiency of the workforce.

He devised a means of detailing a division of labor into time-and-motion studies and a wage system based on performance. Known as "Taylorism" (the grindless repetition of tasks) would become the standard for businesses worldwide in the next decade.[3]

Henry Gantt (1861-1919), studied management techniques specifically in the field of the construction of naval ships in the First World War. As a result, he created the Gantt chart in 1917. It was a system of outlining the sequence and duration of all tasks in a process, reflected by task bars and milestone markers. This in essence was "Time Control."

The Human Relations Movement

This originated in the 1920s, a forerunner to modern human resource management, focused on employee feelings and group behavior. It was influenced by the Hawthorne Studies[4] which measured the relationship between productivity and the working environment and showed that changes in the environment such as lighting did affect productivity. This was a significant break from Taylor's theories in that the workers were not solely motivated by self interest. Further research led to the understanding that workplaces are more than machine like environments in that there are social environments and human emotions that require consideration.

As unions began challenging the fairness and validity of Taylor's scientific management theories, new personnel roles emerged as employers recognized that professionals could play a middle role between employees and employers, and this set the stage for investments in the role of effective human resource management.[5]

Personnel Program

In the 1930s personnel programs emerged mainly to increase productivity. These included safety programs, medical aid and sick benefits, holidays, housing allowances and other benefits.

Workplace Safety

The beginning of the 20th Century saw a move to improve worker security through a series of workplace safety acts and worker's compensation laws. Other changes saw the introduction of health and safety into the food industry, through the Meat Inspection and Pure Food and Drug Acts of 1906.

Early Mass Communication

The modern field of communications traces its lineage through business information, business communication, and early mass communication studies published in the 1930s.

Public Relations

The field of Public Relations broadly interpreted to include corporate and government communications, non-advertising publicity, media relations, political campaigning, health communications and other methods that seek to persuade "below the line" and to create dialogue.

Personal Risk

Other major changes in the 20ᵗʰ Century were with personal risk. In the U.S. this was enacted by Medicare and Medicaid and corporate pension reform, greatly extending consumer protection which moved to safeguard the environment.[6]

The Evolution of Quality Management

The beginning of the 20ᵗʰ Century saw a formulization of Quality Management as various techniques and practices were developed. Frederick Taylor defined the *Principles of Scientific Management* (1911) which devised a means of detailing a division of labor in time-and-motion studies and a wage system based on performance. Taylor's gospel also known as "Taylorism" (the grind less repetition of tasks) would become the standard for businesses worldwide.[7]

In the 1920s Dr. Walter A. Shewhart of Bells Labs, a physicist, introduced in 1924 the Statistical Process Control and the "plan-do-study-act" (PDSA) cycle.[8] The methods were based on continual monitoring of process variation. The concepts of "common cause" and "assignable cause" variability, and "a state of statistical control" were introduced. It was the first use of statistically-based tools and techniques for the management and improvement of processes. It was the basis for the modern quality movement, including Six Sigma. Shewhart's work was further developed by Deming and the early work of Shewhart, Deming, Dodge and Romig constitutes much of what today comprises the theory of statistical process control (SPC). However, there was little use of these techniques in manufacturing companies until the late 1940s.

In the 1930s - Dodge and Romig introduced the Acceptance Sampling Methods. This was probabilistic approach to predicting lot acceptability based on sampling results. It centered on defect detection; the concept of acceptable quality level (AQL)

Projects

Scheduling further evolved with Gantt's work in 1920s where activities could be laid out visually on a chart to indicate progress. These were extensively used in the Hoover Dam project to overlap the project phases.

Financial Tools

The growth of capitalism in the 16th Century (which saw the first limited company and national bank) evolved the calculation of profit and loss in standard units, and a real world representation through tokens and symbols.

The following were developed in the 20th Century for specific industries, and have come into project management as best practices:

- CBA - Cost Benefit Analysis - Economic accounting is not new. The Frenchman Jules Dupuit, followed by the Englishman Alfred Marshall, developed the concepts in the early 20th Century. In the 1920s, the U.S. Army Corps of Engineers developed a process for economically evaluating which public projects to pursue. Economists adopted the Army's process, and have been formalizing and standardizing the CBA since the 1950s. The CBA provides a repeatable, objective method of measuring if, and by how much, the economics of a business change by pursuing investment options. U.S. Federal Government technology projects actually require, by law, the use of CBA in justifying projects.

- ROI - Return on Investment was originally developed circa 1912 at E.I. du Pont de Nemours and Company (or Du Pont). F. Donaldson Brown developed a new method of measuring financial performance in an efficiency report he authored, and became VP Finance for Du Pont. As chief of financial operations, Pierre du Pont introduced the principle of return on investment. His model has often been referred to as the Dupont Model, and 80 years later, it is currently in vogue. When well detailed, it is among the more comprehensive and practical of the methods used in modern management. Brown's efforts at DuPont led to DuPont taking a financial interest in General Motors (GM), and actually helped GM survive through difficult times in the 1920s. GM went on to become highly successful, and became recognized as a vanguard in the planning process, cost accounting, and in capital productivity.

- EVM - Earned Value Management is a collection of management practices and a structured method to establish a Performance Measurement Baseline, that can be used to measure and analyze performance on all projects. The roots of EVM were in industrial manufacturing at the turn of the 20ᵗʰ Century, but the idea took root in the U.S. Department of Defense in the 1960s. The original concept was called PERT/COST.

Significant projects

Panama Canal Project (1904-1914)

Background

The idea of connecting the Atlantic and Pacific Ocean through Central America dates back to the 16ᵗʰ Century when a survey for a route through Panama was made. It was situated in one of the most difficult and inhospitable climates on earth with a tropical terrain of mountains and hostile jungles. A railroad was built across Panama, opening in 1855, by a publicly traded corporation based in New York City, the Panama Rail Road Company, chartered in 1849. It was financed by mainly U.S. private companies.

The project costs were $8 million about eight times the initial estimate in 1850. The project presented considerable engineering challenges, going over mountains and through swamps. Over 300 bridges and culverts were built along the route. Post project the ongoing work proved just as challenging in upgrading and making the railroad permanent. In the tropical climate wood decayed rapidly so bridges were replaced with Iron Bridges, and trestles with gravel embankments. The lifespan of a tie was a year so these were replaced with lignum vitae ties (tropical hardwood).

An estimated 5,000 to 10,000 people died during the project although the Panama Railroad company kept no records. This was indicative of the serious challenges facing a canal project.

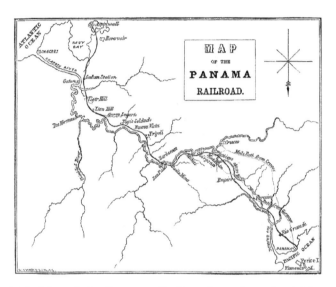

Figure 14.1: Map of the Panama Railroad at a length of 80 kilometers (47.61 miles), 1861.[9]

Initiating

The U.S. was interested in getting the canal constructed. The business case was lucrative. The journey for a ship sailing from New York to San Francisco via the canal cut the travel distance to 9,500 kilometers (6,000 miles), from the 30,000 kilometers (18,000 miles) route around Cape Horn.

The Panama railroad was very successful financially, and by the time the line was officially completed (January 28, 1855), more than 33% of the $8 million cost had already been paid for by eager fare paying passengers and companies paying freight tariffs. Engineering and medical difficulties made it the most expensive railroad project ever completed (per unit length of track). Likewise the on-going maintenance and upgrades rates were very expensive. But the (first class passage fare $25 each way) very high fares and freight costs made it one of the most profitable in the world. The Panama Rail Road Company stock became one of the most highly valued of the era. The railroad was successful but an all-water route between the oceans was still very much desired. It was assumed that it would likely have a very positive return in the long run like the railroad.

Figure 14.2: Panama Railroad and the cut at Basaltic cliff, 1861. From "Illustrated History of the Panama Railroad" by Fessenden Nott Otis, Harper & Brothers, New York, 1861.[10]

The U.S. had been negotiating a treaty with the Columbian Government since 1855 granting the U.S. the exclusive rights and privileges to build a canal through the territory of Nicaragua. In 1880, two Frenchmen, Commander Lucien Napoleon Bonaparte Wyse and General Turr, organized the Panama Canal Company and obtained a concession from the Columbian Government. The success of the Suez Canal (1869 see page 425) buoyed the French and they began construction of the

Panama Canal immediately in 1880. But the French abandoned the project in failure after they had spent over 20 years and $260 million on it.[11]

For the U.S. the project was vastly more strategic for economic, political, and military reasons. Panama was part of Colombia, and with U.S. backing, Panama seceded from Colombia in 1903 and promptly signed a treaty with the U.S. allowing for the construction of a canal and U.S. sovereignty over a strip of land on either side of the structure (the Panama Canal Zone).

Planning

To be successful the U.S. project had to learn from mistakes made by the French. The first French Canal Company inaugurated the undertaking with an exclusive concession obtained from Colombia in May, 1876.

Ferdinand de Lesseps, responsible for the completion of the very successful Suez Canal project, was placed at the head of the new project. It was thought he had credibility and would secure investor's confidence as the company sold shares in the project.

His belief that the project could be accomplished more quickly and easily than the Suez was very misguided. He was not an engineer and relied on a survey that was only two-thirds complete. The original cost and time estimates were $168 million and 7 years. In accordance with the terms of the concession the plans and estimates were submitted to an International Engineering Congress in Paris, in 1879 that he organized to discuss several schemes for constructing a ship canal. De Lesseps opted for a sea-level canal based on his experience with the construction of the Suez Canal. From the 136 congress delegates only 42 were engineers, the remainder were promoters, politicians, speculators, and personal friends of De Lesseps. The engineering congress re-estimated the cost and time at $214 million and 7-8 years. They approved and passed the plans.

De Lesseps made 2 visits to the Isthmus, in 1879 and 1886, staying about 2 months on each occasion. He was accompanied by an engineering commission who came up with the revised estimate in 1879. De Lesseps further reduced this estimate to $131 million, and then without apparent reason, to $120 million.[12]

The total subscriptions, stocks and bond issues, reached $394 million and the shareholders numbered 200,000. In the U.S. De Lesseps was unable to dispose of the reserved stock as Americans were interested in a canal but not under French control.

The excavation of the Culebra Cut was started in 1882 but progress was slow with no tracks to remove the spoil that the excavators were producing. When the problem was resolved, the highest peaks of the cut were tackled but, the worry of landslides and what slope should be taken became a major concern.

In 1883 there was a realization that the tidal range of the Pacific to Atlantic differed in levels by up to 6 meters (20 to 1 feet) and this would be a danger to navigation. A tidal lock was proposed at Panama to preserve the level from there to Colon and this saved about 10 million meters^3 of excavation. It would also prevent landslides.

In 1884 there was a very high sick and death rate among the workforce through cholera, malaria and yellow fever. The hospitals were well equipped, with excellent doctors and surgeons, and supplied with the best medicines and instruments of the time, but they were poorly managed. They were handled under contract, and the administration was left almost entirely to French Sisters of Charity, who were devoted and religious women, but were not trained nurses. Workforce turnover became a huge problem as inadequate medical care was available and workers returned to France.

In 1885 the project ran into financial difficulties and applied to the French Government to issue lottery bonds. This was done successfully during the Suez Canal project.

In 1886 construction equipment was in short supply with too few dredgers, and too much work was being done by hand.

The crash came in December, 1888. At this point $157 million had been spent in Panama, and $78 million in Paris for a total of $235 million. This vast sum is said to have been "one-third expended on the canal work, one-third wasted, and one-third stolen." The project costs were as follows: salaries and expenses of management aggregated $16.5 million; rents and maintenance of leased property, $3.3 million; material and supplies, $29.7 million; buildings, $15.4 million; construction and engineering expenses, $89.4 million; land purchases, $950,655; and medical and religious attendance, $1.8 million.

In view of the various forms of graft, extravagance and waste, it is not surprising that there was so little to show for in actual work accomplished. At the end of eight years the work was about two-fifths completed. About 59.75 million meters^3 (200 million feet^3) had been excavated, a total which included 14.255 million meters^3 (46 million feet^3) from the Culebra Cut. This lowered the peak by 102 meters (336 feet). The overall value of work was about $25 million.

The work was done by contractors, few had met their contracts, and many made small fortunes. Those who were entrusted with the work of excavation were paid for the amount of spoil which they took from the canal prism. With no data available on the cost of such work, it was impossible to even estimate what the charge should be. In many cases the contractors took out what was most easily excavated and avoided the hard spots.

After the project was shut down M. De Lesseps and his son Charles were sentenced to five years imprisonment, and similar sentences were imposed upon several other of their associates. It also emerged that 150 French deputies had accepted bribes and voted for the allocation of financial aid to the Panama Canal Company.

The French project carried a horrendous death toll of an estimated 22,500 workmen as the local workforce toiled in extreme conditions and died from disease (malaria and yellow fever) and in landslides.

In summary, the reasons behind the French project failure were due to mismanagement of finances, corruption, poor initiation, lack or planning for the workforce and its welfare in a tropical climate, inadequate equipment, poor contracts, and a lack of good project management.

Executing

When the French left, they left behind a considerable amount of machinery housing and a hospital. The U.S. paid the French Canal Company $40 million for the Panama railroad stock, plant and materials (some was used and some sold for scrap), buildings, surveys, plans, maps, and records, land, clearings, roads, etc. They also paid for 4 years' use of the ship channel in Panama Bay.

The U.S. had to improve the living standards of the project workforce and deal with the health issues. Engineers focused on building the infrastructure necessary to complete the canal including proper housing for canal workers. Significant investments were made in eliminating yellow fever and malaria from the area through extensive sanitation and mosquito-control programs.

Starting in 1904 the railroad had to be massively upgraded with heavy duty rails to accommodate all the new rolling stock, steam engines and shovels brought in from the U.S. and elsewhere. The new railroad closely paralleled the canal where it could. In many places the new Lake Gatun flooded over the original rail line and a new rail line had to be raised by fill above the water. There were also considerable additions made to the rail system. The new steam shovel technology, many times larger than the original, allowed massive cuts and fills.[13]

The railroad facilitated the building of the canal by providing a ready supply route of equipment, supplies and the project workforce. More than 4,000 wagons were used for the removal of the tens of millions of meters^3 of excavated material from the canal cuts. Each wagon was capable of carrying 15 meters^3 loaded by steam shovel. Up to 160 locomotives pulled the wagons which were unloaded by 30 Lidgerwood unloaders that rapidly discharged the excavated material. Techniques were developed to pick up large sections of track by steam powered cranes and relocate them without rebuilding them. This allowed the track to precede the railroad mounted steam shovels where ever they needed to go. Massive scrapers scraped the dirt cars and allowed them to be unloaded rapidly. The rebuilt, much improved and often rerouted Panama Railroad continued alongside the new canal and across the Gatun Lake. It was completed in 1912 at a cost of $9 million.

The project was challenged by the elevation of 26 meters (85 feet) above sea level, which required a complex system of locks. In 1904 a survey of the area was started for the largest dam along the canal. The first American steam shovels started work on the Culebra Cut with 2,600 men. Both the Atlantic and Pacific portions of the canal were dredged. In 1906 a decision was made to switch to a lock canal, to reduce the volume of material. It required the creation of a lake from the Chagres River to supply water to the canal.

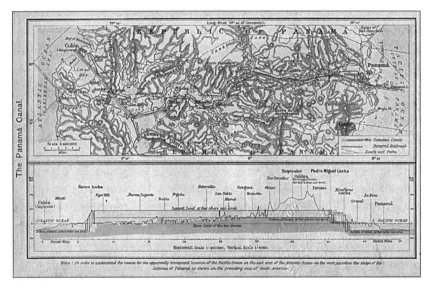

Figure 14.3: Map of the Panama Canal, and the elevations required.[14]

In the first three months of 1907 the total workforce was up to 39,000. Up to 4,535,000 kilograms of dynamite were used every year.

Figure 14.4: Work on the Culebra Cut in 1907 excavated 512,500 meters³ of material. Over 100 steam shovels were used, each excavated 920 cubic meters in an eight-hour day.[15]

Altogether over 268 million meters³ of material were excavated to form the canal and relocated to various areas throughout Panama. Some created a 500 acre town along the Pacific Ocean called Balboa.

Figure 14.5: The Panama Canal locks under construction, in 1910. The partly-constructed middle wall is shown; the large pipe near the bottom is the culvert used to carry water into the locks. Three sets of locks were built, Gatun, Pedro Miguel and Miraflores.[16]

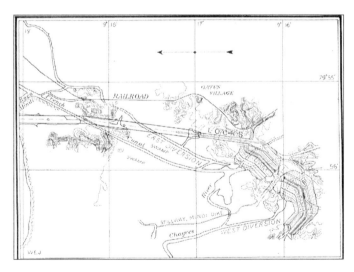

Figure 14.6: Map of the Gatun Dam[17] over the Chagres River created the Gatun Lake 2.4 kilometers (1.5 miles) in length and nearly 0.8 kilometers (0.5 mile) wide at its base.

Closing

The canal was 80 kilometers (50 miles) long and featured 17 artificial lakes, several channels, and 3 sets of locks which raised the water level a total of 42.4 meters (139 feet), where each one took about 8 minutes to fill.

The project workforce fought through the humid weather, 2,667 millimeters (105 inches) of annual rain, mountains, and malaria/yellow fever epidemics in order to finish the canal in 1914, when the first ship sailed through the canal.

In total, 5,609 workmen died during the U.S. construction effort and the overall death toll was a total of 27,500 in both the French and American efforts from disease (particularly malaria and yellow fever) and landslides.[18] It was one of the largest and most arduous construction projects undertaken and cost $380 million to complete.[19]

Between the World Wars and Business Management

Between the two wars new disciplines were added to the study of business management notably, human relationships (between employer and employee), an evolution in marketing (and its importance) and industrial human relations school of management arose to deal with the practical problems caused by Taylorism and the mindless repetition of tasks.

Project engineers developed or adapted coordination techniques that gave the managers control over the progress of the project but did not attempt to dictate to specialized experts how to do their work. MIT professor Erwin Schell articulated this philosophy, telling students in the 1930s, "The work of the engineers in most departments is not sufficiently routinized to allow process control. The most satisfactory policy appears to be that of employing competent men and then holding them [responsible] for results in terms of the erection schedule, leaving ways and means largely to their individual discretion."

Dr. Walter A. Shewart in 1924 developed the first Control Chart, which launched statistical process control and quality improvement.

Ronald A. Fisher (British geneticist and statistician) in 1925 developed Statistical Methods for Research Workers

In 1924 Mary Parker Follett analyzed the dynamic aspects of the individual in relationship to others. Follett advocated the principle of integration, "power sharing." Her ideas on negotiation, power, and employee participation were influential in the development of organizational studies.

Hoover Dam Project (1931-1935)

Background

The driving factors behind the project were first to meet the growth of Las Vegas and Southern California, as these metropolitan centers required the availability of cheap water and power. The second was to control the Colorado River and the cycles of drought and flood in the southwest which slowed the growth of the agricultural industry. By providing hydroelectric power the dam would self-fund the project.

Initiating

In 1927 the Swing-Johnson bill detailing the Hoover Dam project was passed in Congress. The Federal Government had federal land above Black Canyon where Hoover Dam could be built. Up to 1925 the Department of Reclamation had completed such projects, but in 1925 the government began contracting these projects out. The bill started the procurement process. Around the country construction companies looked over the request for proposals. However, many found the plan very ambitious, the landscape was unforgiving, and there were questions as to whether the technology was advanced enough for a dam of that size.[20]

Planning

The bid process started in March of 1931, as five bids were made for the project. It was obvious that the scope of the project was beyond a single company and even the largest construction companies didn't have the workforce or the capital for such a project. The company had to provide a $5 million bond.

The leading dam builder in the U.S. was Frank T. Crowe, a Department of Reclamation superintendent, who built a string of dams in Idaho, Wyoming, and Washington. He wanted to work on the Hoover Dam but under the new rules could not, so he left Reclamation Services and joined one of the smaller companies (Morrison-Knudson Co.) bidding on the dam. He realized only a consortium of construction companies in a joint venture had any chance of winning the bid. Crowe persuaded Morrison to go down this path and he got six smaller companies to band together to form Six Companies Incorporated. Morrison made Crowe construction superintendent.

The consortium brought different expertise to the project. For example, the Wattis Brothers in railroads, the JF Shea Company in tunnel building, the Pacific Bridge Company in bridges, MacDonald and Kahn in large buildings, and Henry Kaiser and Warren Bechtel in road building.

Crowe had much experience on the project's cost estimate and knew what went into the calculations the government used. He had aided developing a cost estimate for a dam on the lower Colorado River back in 1919 and also with the preliminary design in 1924. Not surprisingly, under Crowe's leadership, the conglomerate won the contract in 1931. The Six Companies bid of $48.9 million for the project was only $24,000 above the Department of Reclamation estimate and $10 million lower than the next lowest bid. This was the largest single contract the U.S. Government had ever awarded.[21]

The project was organized with bonuses or penalties attached to the delivery schedule depending on whether the project was early or late. This provided an incentive to begin a very aggressive pace of construction.

The plan required an infrastructure to be completed first prior to the building of the dam. First, as the dam site was very remote, roads and railroad lines had to be constructed, for transporting materials, equipment and the workforce. Second, the Colorado River had to be diverted by building tunnels through the canyon walls.

Executing

The project was run out of the Six Companies office in Las Vegas, and this kick started the hiring of a large workforce. With the Great Depression there was a massive migration of the unemployed to Las Vegas in the hopes of landing jobs with the project, bringing their families and their possessions.

The first deliverables were the railroads and highways to the site. A substantial camp was required to house and feed the workforce. At site a number of plants were built including gravel-screening, concrete-mixing, air compressor, a plate steel fabricating. Without these the project would have been a logistical nightmare.

The next activity was to divert the Colorado River away from the project's foundation site, a monumental task, and this was seasonally dependent as it could be best done during the low-water season in the winter of 1932-33. Work on the 4 tunnels began in May 1931 at a rapid pace of 24 hours a day, 7 days a week, two on each side, were cut through the bedrock walls of the canyon. At 1,200 meters (4,000 feet) long, 18 meters (56 feet) in diameter, and lined with three feet of concrete, these were the second largest tunnels ever built.

The tunnel work was brutal as inside temperatures reached 60°C (140°F) coupled with carbon monoxide from gasoline-powered trucks, dust and fumes from the blasting. Up to 4 workers died from heat exhaustion each week. Crowe had conceived of many new inventions during his dam projects and came up with a "drilling jumbo," 30 x 60 kilogram (144 pound) rock drills attached to four platforms and welded to a truck. The machine bored into the rock face at the same time to speed up drilling and this decreased the overall tunnel construction time a full year ahead of schedule. Work shifted to the temporary earthen cofferdams which were constructed above and below the dam to divert the river into the tunnels. Tunnels and earthen dams were completed by April 1932.

During the project, living conditions were made more difficult as workers poured in looking for work, more than 5,000 in all. Many brought their wives and children and lived in tents. With poor sanitation, little access to clean water, 48°C (119°F) heat and no utilities, this tent shantytown community known as Ragtown was a living hell. Six Companies realized that this community could be there for years so Boulder City, Nevada was completed in 1932. Barracks, housing, post office, library, stores, newspaper and general public welfare with churches and schools for the children were built, and electricity was brought in.

With the river diverted the site was de-watered and dredged down to bedrock so excavation of the dam and power plant could begin. This activity was critical and the base had to be built correctly otherwise there would be potential problems with the dam. Workers used power shovels to dig through to the bedrock 12 meters (40 feet) below ground. In the process they shifted more than a half-million meters³ of river bottom mud. Along the canyon walls high-scalers were suspended and laboriously chipped and shaved away with jackhammers to create a smooth surface so that the dam's walls would adhere. This was a high risk job and the scalers earned $5.60 a day, the highest paid *workers on the project.*

> *"The problem was to set up the right sequence of jobs so the workers wouldn't kill each other off."*
>
> —Frank Crowe, manager of the project

In just two years of winning the contract, June 6, 1933, concrete was poured for the dam's base 18 months ahead of schedule. The concrete was poured into rows and columns of 230 individual blocks in the base so it could dry properly and not crack. Two concrete plants on site produced the concrete which was transported on railcars in large four and eight cubic yard buckets. Crowe had also developed an overhead cable-way system of delivering concrete and moving equipment that was far more advanced than any other system of its time. At its peak 1 bucket was delivered about every 78 seconds.

Figure 14.7: The dam was poured in rows and columns of blocks. Enough concrete was used to pave a highway 5 meters (16 feet) wide from New York to San Francisco.[22]

The cooling of the concrete was a major problem. It would have taken over one hundred years to cool and properly cure the massive amount of concrete. The chemical heat generated by concrete setting was dissipated by imbedding over 937 kilometers (582 miles) of one-inch steel pipe that circulated ice water. The pipe was placed in the interconnecting concrete blocks as it was poured. An ammonia refrigeration plant that cooled the water was capable of creating a gigantic 450 kilogram (1,000 pound) ice block every day. Once the concrete cooled the pipes were then filled with concrete to create a true solid structure, one piece.

The next activity was to build the four 120 meters (395 feet) intake towers and the powerhouse's two 70 meter (230 feet) high wings that were designed to house the 17 generating units. The two outer tunnels would be outlets for huge spillways. The inner tunnels would convey water from intake towers in the reservoir to the power plant.

Altogether the Hoover Dam required 3.1 million meters³ (11 million feet³) of concrete and another million for the power plant, intake towers and other support structures. For the consortium to recover its initial $5 million investment, it gave high prices for the work done in the diversion phase and lower prices for subsequent work. For this to work, Frank Crowe placed the concrete for only $0.90 per foot, a price that was 35% lower than the price of the second lowest bidder, because of the delivery system.

In terms of the project work force in April 1931 there were about 800 men working around the dam site at a common wage of 50¢ per hour. By June, there were 2,000 workers. By November 1932, the project payroll hit $500,000 a month, and at the height of the project in June 1934, 5,128 men were working 7 days a week.

Figure 14.8: the construction continued into the night so the project would stay ahead of schedule and avoid penalties.[23]

Closing

The project saw the development of new techniques and devices which helped bring the project in two years ahead of schedule. For example, the motor-driven jumbo drill, the on-site fabrication of steel pipe, and the overhead cable-way system of delivering concrete. Other developments saw the use of hard-hats (cloth hats dipped into tar and hardened into a tough shell), and dishpan reflectors to light the diversion tunnels.

In 1935 the project was completed at a cost of $165 million. The diversionary tunnels were closed so Lake Mead could be filled. In 21 months, 5,000 men had built a structure with a 183 meter (600 foot) base and 213 meter (700 foot) walls. The official death toll of 112 seems low considering the conditions and the number of men (5,000) who worked on the dam from 1931 through 1935. In 1936 power generation began more than two years ahead of schedule and turbines continued to be added until 1961, when the last one went on line.

Empire State Building Project (1929-1931)

Background

The twenties saw world record fever grip New York in topping the skyline. The investors (sponsors) for the Empire State Building project wanted to beat the Chrysler Building project which was slightly ahead in its height. It had the advantage of breaking ground earlier on September 19, 1928.

Completion Date	Height	Sky Scrapper
1930	283 meters (927 feet)	Bank of Manhattan Building
1930	319 meters (1,047 feet)	Chrysler Building
1931	443 meters (1453 feet)	Empire State Building

Table 14.1: World Record Fever Led To A Race Between Corporations.

Initiating

A contract was put out for companies to compete on. The general contractors Starrett brothers & Eken Inc., known as the premier "skyline builders" of the 1920s, made a bold bid to win the job. They were sure that their competitors had assured the client they had plenty of equipment and anything needed they would rent. The Starrett Brothers approached this differently and told the client they had no useful construction equipment for this project. They explained the size and scope of the project required the design and purchase of new, custom pieces and equipment. These could be sold after and credited back to the project, and this would cost less than renting second hand equipment. The client agreed and they won the contract.

Planning

The schedule on this project was aggressive and planned to be completed in only eighteen months. The contractor planned the project down to the minutest details. In running the project in a crowded Manhattan there were some major risks. First, there was the possibility of serious accidents with the movement and placement of heavy equipment to the project site. Second, the busy city traffic for 12 hours a day meant materials to the site could be severely delayed. This also interrupted the movement of people.

The contractor determined that first, with the immense project scope many different trades people (up to 60) would be required, and second that most supplies had to be ordered to specification in plants as close to a finished state as possible, to minimize preparatory work needed at the site. The suppliers had to be dependable, provide quality work, and meet the timetable. The project schedule was developed such that the phases overlapped. For this Gantt charts were used for the first time to complete this type of analysis.

The project was the first (commercial) to employ fast-track construction, where construction started before the designs were fully completed to reduce delays and manage costs. The structural engineer made a schematic design from the architect's sketches to include the materials (reinforced concrete or steel), types of floors and column spacing.

Executing

Even before the Waldorf-Astoria Hotel was demolished, the site's previous occupant began excavation of the new building in January 1930, as two shifts of 300 workers dug through the rock to create the foundation, in a 24 hour operation.

For the project huge volumes of construction materials were going to be needed: 60,000 tons of steel, 56,000 meters^3 (184,000 feet^3) of concrete, 10 million bricks, 61,000 meters^3 (201,000 feet^3) of Indiana limestone and granite (for the exterior), 3,000 meters^2 (10,000 feet^2) of marble for external use, 6,500 windows, and 100,000 meters^2 (300,000 feet^2) of marble for the elevator lobbies and corridors. The site was not that big and the logistics required solutions to receive, handle and distribute all these materials quickly and efficiently.

In March 1930 the modular assembly of the steel building frame skeleton began. In this type of construction whole sections were pre-fabricated off-site, transported to site and then lifted into position. By pre-marking the steel posts and beams with their place in the framework and the number of the derrick to hoist them, workers could quickly swing the girders into place and rivet them. These could not be raised more than 30 stories at a time, so several large derricks were used to pass the girders up to the higher floors.

The 60,000 tons of steel were manufactured in Pittsburgh forging plants and transported immediately to New York via train, barge and truck, as quickly as 80 hours after coming out of the furnace and off the roller. Many of the steel beams arrived too hot to touch with bare hands.

Figure 14.9: One of the 500 Mohawk iron workers that built the steel frame.

With limited site space concrete was mixed in the basement, and then hoisted to the floors in carts. A small gauge railroad, built on each floor during construction, allowed the carts to be pushed around to where needed. Each cart held eight times more than a wheelbarrow and could be moved with far less effort.

Bricks were another problem and a common practice was to close off the street, deliver and dump piles in the street for bricklayers to wheelbarrow away. This was backbreaking and inefficient. Instead, the bricks were dumped down a chute into a hopper in the basement. The bricks were released from the hopper and dropped into carts, which were then hoisted up to the appropriate floor.

The building became a vertical construction site where men and materials were raised by hoists to upper floors, under construction. Construction was going on simultaneously inside and outside of the building. As electricians and plumbers installed the infrastructure inside while other trades followed in a finely tuned assembly line fashion.

The contractor organized a timekeeping system were daily work completed was checked in unit costs. A just-in-time delivery system of materials could then be synchronized to meet the day's production for the workforce. These were factory efficiencies introduced into the project.

In the summer and fall of 1930, the contractor organized 5 high-class lunch counters on site (on the 3ʳᵈ, 9ᵗʰ, 24ᵗʰ, 47ᵗʰ, and 64ᵗʰ floors) operated at lower-than-average prices. Construction sites to date generally operated without food service with a few having a basic cafeteria.

The food was described as "substantial" and "economical." The lunch counters reduced the number of elevator trips by the workforce. In a project were "time is money" this translated into major cash. The contractor recorded that "not more than 15% of the men left the building during the lunch hour," That gave the workforce more time for lunch and a more productive work force.

At its peak the project workforce reached 3,500 men, who recorded seven million man-hours including Sundays and holidays. The workers were paid an excellent rate of pay ($15 a day) for the day.

As a result, the project build rate was more than a storey a day, incomparable with any other project at the time. This can be attributed to a number of initiatives like effective project logistics, a skilled and organized workforce, and a level of detail that was unknown in its time. Six months after the structural columns were set the steel frame reached the 86th floor.

The speed of construction was amazing as the project was completed ahead of schedule (an astonishing 11 months, in March 1931) and under half of the projected $50 million budget, ($25 million about $400 million today).

Closing

The 86 storey, 58,000 ton building, located on Fifth Avenue in New York, ranks among the world's most beautiful and was promoted as the eighth wonder of the world.[24] It was the world's tallest building for more than 40 years (until the World Trade Center twin towers were erected in 1972).

Figure 14.10: "The Empire State Building, New York, New York." January 8, 1934.[25] The 443 meters (1,453 feet) 103 storey structure (including the mooring mast) was built in just over 13 months.

Other Notable Projects

Titanic Project (1907-1912)

The project objective was to build 3 Olympic Class Ships. It saw an increase in the size of liners (tonnage) by over 40%. The project had a solid business case with a short payback of a few years of operation for each ship. As a result, the latest in emerging technologies was incorporated in particular to provide safety features. For example, from the 16 bulkheads, to the double skin, electric transition doors for the crew, complete electrification of the ship, a 50 phone exchange, and an automatic fog warning system.

Race to South Pole Project (1910-1912)

This was a race between the British and Norwegian expedition teams who were pushing men and technology to the limits in the most hostile environment on the planet. Careful time management was critical for both teams as they had provisioned their route back to ensure that they had enough supplies and provisions as they ran out. Getting back to these buried food caches was a question of survival.

Golden Gate Bridge Project (1933-1937)

In the 20ᵗʰ Century public works projects emerged to address growth in unemployment like the Hoover Dam, Golden Gate Bridge in the U.S., Autobahns in Germany, Ocean liners (Queen Mary and Queen Elizabeth) in the UK, and the Maginot Line in France.

Background

The principal business driver behind the project was growth as San Francisco was the largest American city still served primarily by ferry boats. Without a permanent link with communities around the bay, the city's growth rate was limited.

Initiating

The risks in building a bridge across the bay were very high, with strong, swirling tides and currents, with water 102 meters (335 feet) in depth at the center of the channel, and almost constant winds of 60 mph (97 kilometer per hour). Experts thought that very high winds and blinding fogs would prevent construction and operation of a bridge across the 2,042 meters (6,700 feet) strait.

Planning

The design incorporated a thin and flexible roadway for lower wind resistance and the ability to readily flex. The suspension cables helped transmit these forces to the bridge towers to reduce the overall stress. The steel for the towers and cables was shipped from the East Coast via the Panama Canal.

Executing

The construction was fraught with risk, gusts that could blow off workers. The employer Joseph Straw made a concerted effort to introduce safety. For example, hard hats (made of leather) and safety lines were mandatory, where employees could get fired for not wearing these. Sunscreen and sunglasses were also provided. The employer also installed a safety net for $137,000 which saved 19 men (known as half way to hell club) and became a morale booster for the project.

Closing

The project was finished by April 1937, $1.3 million under budget and six months ahead of schedule, and had the longest suspension span in the world. It was built in a hostile environment perceived very difficult to bridge. The bridge was innovative in making safety a high priority.

Figure 14.11: Golden Gate Bridge, under construction, ca. 1935.[26]

Key Players

> *"Genius is 1% inspiration and 99% perspiration. Accordingly a genius is often merely a talented person who has done all of his or her homework."*

—Thomas Edison

Frederick Winslow Taylor (1856-1915)

Taylor scientifically analyzed each part of an individual's work and trained to perform it in a certain way with co-operation of managers and workers. Although somewhat discredited today he had a significant influence in his time.

Henry Gantt (1861–1919)

He studied management techniques specifically in the field of naval ship construction during the First World War and created the Gantt chart around 1917. These are a system of outlining the sequence and duration of all tasks in a process, reflected by task bars and milestone markers. This in essence was "Time Control."

> *"For the past hundred years Gantt charts have remained little unchanged and are a proven analytical tool for projects was used on major infrastructure projects including the Hoover Dam (1931-36) and Interstate highway system (1956)."*
>
> —Source: http://www.ganttchart.com/Evolution.html

Frank T. Crowe (1882–1946)

A construction engineer he left the United States Bureau of Reclamation after a 20 year career to join the construction firm of Morrison-Knudsen. He was instrumental in creating the Six Companies consortium. He pioneered critical practices in the construction of large dams notably a pneumatic delivery system for transporting and delivering concrete with overhead cables. Known as "Hurry up Crowe" he pushed his project workforce to the limits and just within safety margins. He instilled competition between the various work crews. He built some of the largest dams in the U.S. the Hoover, Shasta and 18 other dams.

Chapter Wrap-up

Conclusions

The success of the Panama Canal project should be contrasted with the disastrous French project. The success of the Suez Canal project created an over confidence for the French. The U.S. project was considered very strategic so it was highly organized, had strong sponsorship, and was driven by a very determined team. The project ran into its challenges but the lessons from the previous French project were taken on board and followed.

The period saw an improved accuracy in project estimation with the Empire State Building and the Hoover Dam projects. Both projects came 11 and 24 months ahead of schedule because of a concerted effort, and focus on aggressively approaching the schedule. Both projects incorporated many technological innovations which improved productivity, based on concepts derived from the Industrial Revolution.

The period saw a greater regard for safety conditions on projects, the use of safety nets, harnesses, and hard hats on the Golden Gate Bridge and Hoover Dam projects. These were all firsts and widely copied by other projects. They served to improve the project stakeholder and workforce relationship and confidence in the project.

The success of these four mega projects highlighted the U.S. prowess at delivering projects in a century which the U.S. dominated as a superpower. In particular, the Empire State Building project delivered substantially under budget and ahead of time.

Key Lessons

PMBoK® Guide Knowledge Areas:

- Integration Management

  - The U.S. approach to the Panama Canal project was innovative and had flexible change control. Based on the disastrous French project earlier, changes were made to the project that improved the governance, and provided greater accountability. Changes were also made to the project approach, and this included the creation of locks to reduce the hazardous landslides.

  - When the U.S. took over the project it was completed at enormous cost because the political will was there as the strategic importance of a canal drove the business justification. A canal would effectively allow for a reduction in a two coast navy.

  - The U.S. project recognized that the sheer scale of the project required substantial project planning, and the coordination to site of vast quantities of equipment, materials, and supplies. This was done to avoid repeating the mistakes of the French effort which had relied on the workforce to fill the gaps when the equipment failed to deliver.

  - One of the contributing success factors to the Hoover Dam project was having a consortium with a diversity of expertise and specialization in different areas of the project. The right skills were then available as required.

- ○ The project was a public works project at a critical time with unemployment rising.

- ○ The Empire State Building project was initiated in 1929 before the Wall Street crash. When it was completed in 1930 the Great Depression reduced demand for office space and the building become known as the "empty state building." It stood practically empty until the 1950s. The business justification was superseded by changes in the environment and the marketplace.

- Scope Management

  - ○ The scope of the Panama Canal project was reduced considerably by creating Lake Gatun. A pragmatic approach was taken to getting the canal finished and operational.

  - ○ The scope of the Hoover Dam project precluded a single company to take it on and could only be done with a consortium of companies offering expertise across the scope of the project. The project scope included a massive supply chain of materials which required the creation of roads and rail to the site.

  - ○ The scope of the Empire State Building was increased by constructing in a crowded Manhattan, with 12 hour traffic, and a limited construction site space to store materials and perform pre-assembly. This required the careful planning of the supply of materials in a "just-in-time" delivery which proved to be highly innovative for its time.

- Time Management

  - ○ The Panama Canal project with the U.S. effort took a very pragmatic approach to minimize the work effort and keep the project on schedule. For example, the creation of Lake Gatun minimized the digging effort, as did the availability of equipment like steam shovels in quantity.

  - ○ The Hoover Dam project had to stay on schedule otherwise the consortium had to pay a $3,000 per day penalty for every day the project went over. This

made it imperative for everything to go according
to plan. The project came in two years ahead of
schedule.

○ The most time dependent activity was the diversion
of the Colorado River during the low-water season.
Its success allowed rapid progress with the schedule.

○ The Empire State Building project took the approach
of fast-tracking and overlapping phases. For example,
construction started before the designs were fully
completed, and the foundation was created before the
site was cleared. This approach was necessary as the
Empire State Building project got into a race with
the Chrysler Building project. However, as the Great
Depression set in the market economics changed.
Once the Empire State Building project was complete
the original business case was not met as the building
struggled to find tenants, and this continued for 20
years.

- Cost Management

  ○ The accuracy of the Panama Canal project budget
  estimation process was improved with the U.S. effort
  based on the failure of the French effort. The budget
  was also refocused on essential project areas like
  the equipment and railroad which were massively
  upgraded with heavy duty rails, all new rolling stock,
  steam engines and shovels.

  ○ With the Hoover Dam project Frank Crowe
  understood the estimation process for a dam very well
  and was able to come in with very accurate estimates.
  These came very close to the Bureau of Reclamation
  estimates.

  ○ Crowe priced work in the diversion phase at a higher
  rate, and subsequent work lower knowing that
  the introduction of innovative equipment (like a
  drilling jumbo and an overhead cable way system of
  delivering concrete) and techniques would impact the
  cost of the former. To further lower transportation
  costs a number of plants were built close by including

gravel-screening, concrete-mixing, air compressor, and a plate steel fabricating plant.

- ○ The Empire State Building project was based on a 50 year history of steel structure project experience, and this was critical to making parametric estimates.

- ○ The project aggressively managed the project budget and proactively reduced costs by introducing just-in-time logistics, and innovations like the mini railroads on each floor, and the ability to hoists carts between floors. Costs were also reduced by prefabricating off site.

- ○ The schedule of delivery from suppliers for steel and brick had to stay on track otherwise it would throw the time line off.

- Human Resource Management

  - ○ In this period there was a shift in the concerns for the welfare of the project workforce. For example, in the Panama Canal project the U.S. effort starting in 1904 was very much focused on the welfare of the workforce. Significant investments were made and engineers built the infrastructure to complete the canal including worker housing and extensive sanitation and mosquito-control programs. In contrast the death toll was 20% of the total of workmen that died in the French effort.

  - ○ Similarly, with the Hoover Dam project a large transitory workforce had to be taken care of, including transportation to site, living quarters, and kitchens. As workers poured in looking for work, a tent shantytown known as Ragtown grew up. This had to be resolved by providing basic essentials and utilities, that saw the development of Boulder City.

  - ○ Frank Crowe created a competitive environment were teams vied to pour the most concrete blocks.

  - ○ The consortium was made up of companies that specialized in specific areas of the project.

- Increased concern for the project workforce saw the Empire State Building project set up high-class lunch counters on site operated at lower-than-average prices. This coupled with an excellent rate of pay created a more satisfied and productive workforce that translated into cash were time was money.

- The Empire State Building project also brought in 500 Mohawk "Iron Workers" who had a tradition of iron working since the mid-1880s. This important experience in assembling complex iron structures very much helped complete one of the most critical and very dangerous activities in the project. Assembling girders at the required heights had with it safety issues and was not sought after work.

- In all of these projects the increased concern for the welfare and safety of the project workforce improved the relationship of the project workforce, management, and stakeholders.

- Quality Management

  - For the Panama Canal project the complexity in canals, dams, locks, and lakes required close control and supervision of the surveying.

  - With the Hoover Dam project certain aspects had to have quality control like the laying of concrete. Without strict controls over the volumes poured (blocks were used), and the use of refrigeration pipes, the concrete would not cool and properly cure.

  - For the Empire State Building project the rate of construction of one floor per day required due diligence in the assembly of the structure. The key dependencies were with the suppliers supplying quality materials, and the sequencing of these for the correct order of assembly.

  - Suppliers were selected on their dependability and quality of work.

- Communication Management

- The Panama Canal project required a substantial budget which required substantial project support that could only be realized through effective communication management. The project was promoted as that of significant national importance.

- The high national visibility of the Hoover Dam, as a public works project, forced the project into taking a proactive approach in improving the living conditions of the large transitory workforce to maintain a positive image with the public.

- The Empire State Building project entered a skyscraper race with the Chrysler building which drew massive press attention and public interest. Both these had to be carefully managed.

- Risk Management

  - The Panama Canal project was fraught with many risks the most serious were the immediate dangers to the workforce, through tropical diseases and poor working and living conditions.

  - These contributed significantly to the French project failure. As a result, the U.S. project made these risks a priority.

  - The project reduced the threat of hazardous landslides through the creation of locks.

  - For the project financial risks were accepted more because the project was of national interest and security.

  - The Golden Gate project had a greater regard for safety with a safety net fitted at the employer's expense that was a great morale booster for the project workforce.

  - It was mandatory for the project workforce to wear hard hats made of leather and use safety lines. Workers were fired for not wearing these.

  - The bridge's design also mitigated the risk of very high winds.

- ○ For the Hoover Dam project the Six Companies consortium minimized the financial risk ($5 million bond) by bringing companies together with proven expertise and experience in different areas of the project. Other areas of risk to the project related to health and safety issues particularly with the tunneling, and the high wall scalers.

- ○ In a setting familiar to today the Empire State Building project carried several project risks caused by the limitations of the site, its size and location. First, the movement and placement of heavy equipment in a crowded Manhattan was a potential for serious accidents. Second, materials to the site could be delayed by the busy city traffic, 12 hours a day. Also the materials could interrupt the movement of people.

- ○ The solution to these was a modular assembly construction. Sections of the steel building frame were pre-fabricated off-site, transported to the site and then lifted into position. This approach became a best practice on future inner city skyscraper projects.

- Procurement Management

  - ○ A lack of diligence with this discipline was a significant factor for the problems encountered in the French attempt of the Panama Canal project. The effort had been hampered by poor management which failed to prevent suppliers and contractors exploiting the project. This wastage contributed to bringing the project to a halt.

  - ○ The successful completion of the Panama project, encouraged contractor driven mega-projects like the Hoover Dam, Empire State Building. Both were very tightly managed through contracts and very successful in finishing ahead of time and on budget.

  - ○ The contract for the Hoover Dam was driven by bonuses and penalties attached to the delivery. This refocused the approach dramatically and increased the pace of the project.

    ◦   The Empire State Building project required a high degree of cooperation between the suppliers who were shipping steel beams, and bricks to site. This was a just-in-time project were there was very little room to move on site so everything had to be well coordinated.

Educators

- Discuss in the context of today's projects the factors behind the improved accuracy in project estimation with the Empire State Building and the Hoover Dam projects.

- Discuss the risks taken with the Empire State Building and the Hoover Dam projects.

- Discuss the difference in the French and U.S. approaches to the Panama Canal project. Where adequate lessons taken from the Panama Rail Road project?

CHAPTER 15

Brief Summary of Projects

This chapter provides an opportunity to step back and look at all the projects holistically and sequentially. It further reinforces why these projects were selected in the first place.

Comparing Projects

The following list outlines the projects according to different characteristics:

- Most technical challenges, and the greatest degree of difficulty encountered

    ○ The Giza Pyramid project created a perfectly level base, erected a burial chamber out of 40 to 60 ton granite blocks, and created a ramp up to one kilometer (half a mile) in length.

    ○ The Transatlantic Cable project payed out a 4,000 kilometers (2,500 miles) cable at sea, in all weathers, to a depth of up to four kilometers (2.5 miles). It also recovered the broken cable, several times, by grappling the ocean floor.

    ○ The Hoover Dam project diverted the very large Colorado River through tunnels, and created an extremely large structure (183 meters (600 feet) base and 213 meters (700 feet) walls).

- For the Florence Cathedral project Brunelleschi created a dome with a diameter of 42 meters (130 feet) that even today would be extremely difficult to recreate.

- Most logistical challenges in getting construction materials and equipment to site

  - The Transcontinental Railroad project had an extensive 30,000 kilometers (18,000 miles) long project supply chain around the Cape Horn.

  - The Giza Pyramid project transported 40 to 60 ton granite blocks 800 kilometers (500 miles).

  - The Panama Canal project had areas that were very difficult to get to and required upgrading the Panama railway for this access.

  - The Empire State Building project operated within a small site (city block), in a downtown locale with continuous traffic in a 12 hour window.

- Most complex equipment required

  - The Colosseum project required vast quantities of essential equipment including captsans, windlasses, gins, cranes, as well as sleds and wagons.

  - The Panama Canal project required much heavy equipment (locomotives, steam shovels, cranes) to literally shift mountains of excavated materials.

  - The Hoover Dam project required complex cooling (800 kilometers (500 miles) of piping), drilling, and concrete delivery systems.

  - The Empire State Building project had innovations like the mini railroads on each floor, and the ability to hoist materials in carts to each floor.

- Most pressing time requirements

  - The Giza Pyramid project had to be finished within the life time of the pharaoh (20 years).

  - The Colosseum project was a gift to the people and was under time pressure to be completed by four contractors working under a penalty and reward contract.

- ○ Both Columbus and Magellan pushed their crews to their limits, as they faced a threat of a mutiny if land was not found within an expected time frame.

- ○ The Crystal Palace project had to be built in 8.5 months in time for the Great Exhibition of 1851. The world exhibition could not have proceeded without it.

- ○ The London Sewer project solved a pressing public health issue where outbreaks of cholera had caused a significant death toll.

- Most stringent quality requirements

  - ○ The Giza Pyramid project required substantially perfect leveling of the base otherwise being off one inch would mean yards at the top.

  - ○ The Parthenon project built slightly curving long steps that made the line look perfectly straight horizontally, a phenomenon known as line sag.

  - ○ The Pantheon project required a carefully calculated and variant mix of concrete to hold up a dome with a large 8 meters (25 feet) opening.

  - ○ The Transcontinental Railroad project took the innovative but pragmatic approach of building first and improving later.

- Most human resource challenges to the project workforce

  - ○ The Transcontinental Railroad project struggled in finding enough workers because the gold rush had created a severe labor shortage in California.

  - ○ The Panama Canal project had to face the threat of tropical diseases and the health of the workforce. The French project had appalling losses of life.

  - ○ The Golden Gate Bridge project was built at heights which required far greater safety for the workforce.

- Largest project workforce engaged in a single project

- ○ The Great Wall of China project at some points had a recruitment of 1.8 million workers.

- ○ The Palace of Versailles project had a peak project workforce of 36,000.

- ○ The Giza Pyramid project had a peak project workforce of up to 40,000.

- Most influential project that changed the course or direction of history

  - ○ The Colosseum project helped stabilize the Roman Empire.

  - ○ The Columbus project influenced and accelerated exploration broadly which developed into a European race.

  - ○ The London Sewer project was the first that was initiated because of statistical evidence.

  - ○ The St Peter's Basilica project was a catalyst for the Protestant Reformation movement.

  - ○ The Palace of Versailles project was one of several root causes of the French revolution.

  - ○ The Transcontinental Railroad project had a significant impact on opening up the American West of the continent to settlement.

- Most influential, refined, and exquisite final deliverable

  - ○ The Parthenon project was widely copied through history.

  - ○ The Pantheon project created a completely unique structure that was not emulated until 1,300 years later.

  - ○ The Hagia Sophia project influenced Mosques for thousands of years.

  - ○ The Chartres Cathedral project was probably the greatest project in the medieval building boom.

  - ○ The Florence Duomo project is still one of the most unique standing domes.

- Most important in generating and influencing other projects
  - The First Railway project (Stockton Darlington Railway) initiated a land transportation revolution, 1825-1860, where within 35 years projects had gone from laying 20 to 2,000 miles of track.
  - The Transcontinental Railroad project showed that industrial megaprojects could be delivered under time and budget. It also greatly influenced the building of national railroads in Canada and Russia.
  - The Crystal Palace project, with its ground breaking glass and iron structure, was a clear sign of the future, and what could be created with these materials. This was a prototype of sky-scrapers of the 20th Century.
  - Both the Hoover Dam and Empire State Building projects incorporated over-lapping phases and building ahead of schedule. This led to the projects coming in significantly under time and budget.
- Most expensive
  - The Palace of Versailles project (cost $6 billion) was a contributing factor of the French Revolution.
  - The St Peter's Basilica project (cost $48,000,000) was a catalyst for the Protestant Reformation.
  - For the Panama Canal projects, French and U.S., the combined costs of both efforts was $600 million.
  - The Hagia Sophia cost 20,000 pounds of gold, an immense fortune.
- Best organized project workforce
  - The Colosseum was built by four contractors who engaged workers from the Roman guild system.
  - The Empire State Building project put inordinate amounts of time into planning all the activities and the assignment of the project workforce to them.

- ◦ The Hoover Dam project was under the consortium of Six Companies Incorporated to provide a well rounded workforce in skill sets for the project.

- • Most under time and budget (efficient project delivery)

  - ◦ The Transcontinental Railroad project was completed eight years ahead of schedule.

  - ◦ The Hoover Dam project was completed more than two years ahead of schedule.

  - ◦ The Empire State Building project was completed eleven months ahead of schedule.

- • Most persistent projects

  - ◦ The Gothic cathedral projects typically spanned decades, some centuries, but they were still completed.

  - ◦ The Transatlantic Cable project required five separate attempts and Atlantic crossings to get a fully working cable link.

- • Most risk encountered by a project

  - ◦ In terms of threats to human life several projects like the Panama Canal (French attempt), Hoover Dam, and Magellan's, all experienced the loss of life.

  - ◦ In financial terms, Magellan's project, the Gothic cathedral projects, the Transatlantic Cable project, Transcontinental Railroad project, and Panama Canal project.

- • Most ahead of its time a project pushing technology and people to limits

  - ◦ Magellan's project in terms of the limited capability of ships, maps, and navigational equipment.

  - ◦ The Transcontinental Railroad project, considered the greatest American technological feat of the 19th Century.

  - ◦ The Transatlantic Cable project in terms of the depths (4,400 meters, 3 miles) that the cable was laid at, and the number of major setbacks that occurred through the project.

- Most inhospitable environment for a project

  ○ The Panama Canal project situated in one of the most difficult and inhospitable climates on Earth with a tropical terrain of mountains and hostile jungles.

  ○ The Transcontinental Railroad project was built across an extensive mountain range and desert.

- Most out of control project

  ○ The Palace of Versailles project lacked change control and the principal stakeholder did not pay attention to the budget.

  ○ The Panama Canal project (French attempt) had out of control procurement, and very poorly planned execution.

- Most over the top project, lavish and beyond functional requirements.

  ○ The St Peter's Basilica project was built to be the center of Christianity.

  ○ The Palace of Versailles project had to over shadow any palace already built.

- Similarities in projects

  ○ The Crystal Palace project delivered 24-foot bays or modules, and repeated this 77 times. The approach very similar to the repeating arches of the Colosseum project. This allowed relatively unskilled workers to construct quickly, and with a high level of quality control.

The Influence and Evolution of Selected Project

There is little question that projects influence other projects. For example, ancient masterpieces like the Giza Pyramid, and the Parthenon had huge impacts on projects and architecture. The Giza Pyramid was stupendous in every aspect and was a catalyst for millennia to follow. The Parthenon project showed it was possible to deliver esthetically pleasing buildings within the scope of a modest project budget, great beauty at an affordable price. The Colosseum project delivered a beautiful building but also a practical one of great importance in daily Roman life. It symbolized the practicalities of Roman project management using the arch (process – repeated through the design), concrete with a veneer of marble (technology), and the system of guilds (people). The Pantheon project was influenced by the Parthenon project. It then influenced many other significant projects because it created a dome that was inspirational. It was copied by the Emperor Justinian with the Hagia Sophia project, and by Brunelleschi with the Florence Duomo project.

Magellan was influenced by Columbus and both their projects were influenced by Marco Polo and other explorers like Alexander, or Leif Ericsson. Hardwick Hall's use of glass influenced the Palace of Versailles project which was a catastrophe for France but remained widely copied across Europe.

With the Industrial Revolution new business models had to arise to support projects like the Iron Bridge project or the First Railway project which integrated many pieces of technology together (steam engines, rails, wheel systems, locomotives, bridges). The Transcontinental Railroad project, influenced by these, showed how a mega industrial project could be delivered under schedule and under budget. This influenced other national railways worldwide but also other mega projects like the Panama Canal, Hoover Dam, and the Empire State Building.

Great Project Managers through History

Through the book a number of key individuals are named as project leads that assumed or took on the role of the project manager.

Project Managers and Sponsors

The following examines the characteristics or traits of these individuals.

- Hemienu (architect of Giza and overseer of work), as a royal prince his duty was to deliver a project that was going to take him the rest of his life. For this he required great patience and a depth of understanding of the project. He fully understood the totality of the project, the logistics and technical challenges in the building of the ramps, and the complexity of king's burial chamber at the core of the structure.

- Pericles was a victorious Greek general who took the initiative to restore the destroyed parts of the City of Athens. He provided a vision to rebuild the city and created a public works project, the Parthenon, which not only created employment but reinvigorated a peace time economy.

- Although the Colosseum architect is unknown the Roman architect Vitruvius provides a vivid description of traits required by an architect in his book (Treatise on Architecture). This included understanding the complex procurement process that was an integral part of the Colosseum project.

- Hagia Sophia was built by the Emperor Justinian to put the Eastern Roman Empire on the map for which he employed two architects (Anthemius and Isidorus). Both were academics the equivalent of university professors with strong theoretical backgrounds.

- Villard de Honnecourt provided insights on a wide range of topics that were part of a tool kit of a cathedral master-builder. Together with the chapter the master-builder ran the Gothic cathedral project. Some of the required traits to be successful with these projects included sheer persistence, faith in completion, and resoluteness to cause. It was common to have a turnover of master-builders through the life time of the project.

- The Renaissance Architect, as exemplified by Brunelleschi, pushed materials technology to its limits, and recreated classic architecture 1300 years old. He had to continuously manage different and difficult stakeholders, in this challenging project.

- The Renaissance explorer, as exemplified by Columbus or Magellan, had to bide their time and skillfully manipulate their projects to get the all important sponsorship so to get the projects started. They personally sacrificed everything to get the projects off the ground.

- The Victorian (Industrial age) engineer, as exemplified by Darby or Stephenson, took great personal financial risks. They had to personally underwrite their projects to prove the value of the new technologies that they were developing.

- Edward Pease and Cyrus Field were two Victorian sponsors, directors, and entrepreneurs who played a significant role not just in the funding but organizing the momentum behind their projects, the First Railway (Stockton and Darlington Railway), and the Transatlantic Cable.

- Theodore Judah, as the Chief Engineer at the Central Pacific Railroad, was very persistent in the Transcontinental Railroad project. He was successful in lobbying the Federal Government to pass the act, and then finding sponsors for the project.

- Frank Crowe had a very good understanding of the estimation process for a dam and was able to very accurately come up with estimates for the Hoover Dam. He also developed mechanical devices and technologies that when used accelerated certain activities in the project. As a result, the project came in two years ahead of its schedule.

Advent of Knowledge Areas

The book presents the idea that the nine *PMBoK® Guide* Knowledge Areas (KAs) were instictively known and followed in the earliest ancient projects. The notion that these concepts were not practiced until modern times does not hold up well. For example, ancient projects, exemplified by the Giza Pyramid, or Colosseum, were pushing the limits of:

- The heights of the structures and their stability.

- The available technologies.

- The engineering complexity of the design.

- The volume of materials, provisions, tools, and equipment in the supply chain.

- The size of the vast project workforce that had to be organized.

Across history architects or master-builders required a comprehensive grasp of the totality of their projects, including all the major activities and the nine Knowledge Areas.

Integration Management

This was all important to ancient projects. The concept of a written project plan was probably not developed until the 20th Century although through history the master-builder would have been very familiar with the totality of the project. Many projects (Giza Pyramid and Colosseum) were targeting a project end date, often for political reasons. So, the concept of a simple schedule and plan was required and this ties back to the military where operations were carefully planned and finalized to a closure through a confrontation or battle. As one of the oldest professions the military has always dealt with major scope, complexity, and a great number of dependencies. The aspects of the military that have come into project management include:

- Planning.
- Execution.
- Scheduling.
- Logistics[1] and supplies.
- Organizing vast numbers of people.

The other aspect of Integration Management, change control, manifests itself through most of the significant historical projects. In construction projects it was in the hands of the overseer of works to monitor the project and control the budget. Project overruns, and overspends were as problematic then as they are today.

Although cost-benefit analysis is perceived to be new to projects, throughout history projects either had to have a sponsor with the political will to see it through to completion, or at least to see a return on investment. For example, the Gothic cathedrals drew in pilgrims into the towns and boosted the economy; the Parthenon project costs were high but the business justification was based on stimulating the local economy and incorporating local small businesses and unemployed workers; the voyages of exploration were based on attaining a significant financial return based on valuable cargo (spices) brought back.

Through the centuries, projects have become somewhat more complex with the introduction of new materials, or techniques, and therefore have required more discipline, with time sensitive impacts becoming more predominant and more costly. The ability to manage these became essential through the 19th Century with the Industrial Revolution.

Projects using new and untested technologies, like the rail-roads, tended to justify the investment through similar business models like the canal. Although a transportation infrastructure this proved to be a completely inadequate model in execution. Railroads could meet both freight and passenger commerce.

To successfully deliver a project in a short time frame required a project charter, or clear project objective. For example, the Hagia Sophia project took 5 years. Even with a project stakeholder who had access to wealth and power, a business justification for the project was important no matter how self serving the project was to the stakeholder. This project was an important symbol of power of the emerging Eastern Roman Empire. In addition, a project plan may not have been formally defined but many of the ancient projects followed a simple work breakdown structure (see Appendix E).

Scope Management

From the earliest ancient projects scope management had to have been understood and used then as it is used today, with little difference. The notion that scope management was a concept not understood until the 17th Century does not hold up very well. There is strong evidence that the master-builders on ancient projects (the Giza Pyramid, the Colosseum, and the

Gothic cathedrals) had a comprehensive and intuitive grasp of the totality of their projects, the entire building operation, including all the major activities, and the concept of a work breakdown structure. From the earliest ancient projects, scope had to have been understood and practiced with very little difference from how we practice it today. Otherwise projects would have been very difficult to manage and deliver in an acceptable time frame. The main exceptions to this were projects which were clearly out of control, like the Palace of Versailles where scope changed continuously and dramatically when whole buildings were torn down right after construction.

Ancient projects followed a simple work breakdown structure where the sequence of activities was well understood by the master-builder. It would not be that different to the example in Appendix A for the Colosseum. This would have been very intuitive to the master-builder.

Ancient projects did not just consist of assembly and erection activities. A great deal of planning had to take place and this is not very visible when examining many ancient projects with sketchy details. For example, the ancient projects had to go through extensive planning with preparation and logistics, but little evidence survives today to show this.

Preparation requires the physical planning of the site including surveys, site selection, layout, and initial steps in the planning process. Surveying skills were required, and the Egyptians had established these due to the annual flooding (inundations) of the Nile. For the Romans, land surveying was fundamental to the creation of the vast empire and the building of the road network and cities.

The Egyptians tracked building projects and created a very formulated approach to new projects which listed complicated and ever-present ritualistic formulas of tasks. Conceptually, this was similar to a work breakdown structure. They had several centuries of experience with pyramids before the Giza Pyramid project. Similarly, by the time of the Colosseum project, the erection sequence of construction and listed tasks was well known for projects of this type but not size. In parallel, the Romans built up and refined estimation techniques. For example, the Colosseum project was a scale up of other amphitheaters like Capua (built under Augustus 63 BCE to 14), Pompey (built in 14), and Verona (built in 30).

The complexity of many of these ancient construction projects provides strong evidence that through a good understanding of the scope, a work breakdown structure was followed through the project phases by the Egyptians at Giza, the Romans with the Colosseum, and the master-builders of the Gothic cathedral projects.

The long term nature of the Gothic cathedral projects required paying special attention to Scope Management. Villard de Honnecourt kept a model book which included his plans for defining project scope. The unparalleled commitment and the extremely lengthy time frames measured in decades and centuries required a very well defined scope of what was being built as it was passed through the ever-changing project team. They could not afford too many deviations from the initial design. Also, with the lengthy time spans, there was no single master-builder responsible for a project from beginning to end, so new masters had to be introduced in each new generation. They had to adapt their plans to what was already constructed. At Chartres, up to nine different master-builders were employed through the term of the 75-year project.

Sometimes the scope of the project changed dramatically and unexpectedly. Like for Columbus, when one of the ships was lost which made the return much more difficult. Men were forced into making a landed settlement which they were not prepared for and this ultimately failed. Although the initial scope seemed straightforward, the inaccuracies in the estimates could have been catastrophic had the American Continent been further away.

The Palace of Versailles project is a good example were scope continuously changed and was not effectively managed. The main problem was the project was under the control of the principal stakeholder, the King of France. The schedule and costs were impacted dramatically but not much could be done about the King's changing whims.

Time Management

This Knowledge Area was important with ancient projects as time costs money, the same then as it does today. Ancient projects did not have inordinate amounts of time, and the time pressures were very similar in keeping to a deadline. For example, the Roman Colosseum was built by the government as a political gesture to highlight its generosity back to the people. Similarly, the Parthenon project was a high-profile project and had to be completed in a single lifetime as it was a showcase building built for political purposes, and to put Athens on the map as the center of an empire. Any delays would have been problematic. The time pressures were very similar to today. Even the Gothic cathedral projects were under the same constraints, as longer project cycles cost more, and only when completed would attract the pilgrims who would bring prosperity to the town.

The leaders on ancient project required a comprehensive grasp of the complete project for all the major activities undertaken by the various guilds, and all their ramifying complexities. With this understanding they could plan and coordinate the sequence of activities most efficiently and with the least disruption. A good example was the quarrying of the granite for Giza Pyramid and scheduling its delivery.

For many projects the scheduling of activities in a substantial building was significant. The order and sequence of activities in the erection had to be worked out and adhered to. Constraints like the physical limitations of the area of site, and the amount of available space dictated the access to materials to work with. For example, with the Gothic cathedral projects only small hand carts could be used to move materials around in the nave as the width between the bases of the pillars was as narrow as 1.2 meters (4 feet), insufficient for large loads or even the ordinary traffic of a building.[2] These tight confines typical to Gothic cathedrals demanded a highly orchestrated schedule of activities which had to be carefully thought through.

Gothic cathedral projects are a very good example of long-range planning and scheduling. The final project output built with painstaking precision and incredibly complex detail, in some cases took over a hundred years to complete.

Logistics is a military term for the procurement, distribution, maintenance, and replacement of materials and personnel. In a building project, equipment, materials, and subassemblies required shipment to a construction site. The timing of these activities needed to be carefully planned and coordinated through the project phases.[3]

The voyages of exploration were wholly dependent on carrying adequate supplies of food and water on board the vessels. As a result, time overrode both the cost and scope on these projects in importance. The challenge for Columbus and Magellan was that the scope (the destination point) was fixed. The only flexibility they had was turning back before the objective was reached. As a result, they had to accurately measure distance so that they could balance the time and the remaining supplies to survive the return leg of the journey.

In many projects the completion of the project would dramatically enhance the sponsors as in the case of the Roman Emperors responsible for the Colosseum, Pantheon, and Hagia Sophia. The main exceptions were projects that were clearly out of control, like the Palace of Versailles and St Peter's Basilica. The Gothic cathedral projects had extremely long time-lines, but this was planned for and the completed building segments were used as soon as they were built. In the industrial age delivering a project in a specific time frame became more critical as technology was rapidly evolving and any advantages would be lost to a competitor. For example, with the First Railway project there was much competition to be the first with a working railway.

The Transcontinental Railroad project approach was to build first and improve later, as the railroad became operational. This approach established a revenue flow early and kept the project on schedule by focusing on the critical path activities. Quite remarkably the project was completed eight years ahead of schedule.

Cost Management

This was significant for ancient projects as they relied on investments and loans as do projects today. The workforce required some sort of a regular payment. Finding additional funds for failing projects was no different to today.

There were some exceptions, projects that were almost oblivious to cost overruns. For example, St. Peter's Basilica (Rome) was to be the largest church ever built and the center of the Catholic Church unrivaled and unparalleled by anything. It also marked the return of the Papacy to Rome from its exile in Avignon. It was ostentatious in every detail and cost a staggering $48,000,000. It was also a likely contributing factor to the Protestant Reformation movement.

The challenge of funding mega projects is exemplified by the Transatlantic Cable and Transcontinental Railroad. Both incorporated new technologies and materials, and required vast funding, that could only be supported by a share issue, government bonds, and land grants. Little different to Julius Caesar's initiative for funding massive construction projects when he took back from money changers the power to coin money, and then created a plentiful supply of money.

Ancient projects were very much aware of the impact of costs on projects and effectively managed costs by employing different strategies. For example, the Romans exploited local materials (tufa, bricks, stones), and then used varying types of concrete for the core, over which a veneer of marble was attached. This approach provided a low cost way of constructing buildings of high quality. Other examples were the incorporation of labor saving devices (cranes), or material saving techniques like the arch.

The Greeks and Romans evolved their systems for keeping accounts which had developed from simple systems used for keeping family records, and making public transactions visible. They used these in managing project budgets.

The Romans refined their estimation techniques with experience. For example, with the Colosseum a level of parametric estimating was available aggregated from previous amphitheater projects (like Capua, Pompey, and Verona).

Generally, the projects that successfully managed and controlled their budgets had very effective governance structures in place. For example, this was very much a part of the Gothic cathedral projects with chapters and overseers of works.

The development of machines (based on the six simple machines defined by the Renaissance scientists as the lever, wheel and axel, pulley, inclined plane, wedge, and screw) reduced the unskilled labor required as did the introduction of animal muscle power. For example, the Egyptians used wooden sleds on tracks, cranes and lifting devices; the Greeks used yardarms of sailing vessels with pulleys and ropes; the Romans used a great wheel, with a treadmill within, as a hoisting device that could raise loads of considerable weight. The use of beasts of burden evolved with the Gothic cathedral projects which used ox and cart to transport stone from quarry to site. In Laon (in the Picardy region of France) ox were considered so important they were immortalized as sculptures in the towers.

By the 20th Century projects like the Empire State Building aggressively managed the project budget and pro-actively reduced costs by introducing just-in-time logistics, and innovative new technologies like the mini railroads and carts on each floor, and the ability to hoist these carts to each floor.

Quality Management

Ancient projects relied on quality management as do projects today. The theories of Quality Management may not have been developed and formulized until the mid 20th Century, but neither had mathematics and engineering until the 18th Century when calculating stresses and loads on structures was introduced into projects. All ancient projects took a non-empirical approach and as a result far greater attention had to be paid during the project to the integrity of the construction and structure. This did not restrict ancient projects in their approach to quality management. There is much evidence that many elements of today's quality management existed in ancient projects. The first great project the Giza Pyramid required incredible accuracy over distance where if the base was off by one inch and not perfectly level it would mean being off by yards at the top, a lesson they learned from the Bent Pyramid. Quality planning had to be at the forefront to avoid costly mistakes later on.

"...the dimensions of the pyramid are extremely accurate and the site was leveled within a fraction of an inch over the entire 13.1-acre (5.3 hectares) base. This is comparable to the accuracy possible with modern construction methods and laser

leveling. That's astounding. With their rudimentary tools, the pyramid builders of ancient Egypt were about as accurate as we are today with 20th Century technology!"

—Craig B. Smith, P.E., Ph.D., Chief Operating Officer, Daniel, Mann, Johnson & Mendenhall (DMJM)

In the ancient building projects quality management played a significant role in a number of ways driven primarily by safety. First, the quality of materials selected for a project had to pass an inspection. For example, knots and shakes in the timber, and fissures or pockets of weakness in stone, could cause premature disintegration. Stone was only supplied from local quarries after it had been inspected and passed by the master-builder. Second, the precision with which the building materials were assembled. For example, the Greeks used a number of tools like red clay, in copying stone pieces, the Pantograph in recording the shape of an object, and ingenious axial dowels to align the massive column drums of the Parthenon. Near perfect alignment was not just there for esthetic purposes but safety as well. Accuracy of structural judgment was integral were thrust had to be balanced by opposing thrust. With the ancient projects, master-builders lacked the scientific testing of materials and structures. They had to rely on experience and judgment, to determine the level of precision and deviation that was acceptable. They also employed scale models as well as precise assessment of failures in buildings that had collapsed or were in imminent danger of collapse.[4] They also had to track the state of the construction. In these projects quality was incorporated so there was quality control at a local level.

In ancient building projects inspections led to the establishment of building codes in urban civilizations like ancient Egypt, that were adopted by the Romans and later in medieval London that assured good practices in building construction. Poor construction practices were penalized with legal sanctions.[5] Building codes forced projects into securing permits before construction and submitting plans to building inspectors.

There are certain quality patterns that thread through the historical projects of time. The Gothic cathedral projects had much in common with the Parthenon project in the refinement, sophistication, and meeting of specifications of the project so precisely. Likewise the guilds, at the heart of ancient construction projects, (Colosseum, Pantheon, Gothic cathedrals) used their governance frameworks to pass the responsibility of tight quality control with the skilled local workforce to maintain at an individual level. Guilds, going as far back as the Roman era, prided themselves on their work and a system of inspections was predominant through these projects. This was also due to the non-empirical approach to projects were great attention had to be paid to the integrity of the construction and structure.

Similarly, mason's marks were another important technique used in stone construction projects, like the Gothic cathedrals, with requirements for accurately cut and closely fitting blocks. These helped the stone-setters more accurately set individually "hewn to shape" blocks into place without errors. The marks also introduced "traceability" so if a block didn't pass an inspection these would identify the stonecutter so he could be held accountable for its craftsmanship.[6]

Guilds existed as far back as the Roman era. During the Middle Ages, these guilds took the responsibility for quality control, policed the quality of their member's workmanship, and localized this responsibility. There is also evidence that guilds in the 15th Century started to standardize.

> *"In 1459 … master stonecutters from such cities as Strasbourg, Vienna and Salzburg met at Regensburg to standardize the statutes of their lodges. Among other things, they declared that no one ("no workman, no master, no journeyman") should reveal to the outside world the art of taking an elevation from a plan."*[7]
>
> —On the Correct Building of Pinnacles, Wenzel Roriczer
> Roriczer (15th Century Architect)

Prior to the extensive division of labor and the mechanization resulting from the Industrial Revolution, it was possible for workers to control the quality of their own products. Working conditions then were more conducive to professional pride.

The Japanese used religious ritual as a form of quality control in their production of steel specifically for its renowned Samurai swords that required the highest of quality standards. The Japanese reliance on this sword was high as they had abandoned firearms for 300 years in favor of these weapons.[8]

The Industrial Revolution incorporated quality improvements that saw the cleanup of industrial processes and the removal of impurities which led to new materials like glass and Bessemer steel. These were catalysts in significant projects like the Crystal Palace project, and the Transcontinental Railroad project. Industrial technologies like the steam engine, and railroads required high tolerances to be effective and efficient.

Human Resource Management

The management of people and the ability to organize a vast workforce has been fundamental to all projects ancient and modern. Practically all the early projects took a note from the military and were organized on these lines and the system of guilds. The Romans who ran a long line of projects, very much exemplify this, and the guilds (defined by Theodosian Code) were fundamental to the success of their projects.

The concept of the guilds continued through into the Middle Ages and underpinned the Gothic cathedral projects. The guilds incorporated quality management, maintained a highly skilled workforce, with a system of apprenticeships, and took care of its members providing meals and lodgings. In ancient projects the numbers involved and the size of the project workforce (in excess of 10,000) far outweighed numbers in modern projects.

Even though slaves existed in ancient times, there were 21 million in the Roman Empire.[9] These were mainly owned by private families and individuals, like household servants or concubines. Major projects tended to contract companies and local workers (citizens) rather than slave labor.

The Giza Pyramid was constructed over a 20 year period entirely by humans and with no animal muscle power. There is little evidence that in the Fourth Egyptian Dynasty hordes of ill fed slaves existed. The precision of the building and the efficiency needed to erect it swiftly required the skills of craftsmen.[10] The workforce was organized according to the skills of the workers. Two hundred years of experience with Pyramid projects helped identify workforce size, and the mix of trades, and skills needed. The skilled workforce was made up of surveyors, stonecutters, mason, mortar makers, and carpenters. They worked all year around either on site or in the quarries.

They benefited from a system of privileges. The unskilled workforce (mostly farmers) worked between July and November when their fields were flooded. At Giza they were organized into gangs of 25 men to transport the stone from quarries to site with a soldier for a foreman.[11] Typically, 10 men would pull one block.

Early projects that required the management of large work forces (human resources) relied on a quasi-military structure. Lessons from the military were used, very evidently with the Romans, whose society was based on an empire driven by military conquest and creating the necessary infrastructure to keep possession of these conquered lands.

In the large scale government projects of the Roman era the main construction force was made up of contractors, who used the guilds for their workforce. Aside from these, the employment of soldiers from the Legion and some government slaves (*Servi Caesaris* were the most renowned[12]) were used. Roman projects came under the authority of the architect who individually had broad training and experience which incorporated the skills of an engineer, artist and craftsman.

With the Gothic cathedrals (the 12th to 14th Century) the more ambitious the project, the further it extended plans into the future and grew more distant from the planners themselves. The commitment and time frames involved are unparalleled in the field of human endeavors. Thus, it required tremendous assurances in the ability to train and pass off the project to the next generation.

The 15th/16th centuries voyages of exploration provided a contrast in styles of management. Magellan took a more heavy handed approach with his crew than did Columbus. Where Columbus avoided mutiny and conflict, Magellan dealt with mutiny heavy handily. They both had serious challenges in getting the funding in place for their voyages, and had to bide their time. They also had to carefully manage their fickle sponsors and stakeholders. Finally, acquiring a team was very difficult because of the perceived dangers, even with a reward system in place. Columbus carefully planned the trip to East Asia right down to determining the different skills needed to complete the project. The sponsors swapped out Magellan's Portuguese crew out for Spaniards a few weeks before the voyage.

Through the Industrial Revolutions a more empathetic view of the worker gradually emerged, as an expanding workforce of thousands needed to be taken care of in terms of housing, health, welfare, and education.

Many of the mega projects of the 19th and 20th Century were contractually structured to incorporate penalty and reward systems. The Transcontinental Railroad project offered parcels of land next to the completed railroad as an incentive for the railroad companies to race each other. The Hoover Dam project was based on a penalty ($3,000 for every day late) and reward system.

With the Hoover Dam project a large transitory workforce had to be taken care of, and this included transportation to site, living quarters, and kitchens. As workers poured in looking for work a tent shantytown known as Ragtown grew up which was eventually resolved through the development of Boulder City.

In the Empire State Building the project's general contractors improved the work-site environment when they recognized the importance of feeding the workforce. Restaurants were located on the 3rd, 9th, 24th, 47th, and 64th floors of the construction site. These were high-class lunch counters that operated at lower-than-average prices, and discouraged the project workforce from leaving the site. A more satisfied workforce translated into a greater return were time is money.

The increased concern for the welfare (and safety conditions) of the project workforce continued into the 20th Century and became more evident with all four U.S. mega-projects. For example, with the Golden Gate project, the bridge had a safety net fitted at the employer's expense which saved the lives of 19 workers, known as the "half way to hell club". This improved the relationship between the project workforce, management, and stakeholders.

Communications Management

Ancient projects with an extremely large workforce (in excess of 10,000) needed very effective communications management across the project site facilities like quarries, workshops, and construction sites, as well as the offices, and the workers villages.

There are several different types of communications[13] to consider in ancient projects, from oral/verbal (face to face), to written (clay tablets, and parchment), to visual (Heliographs) or audio (horns) signals. Alarm signals were available and used when there was danger, an emergency, or a threat. With the scale of the projects there were so many different things going on around the construction site, it was important that all communications were necessary (not a distraction), clear, succinct, accurate, and addressed to the relevant persons.

The first recorded use of the Heliograph, an instrument for instantaneous optical communication of over 50 kilometers, was in 490 BCE when the ancient Greeks used polished shields to signal in battle. The Romans used mirrors to send messages over considerable distances. For example, the Roman army used these along established lines, like Hadrian's Wall where the forts were located a mile apart.

The hierarchy of the trade guilds and master-builder provided a logical way to organize, and hence communicate along. Project communication tended to be face to face (verbal), primarily to keep knowledge "in house" within the guilds. Other forms communicated ideas and design through templates and models to all project areas. This reduced the need for lots of documentation.

In Europe during the Dark Age the network of churches grew and communications improved through initiatives undertaken by the church. The medieval communities grew into towns and started to trade with each other. The Gothic cathedral projects became the centre pieces to this revitalization.

With certain projects like the Gothic cathedrals communications management was also critical with sponsors, as these projects were dependent on their generosity and continuous donations. Several stakeholders existed first, the town and the inhabitants, second, various benefactors, the bishop and chapter. The chapter (at the centre of the project) had to determine other stakeholders, plan communications, set expectations, distribute information, report performance, and manage these stakeholders.

Communication changed in the 15th Century with Guttenberg's moveable typeface printing press which made books widely available in Europe. This provided subject expertise at the finger tips of the project manager. It was now a lot easier and less expensive to have a book over an expert. With time printing would have a massive impact on communications.

Columbus had great difficulty in acquiring a crew for his voyage of discovery, as there was a fear and reluctance on the part of potential crew-members to go. Columbus used one of his key stakeholders, a well respected and local Captain Pinzon, to communicate and convince a potential crew that the journey could be completed safely. For many it required a leap of faith. Once underway, the crew was very trepid and lacked confidence in the voyage. It was important that Columbus was consistent in his communication about reaching his objective and resolute to the cause of completing it. He could not waver, and had to build up confidence along the journey. He also had to show a lot of empathy to the crew.

In contrast Magellan's voyage around the world fell into trouble because he mishandled his communications. Magellan had incomplete maps of the Americas, and an inaccurate estimate of the circumference of the earth. He claimed he knew where the passage to the Spice Islands was. He didn't and lied to his sponsor (the King of Spain) and to the crew. During the voyage Magellan didn't communicate constantly or effectively, which was vital in heading off mutinies. When the crew realized that Magellan had lied he had a full scale mutiny on his hands which he only overcame with the use of brute force. In addition, by drawing and quartering the mutineers, Magellan was sending a very powerful message to the rest of his team relative to his limited tolerance.

With the Palace of Versailles, King Louis XIV's hands-on role in the project disrupted the hierarchy of communications flow, so it went through him. There was also a major communications breakdown between the King and his Treasurer Colbert.

In the Industrial Revolution projects grew in scale and scope and were seldom concentrated in one location. For example, the Transcontinental Railroad project had teams working across the continent and project materials were sourced even further out and shipped around South America. So, the communication efforts had to quickly adopt new methods and technologies like the telegraph. This project along with other mega projects, like the Transatlantic Cable, required constant and proactive communications to keep them in the public eye so investors and public funding could be found. In both cases many reputable engineers testified that the projects were impracticable and unfeasible.

With the London Sewer project the publication of statistics pointed to the root causes of the problem. This provided a catalyst in getting interest in the project and was the first step in getting public acceptance.

Risk Management

This is probably the oldest of the *PMBoK® Guide* knowledge areas. The earliest form of risk management was through insurance, primarily used to hedge against the risk of a contingent loss. It was also a financial tool to reduce risk by sharing financial burdens with others, compensated for taking the added risk. This was seen as early as 1800 BCE in its use to help finance ship voyages. The craft guilds in ancient Greece and Rome provided life insurance by trade. During the Middle Ages, as trade expanded, new forms of insurance were used to protect farmers and traders from droughts, floods, and other disasters.

In ancient construction projects risk management played a big part in how the projects were delivered as the primary concern was safety. Injuries and deaths were very bad for project morale. The risks most commonly identified where environmental (weather, earthquakes), defective (poor quality) materials, and the degree of stresses in the structures. All these had to be considered throughout the project.

A simple examination of ancient projects shows that over time civilizations took on projects with increasing amounts of risk, knowingly. For example, Romans over time quickly pushed concrete technology in their buildings to the limit with extensive arches, barrel vaults, and domes as seen in the Colosseum and Pantheon. Project architects became more comfortable with the increased level of risk and more effective in managing it. Ancient projects used a trial and error approach to construction based on experimental construction knowledge that was not theoretical. For example, the Egyptians built with limited mathematics, as they didn't know the structural strength of stone and based their know-how on trial and error.

The ancient construction projects (Giza Pyramid, Parthenon, Colosseum, Gothic cathedrals) were fraught with danger primarily because of the handling of vast quantities of materials, across quarries, workshops, and the construction sites. These projects had extensive supply chains with many dangers and risks to the project workforce because of the quantities of materials being moved with relatively simple equipment (sleds, ropes, levers). Huge blocks had to be pulled out of quarries and up ramps into position in a confined space. Some areas were very dangerous like the king's burial chamber which had to be reinforced with massive 30 ton marble blocks to prevent collapse.

There were also risks in the large size of the workforce, some projects were in excess of 10,000, and the dangerous aspects of the construction work across the project site. Typically, many unskilled laborers were employed with differing experience in projects and needed to be apprised of the dangers. Roman architect Vitruvius laid out in his treatise specific dangers; everything from erecting roof vaulting, to digging wells (working in confined spaces), to working with large construction equipment, and to the dangers of fire to certain materials.[14] Overtime these types of best practices were incorporated into the building codes of municipalities like Rome and medieval London that assured good practices in building construction.[15]

With the Hagia Sophia project some environmental risks were mitigated when the workforce discovered and applied cement with earthquake-resistant properties withstanding earthquakes of up to 7.5 on the Richter scale. They made the building light and flexible using crushed brick in the mortar to give it a high tensile strength. The thick mortar joints, thicker than the bricks made the material more like reinforced concrete. They added windows into the dome to avoid cracking. They created shock absorbers by using lead at the foundation of the major columns carrying the dome. In addition, to secure the building from fire no wood was used in its construction except for the doors.

In terms of managing risks some specific examples involve the Gothic cathedral projects. Medieval towns were gripped in a world record fever as they tried to create the tallest and largest cathedrals. As the choir rose to a stunning height and the ratio of glass to stone increased in the walls there was an increased risk of collapse. The workforce had to be sensitive in every step to the degrees of stress as they worked within stringent structural tolerances. To mitigate the risk new support techniques developed like flying buttresses that transferred the weight off the walls. Pressures were directed and channeled to areas where they could most effectively and securely be received and grounded. The cathedrals were stone skeletons and never before had such accuracy of structural judgment been required in building. In addition, this period also saw the greatest advances in scaffolding practices.[16] The sites were hazardous and at risk were the lives of the project team as well as having to redo years of work. A cautious and safe approach was required daily.

Risk management is based on a repository of knowledge. That is to say the more knowledge that is available the better the ability to solve problems in hand. The term 'risk' came into the English language through the Spanish or Portuguese, where it was used to refer to sailing into uncharted waters. The medieval voyages of exploration carried tremendous risks for the crews. For example, the risk of the unknown (not having maps), or carrying enough food and water to survive the journey, or extreme weather conditions. Columbus estimated the distance to Japan about 2,000 miles which fortuitously was somewhat similar to the distance to the Americas.

The voyages of exploration further evolved risk management in projects as the scale of and breadth of potential risks increased, and the need to identify and respond to these during planning. During these voyages, risks had to be monitored and controlled, so it is little wonder that modern insurance companies like Lloyds of London were started to protect shipping, a century later, as the volume of shipping expanded.

Columbus and Magellan were both proactive with risk management in identifying the potential risks ahead of time and then preparing themselves to deal with them, like losing ships in a storm or a mutiny. They mitigated risks by taking a fleet of ships, a sound contingency. Columbus lost one ship and returned with two, Magellan lost four out of five ships. Columbus also took provisions for one year which gave him a lot more flexibility. In contrast, Magellan took risk management to another level.

By sailing south along the coast of South America in search of a passage to the Pacific he was forced to explore every inlet, river, and bay. The risks included those of the unknown, foul weather and storms, running out of provisions, leading to starvation and mutiny. The unknown included incomplete maps of the Americas, and an inaccurate estimate of the circumference of the earth. When a full blown mutiny broke he heavy handedly quelled it. Magellan was impetuous and when he finally cleared the passage to the Pacific he pushed ahead without taking adequate provisions on board which proved disastrous. Only 18 of the 246 crew returned safely back to Portugal on board two of the five ships that initially set out. Magellan never returned as he was killed in a skirmish in the Philippines.

With the Taj Mahal the architects were aware of environmental risks and took into consideration the possibility of an earthquake. The four Minarets at the four corners of the plinth were deliberately erected leaning outwards so that in the unlikely event of an earthquake in case of collapse they would fall away from the tomb. This is a good example of how architects identified a risk and responded to it with a mitigation strategy.

The Industrial Revolution saw the development of new technologies and materials. This led to far more ambitious and risky projects that increased in scale. It also led to the evolution of limited liability that shielded the board of a company. Risk management had to further mature in all areas with the need to protect the workforce and stakeholders from these risks.

In 1906 the San Francisco earthquake challenged perceptions about risk and its management, and raised many questions about coverage, exclusions, causes of loss, loss adjusting, and loss wordings. Many losses could have been avoided such as the fires that arose. The insurance industry embraced these changes leading to more sophisticated building and risk modeling practices used to protect against the effects of a severe earthquake. Today buildings in San Francisco are largely earthquake resistant.

Procurement Management

This *PMBoK® Guide* knowledge area has been constantly critical from ancient to more recent projects. It is not surprising with the volume of materials required for some of the projects. For example, both the Giza Pyramid and the Colosseum projects were located within or close to quarries as materials had to be procured for the project. The Giza Pyramid project was central to an extensive supply chain of many millions of tons of materials. Ancient projects leaders were very much aware of the impact of costs on projects and effectively managed these by employing different strategies. The Colosseum project was delivered by four contractors.[17] Their contracts detailed specifications of the work, requirements for guarantees, and the methods of payment and time. The Governments of the Roman Empire mandated that public works projects went through a procurement process.

As to construction contracts the Romans employed highly sophisticated oral and written contracts that "were explicit in assigning detailed responsibilities" for all parts of the job, including labor and materials—and there were even arbitration clauses...they possessed the same basic function of our modern contractors.[18]

Procurement played an active part through the Middle Ages with the great cathedral projects (Chartres, Florence, St Peters). The scale of procurement with these projects was so great that they could secure very advantageous arrangements for obtaining materials, like timber and marble. These could be then supplied to other builders in the public or private sectors. Evidence of procurement exists in the Fabric Rolls[19], early public records that give accounts of expenditure on buildings like the English Cathedrals. These list the nature and quantities of materials purchased, who supplied them, what was paid for them, as well as the names and wages assigned to each category, and the grade of the workforce.

Another aspect of procurement is not just products but services. The discovery of gold in California in 1840 created immense labor shortages in the whole region. When the first Transcontinental Railroad project got underway finding a workforce on the West Coast became extremely difficult. The U.S. Government initiated a scheme with the Chinese Government were a substantial Chinese workforce of 10,000 was procured for the project. This became a substantial logistical problem for the project alongside the shipment of materials and equipment. Some materials like the rails were shipped from the

East Coast via Panama or around Cape Horn at the tip of South America a distance of 30,000 kilometers (18,000 miles).

The Industrial Revolution influenced procurement management with its specialist requirements for materials to be prefabricated for assembly. The mega projects required massive supply chains. The most notable and influential was the Transcontinental Railroad project which extended around Cape Horn. In addition, the use of materials, like iron and glass, on such an unprecedented scale required massive share issues to fund these projects. For example, the Crystal Palace Company was formed and issued shares that raised £1.3 million. Similarly, the first Transcontinental Railroad project was funded with U.S. Government issued bonds as was the First Transatlantic Telegraph Cable project.

A lack of diligence with procurement management had been a significant factor for problems encountered with the French attempt of the Panama Canal project. Into the 20th Century procurement management was critical to mega projects, like the Hoover Dam and the Empire State Building, that required an increase in contractors and their specializations.

The contract for the Hoover Dam was tendered out and the eventual winner was a consortium of six companies. The use of contractors in all mega projects required good planning and maturity of procurement processes to execute the project. The contract was driven by bonuses and penalties, attached to the delivery. This refocused the approach dramatically increasing the pace of the project.

The Empire State Building project required a high degree of cooperation between the suppliers who were shipping material notably steel beams, and bricks. The small city-block site, with very little room to move, required a just-in-time approach.

Procurement management was taken to further heights in the UK and U.S. during World War II when the huge military buildup required the rapid procurement of vast amounts of munitions and supplies.

Project Success versus Failure

Two projects covered from a project management perspective appear ruinous namely, St. Peter's Basilica and the Palace of Versailles. The ostentatiousness of St. Peter's Basilica was likely a contributing factor to the Protestant Reformation. Yet the buildings were successfully used and copied, and the Basilica is probably the most famous church in Christian history.

CHAPTER 16

Conclusion to the Book

This book has covered one of most important disciplines that is available to us in today's world. Imagine a world without project management. Most of our modern and material world would not exist or function without it.

There has never been a time in history when project management was not important but it has never been more important than now based on how pervasive it is, and how much of the world's economy depends on it.

Today, project management is everywhere, in every industry and every field, and it is still continually growing into the foreseeable future. Yet, this should not be unexpected as for thousands of years humans have run projects. Projects have been at the heart of human civilization and its progress.

The evidence is overwhelming that today's project management is a steady evolution and built on projects from the past. The basic approach is the same. Only the materials and technologies used in the projects differ significantly.

The main conclusion of the book is that there is little new in the principles of successful project management. The same practices that were prescribed and followed by Vitruvius, Villard de Honnecourt, and Brunelleschi are very relevant in today's modern projects.

Key Findings

Some of the key findings gained in the examination of the 25 significant historical projects are:

- Project management is not a new 20th Century discipline but has existed and been practiced since the beginning of civilization. The main justification for the development and existence of project management is the need to create order and to provide deliverables.

- Supposedly-recent management disciplines, first fully articulated in 1983 as the nine knowledge areas in the *Project Management Body of Knowledge* or *PMBoK® Guide*, were actively used in all of these projects.

- These projects were delivered in similar time frames to today's projects, simply because people initiate projects so they can realize the benefits of the project.

- Project management has continually evolved, adopted and absorbed developments in new materials, technologies, ideas, and practices.

- The history of projects is littered with the repeated use of good ideas, some many centuries apart. For example, Brunelleschi provided the Duomo project workforce with amenities like on-site cafeterias at elevated levels so as to minimize travel at meal breaks. The idea was repeated with the Empire State Building project with great success and this helped boost the morale of the workforce, and improved productivity.

- Projects from the past had very similar characteristics to today's projects. For example, the use of elements we consider modern like repeating patterns, penalties and rewards, standards and measures, procurements and contracts, and an on-site and off-site project workforce organized through the guilds.

- The Gothic cathedral projects were initiated with a very limited budget, and a timeline running into decades, something unheard of in today's projects. There was a faith within the project team that the project governance would carry the project through, even if took a hundred years to complete and the master-builder and project team changed several times over. With a lifespan which is unacceptable by today's standards these projects were executed with a high degree of confidence.

Learning's for Today

What can we learn and take forward from this analysis. One of the most significant learning's is that yesterday's projects were successfully completed without today's technologies, tools, knowledge, or published processes. For example, the scientific and mathematical calculations for determining solutions, like the stresses through a building, were only available in the last 300 hundred years. Before then a purely non-empirical approach was used which was based on observations and experience with materials and construction. Yet, surprisingly the complexity of many of these early historical projects rivals any project today in terms of the enormity of the deliverables, their quality, or speed of execution. Projects like Giza Pyramid, Colosseum, and Gothic cathedrals exemplify this.

Dispelling Myths

The analysis provides an opportunity to dispel some of the common myths related to historical projects. For example, these projects were driven by overbearing leaders, had a slave workforce, unlimited budgets, and extended time lines. Many of these misperceptions were propagated through the work of fiction, books and movies, like those by Cecil B. DeMille. More recent research through the work of Egyptologists like Mark Lehner has painted a far more realistic picture of a well looked after project workforce that had decent food, accommodation, and health care. The workforce had to be motivated, somewhat difficult to do with slaves. Hence the bulk of the workforce came off the farms and villages during the inundation (flooding) of the River Nile. The project had to be delivered in the pharaoh's lifetime of 20 years, and was used as an economic stimulus for unifying the nation, one of the earliest examples in history. Similarly, research work by Ahmet Çakmak, a professor emeritus in earthquake engineering at Princeton University has provided new insights into the project at Hagia Sophia.

What Made Yesterday's Projects So Successful?

Success was based on a few simple things:

- The project started with a clear project vision and an effective project governance structure, best exemplified by those associated with the Roman Colosseum, or Gothic cathedral projects.

- A conducive, surrounding environment, to support and allow the project to proceed successfully without getting caught up in red-tape or morale sapping politics. This was established by leaders, particularly sponsors, who could provide the project a sharp focus like for example Emperor Justinian with the Hagia Sophia project.

- Motivated sponsors and leaders, who had a specific goal to keep the project on track and could work well with their project team. Good examples are the Crystal Palace project, with its 8.5 month time line, and the London Sewer project which had to solve a desperate public health issue.

- A system of guilds, based on quasi military lines, provided an organizational structure for different trades that brought in a localized scrutiny and quality control, and looked after their workforce, best shown by the Parthenon project, or by Roman or medieval projects.

- The simplicity of communication, for example, design ideas were transferred to all areas of the project team through templates and miniature models. This reduced the need for lots of documentation as with the Giza Pyramid and the Gothic cathedral projects.

- A pragmatism within the project for quick and simple solutions so that the project continued, without grinding to a halt. A good example is with the Transcontinental Railroad project which faced very severe challenges. These included an inhospitable environment and terrain (mountain range and desert). It required a continuous 30,000 kilometers (18,000 miles) supply chain of materials, and procuring a large enough workforce throughout the project.

Final statement

The most important thing to gain from this book is that project management has progressively evolved through time and across history.

Where does this lead us to today? It is an error in judgement to dissociate these historical projects from modern project management. Across history, the methods and practices of project management are little different from what is done today. Much can be learned from these historical projects that could be used as a baseline for measuring today's projects and putting them into perspective.

Going Forward and Future Research

For this book many ancient projects were not included as their documentation is sparse and sporadic, or nonexistent, or there are no archaeological records or ethno history. Only the final output of these projects, the notable structure or edifice, is still with us today. Over time astute researchers (like Francis Price 1753, Frederick Schwatska 1883, Thomas Lennox Watson 1901) have turned to sleuthing to assess these structures and derive early building practices and methods. This has provided us with unique insights to compare to modern practices. Similarly, author and architect John Fitchen investigated historical construction practices by synthesizing a large body of literature on past methods scattered and fragmented. He diagnosed, evaluated, and interpreted clues to provide meaningful insights into the Gothic cathedral projects.

Other types of analysis like the use of computer models of significant buildings can produce valuable insights and establish how a building was constructed, and insights on the project. For example, Jean-Pierre Houdin used the computer modeling of the Giza Pyramid to establish the internal ramp theory, now gaining wide acceptance in the archeological community. Internal ramps reduced projections of the workforce size. Similarly, the investigation of the Hagia Sophia Cathedral by Ahmet Çakmak, a professor emeritus in earthquake engineering at Princeton University, provides insights on the project.

This book suggests that all significant historical projects should be reinterpreted again but from a project management perspective using today's knowledge and concepts, and available technologies like Internet based communities and interest groups. This will help dispel the myths surrounding these projects and establish the continuum of project management.

Future Plans

The next planned book to follow this is *The History of Modern Project Management* which covers projects after the start of the Second World War. It looks at the many significant developments from 1940 onward up to present day.

Sample Colosseum Work Breakdown Structure

Project Walkthrough by Karen Frecker

a. Design structure.

b. Measure and lay guidelines for site and structures (dimensions, angles, checking for level).

c. Harvest and transport lumber; prepare scaffolding and centering.

d. Assemble, disassemble, and move cranes around the site.

e. Forge and build metal structures (pipes, iron connecting joints for stones).

f. Plan, administer and manage the building project.

g. Support the workers, for instance the work required to house, clothe and prepare food for them.

h. Excavate the earth for the foundation.

i. Transport the materials.

j. Load and unload materials from carts.

k. Hoist the materials into place.

l. Move the materials around on-site.

m. Quarry the travertine and tufa, based on DeLaine.

n. Produce the bricks.

o. Quarry/excavate the rubble.

a. Excavate the lime.

b. Slake the lime.

c. Quarry the pozzolano.

d. Shore the foundations.

e. Lay the foundations.

f. Lay the brick and core for brick-faced walls; lay the brick for the floors of levels 2, 3 and 4.

g. Mix the mortar.

h. Erect scaffolding.

i. Prepare and erect centering.

j. Load various materials into baskets.

k. Lay vaults.

l. Jimmy and adjust stonework, check for level and plumb.

Bibliography

[1] *Building Construction Before Mechanization*, John Fitchen, ISBN-10: 0-262-56047-X, ISBN-13: 978-0-262-56047-4.

[2] *Technology in World Civilization*, Arnold Pacey.

[3] *Science and Technology in World History*, James E. McClellan and Harold Dorn.

[4] *The Day the Universe Changed*, James Burke, 1985.

[5] *A History of Warfare*, John Keegan, 1994.

[6] *Building Construction Before Mechanization*, John Fitchen, 1986.

[6] *The Secrets of the Great Pyramid*, Bob Brier, Jean-Pierre Houdin.

[7] *The Cathedral Builders*, Jean Gimpel.

[8] *A History of Accounting Thought*, Michael Chatfield, 1977.

[9] *Architecture*, Jonathon Glancey, Eyewitness companions DK.

[10] *Pyramid*, David Macauley, 1975.

[11] *The True Story of Christopher Columbus Called The Great Admiral*, Elbridge S. Brooks.

[12] *The Life of Christopher Columbus From His Own Letters and Journals and Other Documents Of His Time*, Edward Everett Hale, *The Explorers*, Richard Humble (Time-Life Books).

[13] *Treatise on Architecture*, De architectura, known today as The Ten Books on Architecture, Roman architect Vitruvius.

[14] *Empire Express: Building the First Transcontinental Railroad* (Paperback), David Haward Bain, 2000.

Measurement Conversions

Converting inches to centimeters ...

1 = 2.54

2 = 5.08

5 = 12.7

Converting feet to meters ...

1 = 0.3048

2 = 0.6096

5 = 1.524

Converting meters to feet ...

1 = 3.28

2 = 6.56

5 = 16.4

Converting kilometers to miles ...

1 = 0.6214

2 = 1.2428

5 = 3.107

Converting miles to kilometers ...

1 = 1.61

2 = 3.22

5 = 8.05

Some other associated conversions - 1 mile = 5280 feet, 1 yard (yd) = 3 feet

Converting kilometers to yards ...

1 = 1094

2 = 2188

5 = 5470

Converting knots to statute miles/hour ...

1 = 1.151

2 = 2.302

5 = 5.755

10 = 11.51

Converting liters to US gallons ...

1 = 0.26

2 = 0.52

5 = 1.3

Converting tonnes to tons ...

1 = 0.9842

2 = 1.9684

5 = 4.921

Converting degrees Celsius to degrees Fahrenheit ...

1 = 33.8
2 = 35.6
5 = 41

Converting degrees Fahrenheit to degrees Celsius ...

1 = -17.22205

2 = -16.6665

5 = -14.99985

Mapping the Projects to PMBoK® Guide Knowledge Areas

The table below outlines the mapping for the projects to each of nine knowledge areas so as the reader can identify how predominant this knowledge area was to the project and the degree of fit. The table uses the Harvey Ball diagram notation.

Significant Projects	Integration	Scope	Time	Cost	Quality	HR	Comms	Risk	Procurement
Giza Pyramid	●	●	●	◑	◔	●	●	●	●
Parthenon	◔	◔	◑	◑	●	◔	◔	◔	◔
Colosseum	●	◔	●	◑	◔	◔	◔	●	●
The Pantheon	●	◔	◔	◔	●	◔	◔	◔	◔

Cathedral Hagia Sophia								
Gothic Cathedrals								
Florence Cathedral Duomo								
Columbus								
Magellan								
St. Peter's Rome								
Taj Mahal								
Palace of Versailles								
The Iron Bridge								
First Railway								
Crystal Palace								
Trans-continental Railroad								
Transatlantic Cable								
London Sewers								
Mechanical Computer								
Panama Canal								
Hoover Dam								
Empire State								
Golden Gate								

Relationship to Other Project Management Methods

Although the primary focus has been on the *PMBoK*® *Guide*, it is very important to expand this to other leading methods. PRojects IN Controlled Environments (PRINCE2®) is a structured approach to project management and provides a method for managing projects within a clearly defined framework.

Overview of the method

PRINCE2® is organized into the following processes with separate key activities for each (40 in total):

1. Starting up a project

   a. Appointing an executive and a project manager, designing and appointing a project management team, preparing a project brief, defining the project approach, and planning the next stage (initiation).

2. Initiating a project

   a. Planning quality, planning a project, refining the business case and risks, setting up project controls, setting up project files, and assembling a Project Initiation Document.

2. Directing a project

    a. Authorizing initiation, authorizing a project, authorizing a stage or exception plan, giving ad-hoc direction, and confirming project closure.

2. Controlling a stage

    a. Authorizing work package, assessing progress, capturing and examining project issues, reviewing stage status, reporting highlights, taking corrective action, escalating project issues, and receiving a completed work package.

2. Managing stage boundaries

    a. Planning a stage, updating a project plan, updating a project business case, updating the risk log, reporting stage end, and producing an exception plan.

2. Closing a project

    a. Decommissioning a project, identifying follow-on actions, and project evaluation review.

Mapping PRINCE2® to the PMBoK® Guide

PRINCE2® has processes at both the Project Level and Stage Level ("phase-by-phase"). These can be matching up to the *PMBoK® Guide:*

PMBoK®	P2: Project Level	P2: Stage Level ("phase-by-phase")
Initiating	Starting Up; Directing	Managing Stage Boundaries; Directing
Planning	Initiating, Planning	Managing Stage Boundaries; Planning
Executing/ Controlling	Managed stage by stage	Controlling a Stage; Managing Product Delivery; Directing
Closing	Closing a Project	Managing Stage Boundaries

Matching up the Knowledge Areas to Components:

PMBoK® Guide Knowledge Areas	PRINCE2® Components
Integration	Combined Processes and Components, Change Control
Scope, Time, Cost	Plans, Business Case
Quality	Quality, Configuration Management
Risk	Risk
Communications	Controls
HR	Organization (limited)
Procurement	Not covered

Source: http://www.pmiwestchester.org/downloads/
Prince2PMBoK.pdf and thanks to Jay M. Siegelaub PMP, MBA.

Overview of the method

Starting up a Project

In this process the project team is appointed and a project brief (with what the project is attempting to achieve and the business justification) is prepared. The overall approach is determined and the next stage planned. The project board is then asked to authorize the next stage.

Initiating a Project

The project brief is augmented to form a Business case which ensures that quality is agreed with the overall approach to controlling the project itself. Project files are created with the overall project plan, and a plan for the next stage that is put before the project board for authorization.

Directing a Project

This process dictates how the Project Board (the executive and project sponsor) should control the overall project and authorize the initiation. It also dictates how the project board should authorize a stage plan, provide ad hoc direction, and how the project is closed down.

Controlling a Stage

This process dictates how each individual stage is controlled and includes how work packages are authorized and received, and the way progress is monitored and reported to the project board. A means for capturing and assessing project issues is determined with how corrective actions are taken, and how project issues are escalated to the project board.

Managing Stage Boundaries

This dictates what is done towards the end of a stage, how the next stage is planned and the overall project plan, risk log and business case are amended. It also covers what has gone outside the tolerance levels of a stage, and how it is reported.

Closing a Project

This covers the end of a project, the formal de-commissioning (resource free up), follow on actions and formal evaluation of the project.

Project Management Organizations

In 1969, the Project Management Institute (PMI) was formed to serve the interest of the project management industry. The premise of PMI is that the tools and techniques of project management are common even among the widespread application of projects from the software industry to the construction industry. In 1981, the PMI Board of Directors authorized the development of what has become *A Guide to the Project Management Body of Knowledge (PMBoK® Guide)*, containing the standards and guidelines of practice that are widely used throughout the profession.

The International Project Management Association (IPMA), founded in Europe in 1967, has undergone a similar development and instituted the IPMA Competence Baseline (ICB). The focus of the ICB also begins with knowledge as a foundation, and adds considerations about relevant experience, interpersonal skills, and competence. Both organizations are now participating in the development of an ISO project management standard.

End Notes

Chapter One

[1] The Farther Reaches of Human Nature.

[2] *Technics and Civilization*, Lewis Mumford, 1963.

[3] The first North American explorers, Stone Age, http://www.pbs.org/wgbh/nova/transcripts/3116_stoneage.html.

[4] Dates, definitions and examples of the earliest known usage of these words can be found in the Merriam-Webster Dictionary.

[5] According to David Nye, Jacob Bigelow, who was a Harvard professor of medicine, formulated the word 'technology' in 1828. In his Elements of Technology, Bigelow "encouraged the fusion of science and art, which he felt was characteristic of industrial society", Nye, 1994, p.46.

[6] "Upon the request of the poet Coleridge in 1833 Whewell invented the English word "scientist;" before this time the only terms in use were "natural philosopher" and "man of science."" See Snyder, Laura J., "William Whewell", *The Stanford Encyclopedia of Philosophy* (Spring 2001 Edition), Edward N. Zalta (ed.), http://plato.stanford.edu/archives/spr2001/entries/whewell/.

[7] *History of engineering*, http://www.creatingtechnology.org/history.htm#1.

[8] *Paul Allen History of PM*, http://members.aol.com/AllenWeb/history.html.

[9] Based on a portfolio of 33 sheets of parchment with 250 drawings from the 1230s, which is in the Bibliothèque Nationale, Paris (MS Fr 19093).

[10] *Sketchbook of Villard de Honnecourt*. This image (or other media file) is in the public domain because its copyright has expired.

[11] *Technology in World Civilization*, Arnold Pacey, p.42.

[12] *Science and Technology in World History*, James E.McClellan and Harold Dorn, p.118.

[13] *The Day the Universe Changed*, James Burke, 1985, p.113.

[14] Derived from experiment and observation rather than theory.

[15] *Project Lessons from the Great Escape*, Multi-Media Publications Inc..

Chapter Two

[1] Project Management Body of Knowledge.

Chapter Three

[1] *Architecture*, Jonathon Glancey, Eyewitness Companions DK.

[2] http://www.historyworld.net/wrldhis/PlainTextHistories.asp?groupid=1431&HistoryID=ab23.

[3] *The Development of Double Entry Bookkeeping and its Relevance in Today's Business Environment*, Regina Libina, Pace University, 2005.

[4] Based on the six simple machines defined by Renaissance Scientists as the lever, wheel and axel, pulley, inclined plane, wedge, and screw.

[5] *A History of Accounting Thought*, Michael Chatfield, Chapter 6, New York: Dryden Press, 1977.

[6] *The Secret of the Great Pyramid*, Bob Brier and Jean-Pierre Houdin, p.23-24.

[7] This image (or other media file) is in the public domain because its copyright has expired. http://pymd.com/Saqqara-Pyramids-Djoser.htm.

[8] This image (or other media file) is in the public domain because its copyright has expired. http://www.guardians.net/egypt/pyramids/dahshur/Sneferu/TheBentPyramid.htm#about.

[9] *Building Construction Before Mechanization*, John Fitchen, 1986, p.4.

[10] Adult life expectancy was about 35 for men and 30 for women. For a Pharaoh it would have been between 60 to 70. Source: Filer, Joyce (1996). *Disease*, Austin, Texas, University of Texas Press, ISBN 0-292-72498-5.

[11] *Science and Technology in World History*, James E. McClellan and Harold Dorn.

[12] *Pyramid*, David Macauley, 1975, p.13.

[13] http://www.touregypt.net/featurestories/pyramidworkforce.htm.

[14] *A New Pyramid Age*, Philip Coppens.

[15] Who Built the Pyramids? Not slaves. Archeaologist Mark Lehner, digging deeper, discovers a city of privileged workers, by Jonathan Shaw, http://harvardmagazine.com/2003/07/who-built-the-pyramids.html.

[16] A team of construction managers from international architectural, engineering, and construction management firm DMJM performed a forensic analysis to determine the construction methods and management techniques that were employed by the ancient Egyptians in constructing the Great Pyramid. "Project Management, Pyramids." By Craig Smith, *Civil Engineering Magazine,* June 1999.

[17] http://www.smithsonianmag.com/science-nature/pyramid. html?c=y&page=2; http://www.archaeology.org/0705/ etc/pyramid.html; http://www.usatoday.com/tech/ science/2007-05-16-pyramid-theory_N.htm.

[18] This image is a computer generated image in the public domain.

[19] This image is not copyrighted. *Shepp's Photographs of the World,* James W. Shepp, Daniel B. Shepp.

[20] *The Secret of the Great Pyramid,* p.14.

[21] *The Great Wall of China,* William Edgar Geil.

[22] This image (or other media file) is in the public domain because its copyright has expired. http://bygones.ebaypix. net/china1805map.jpg.

[23] This image (or other media file) is in the public domain because its copyright has expired. Photograph of The Great Wall of China from 1907, Herbert George Ponting. http://www.geocities.com/blackinkal4/ RoyalGeographicalSociety_Asia_2.html.

[24] *Technology in World Civilization,* Arnold Pacey p.42.

[25] *Science and Technology in World History,* James E.McClellan and Harold Dorn, p.25.

[26] This image (or other media file) is in the public domain

because its copyright has expired. Photograph of Stonehenge 1910, http://www.oldukphotos.com/wiltshire-salisbury.htm.

[27] http://www.mystae.com/restricted/streams/thera/Phoenicians. html.

[28] *National Geographic: Visual History of the World*, p.37.

[29] This image (or other media file) is in the public domain because its copyright has expired. http://www.crystalinks. com/meso.html.

[30] *The Secrets of the Great Pyramid*, Bob Brier, Jean-Pierre Houdin, p.69.

[31] *The Secrets of the Great Pyramid*, Bob Brier, Jean-Pierre Houdin, p.69.

[32] *Pyramid*, David Macauley 1975.

[33] *Building Construction Before Mechanization*, John Fitchen, 1986, p.4.

[34] http://www.all-art.org/Visual%20History/3.htm.

[35] *Science and Technology in World History*, James E. McClellan and Harold Dorn.

[36] *Egypt the Black Land*, Paul Jordan, 2000, p.151.

[37] *Ancient Labor's Untold Story: Evidence of Workers' Organization from 3000 BCE to 550 CE in the Mediterranean World*, Dr. Charles Micallef, 2008.

[38] *Building Construction Before Mechanization*, John Fitchen p.6.

[39] *The Development of Double Entry Bookkeeping and its Relevance in Today's Business Environment*, Regina Libina, Pace University, 2005.

[40] Book-keeping system in which a person charges themselves

with receipts and credits themselves with payments. Used extensively prior the advent of double-entry book-keeping.

Chapter Four

[1] *Architecture* – Jonathon Glancey, Eyewitness companions DK.

[2] *Battle*, R.R. Grant, 2005.

[3] *A History of Warfare*, John Keegan, 1994, p.169.

[4] This image (or other media file) is in the public domain because its copyright has expired. *The Ten Books on Architecture*, Vitruvius Pollio.

[5] *A History of Accounting Thought*, Michael Chatfield, Chapter 9, New York: Dryden Press, 1977.

[6] Chatfield, Michael. *A History of Accounting Thought*, Chapter 6, New York: Dryden Press, 1977.

[7] A black siliceous stone used to ascertain the purity of gold and silver. The streaks of metal left behind on the touchstone are treated with nitric acid, which dissolves impurities, and thus, when the streaks are compared, the contrast between pure and impure metal is heightened. http://www.britannica.com/EBchecked/topic/600649/touchstone.

[8] *A History of Accounting Thought*, Chatfield, Michael, Chapter 10, New York: Dryden Press, 1977.

[9] This image (or other media file) is in the public domain because its copyright has expired. "*Historical Atlas*", William R. Shepherd, New York, Henry Holt and Company, 1923. *University of Texas Perry-Casteneda Map Collection*, http://www.lib.utexas.edu/maps/historical/history_shepherd_1923.html.

[10] Jeffrey Hurwit a professor of art history at the University of

Oregon and expert on the architecture of the Acropolis,
http://www.pbs.org/wgbh/nova/parthenon/hurwit.html.

[11] *"Pericles", The World Book Encyclopedia,* 1968, volume 15,
p.255. The funding of the project seems to have been
quite complex. Aside from these sources of revenue there
were many others including public moneys, tax revenues,
harbor fees, boards of commissioners and judges
contributing as well.

[12] *The Acropolis in the Age of Pericles,* Jeffrey M. Hurwit,
University of Oregon.

[13] http://faculty.frostburg.edu/phil/forum/Athens.htm.

[14] Chisel marks from about 200 different stonemasons have
been identified. http://www.pbs.org/wgbh/nova/
parthenon.

[15] *Plutarch's Life of Pericles.*

[16] *A Text-Book of the History of Architecture, Seventh Edition,*
revised, Hamlin, A. D. F. (Alfred Dwight Foster), 1855-
1926, Not copyrighted.

[17] http://www.pbs.org/wgbh/nova/parthenon/hurwit.html.

[18] *A Text-Book of the History of Architecture, Seventh Edition,*
revised, Hamlin, A. D. F. (Alfred Dwight Foster), 1855-
1926, Not copyrighted.

[19] This image (or other media file) is in the public domain
because its copyright has expired. http://www.mcah.
columbia.edu/arthum/publicportfolio.cgi?view=1960.

[20] This image (or other media file) is in the public domain
because its copyright has expired. http://www.mcah.
columbia.edu/arthum/publicportfolio.cgi?view=1960.

[21] *Secrets of the Parthenon*, PBS Airdate: January 29, 2008, http://www.pbs.org/wgbh/nova/transcripts/3502_partheno.html.

[22] *Secrets of the Parthenon*, PBS Airdate: January 29, 2008, http://www.pbs.org/wgbh/nova/transcripts/3502_partheno.html.

[23] *Plutarch's Life of Pericles*, http://classics.mit.edu/Plutarch/pericles.html.

Chapter Five

[1] De architectura (Latin: "On architecture").

[2] This image (or other media file) is in the public domain because its copyright has expired. *"Historical Atlas"*, William R. Shepherd, New York, Henry Holt and Company, 1923. *University of Texas Perry-Casteneda Map Collection*, http://www.lib.utexas.edu/maps/historical/history_shepherd_1923.html.

[3] *A History of Accounting Thought*, Chapter 12, Michael Chatfield, New York: Dryden Press, 1977.

[4] Title: *Manners, Custom and Dress During the Middle Ages and During the Renaissance Period*, Paul Lacroix,

[5] *A History of Accounting Thought*, Chapter 13, Michael Chatfield, New York: Dryden Press, 1977.

[6] *Ancient and Medieval Banking and Business in the Roman World*, Jean Andreau, Translated by Janey Lloyd, Cambridge: Cambridge University Press, 1999.

[7] De architectura (Latin: "On architecture").

[8] *The Roman Army*, Pat Southern, p.103.

[9] *A Text-Book of the History of Architecture, Seventh Edition*,

revised, Hamlin, A. D. F. (Alfred Dwight Foster), 1855-1926, Not copyrighted.

[10] *A Text-Book of the History of Architecture, Seventh Edition*, revised, Hamlin, A. D. F. (Alfred Dwight Foster), 1855-1926, Not copyrighted.

[11] *A Text-Book of the History of Architecture, Seventh Edition*, revised, Hamlin, A. D. F. (Alfred Dwight Foster), 1855-1926, Not copyrighted.

[12] The Flavian Amphitheatre, Commonly Called The Colosseum at Rome: Its History and Substructures Compared with Other Amphitheatres, John Henry Parker, C.B. Hon. M.A. Oxon., F.S.A. London.

[13] *Building Construction Before Mechanization*, John Fitchen, 1986, p.51.

[14] *Origins of Modern PM*, Patrick Weever, December 2007, http://www.mosaicprojects.com.au/PDF_Papers/P050_Origins_of_Modern_PM.pdf.

[15] *The Roman Pantheon—The Triumph of Concrete*, written by David Moore, P.E.

[16] *The Common People of Ancient Rome Studies of Roman Life and Literature*, Frank Frost Abbott.

[17] Manumission is the act of freeing a slave, done at the will of the owner. Merriam-Webster Online Dictionary, 2009.

[18] *Ancient Labor's Untold Story: Evidence of Workers' Organization from 3000 BCE to 550 CE in the Mediterranean World*, Dr. Charles Micallef, 2008.

[19] The earliest permanent amphitheatre would appear to be that built c.80 BC by Marcus Porcius at Pompeii, which in plan differs very little from that of the flavian amphitheatre.

[20] *The Upside of Down: Catastrophe, Creativity, and the Renewal*

of Civilization, Thomas Homer-Dixon, and Ms. Karen Frecker, an energy analyst based in Toronto, Ontario, Canada.

[21] The Colosseum measures 48 meters (157 feet / 165 Roman feet) high, 189 meters (615feet) long, and 165 meters (510 feet) wide, and Covers an area of 6 acres.

[22] http://www.roman-colosseum.info/colosseum/index.htm.

[23] This image (or other media file) is in the public domain. Source: Profile of Colosseum with seating areas named in english. Created by Ningyou. http://commons.wikimedia. org/.

[24] *The Colosseum*, Keith Hopkins and Mary Beard, 2006.

[25] *A Text-Book of the History of Architecture, Seventh Edition*, revised, Hamlin, A. D. F. (Alfred Dwight Foster), 1855-1926, Not copyrighted.

[26] This image (or other media file) is in the public domain because its copyright has expired. Postcard circa 1960.

[27] This image (or other media file) is in the public domain because its copyright has expired.

[28] This image is not copyrighted.

[29] *Roman Architecture from Augustus To Hadrian The Colosseum: An Analysis Of The Inherent Political And Architectural Significance.* C.J Lyes.

[30] *A Text-Book of the History of Architecture*, Seventh Edition, revised, Hamlin, A. D. F. (Alfred Dwight Foster), 1855-1926, Not copyrighted.

[31] *The Common People of Ancient Rome Studies of Roman Life and Literature*, Frank Frost Abbott.

[32] *A Text-Book of the History of Architecture, Seventh Edition,*

revised, Hamlin, A. D. F. (Alfred Dwight Foster), 1855-
1926, Not copyrighted.

[33] *Oxford Handbook of Engineering and Technology in the
Classical World*, John Peter Oleson.

[34] If the dome was treated as a series of arch segments, then
the weight of the step rings near the base had a beneficial
structural effect, acting somewhat like a buttress.

[35] *An Elementary History of Art*, N. D'Anvers (New York: Charles
Scribner's Sons, 1895), p. 47, not copyrighted.

[36] This image (or other media file) is in the public domain
because its copyright has expired.18th Century, painted
by Giovanni Paolo Panini.

[37] This image is not copyrighted.

[38] This image (or other media file) is in the public domain
because its copyright has expired. http://
worldheritagesite.org/sites/pontdugard.html.

[39] According to Liber IV.

[40] This image (or other media file) is in the public domain
because its copyright has expired. Painting by John
Soane (1814).

[41] This image (or other media file) is in the public domain
because its copyright has expired. *Encyclopaedia
Britannica*, 11th Edition, Volume 4, Part 3, p.585.

[42] This image (or other media file) is in the public domain
because its copyright has expired. Painting by William
Bell Scott (1857).

[43] This image (or other media file) is in the public domain
because its copyright has expired. *Encyclopaedia
Britannica*, 11th Edition, Volume 4, Part 3, p.586, Fig 3.

[44] http://www.aboutscotland.co.uk/hadrian/wall.html.

[45] *The Roman Pantheon—The Triumph of Concrete*, David Moore, P.E..

[46] *Building Construction Before Mechanization*, John Fitchen, 1986, p.51.

[47] *The Colosseum*, Keith Hopkins and Mary Bend, p.144.

[48] The population of the world circa 1 was 200-300 million people and in the Roman Empire under Augustus about 45 million (15% of the world's population) with 4 million Roman citizens.

[49] *The Roman Army*, Pat Southern, p.229.

[50] *Roman Builders – A Study in Architectural Process*, p.27, Rabun Taylor, 2004, Harvard University.

[51] *Vitruvius The Ten Books On Architecture*, Vitruvius Pollio.

[52] *Building Construction Before Mechanization*, John Fitchen, 1986, p.40.

[53] *The Roman Pantheon—The Triumph of Concrete*, David Moore, P.E..

[54] *Roman Builders – A Study in Architectural Process*, p.28, Rabun Taylor, 2004,Harvard University.

[55] *The Upside of Down: Catastrophe, Creativity, and the Renewal of Civilization* by Thomas Homer-Dixon.

Chapter Six

[1] This image (or other media file) is in the public domain because its copyright has expired. *"Historical Atlas"* by William R. Shepherd, New York, Henry Holt and Company, 1923. *University of Texas Perry-Casteneda Map Collection*, http://www.lib.utexas.edu/maps/historical/history_shepherd_1923.html.

[2] *The History of Money*, http://www.pbs.org/wgbh/nova/moolah/

history.html.

[3] *Nature*, Vol 443, 28 September 2006, by Virginia Hughes. Incorporates the investigation of the Hagia Sophia Project by Ahmet Çakmak, a professor emeritus in earthquake engineering at Princeton University.

[4] *A Text-Book of the History of Architecture*, Seventh Edition, revised, Hamlin, A. D. F. (Alfred Dwight Foster), 1855-1926, this image (or other media file) is in the public domain because its copyright has expired.

[5] *A Text-Book of the History of Architecture, Seventh Edition*, revised, Hamlin, A. D. F. (Alfred Dwight Foster), 1855-1926, this image (or other media file) is in the public domain because its copyright has expired.

[6] *Master-builders of Byzantium*, Robert Ousterhout, University of Pennsylvania Museum of Archaeology and Anthropology.

[7] *Shepp's Photographs of the World*, James W. Shepp, Daniel B. Shepp, not copyrighted.

[8] *Shepp's Photographs of the World*, James W. Shepp, Daniel B. Shepp, not copyrighted.

[9] This image (or other media file) is in the public domain because its copyright has expired. http://www. philographikon.com/printsmexico.html.

[10] *Building Construction Before Mechanization*, John Fitchen, 1986, p.49.

[11] This image (or other media file) is in the public domain because its copyright has expired. University of Texas Libraries, http://www.lib.utexas.edu/maps/historical/ history_europe.html.

[12] *Revival and Decline of Greek Mathematics*, Carl Benjamin Boyer, 1991.

[13] Research using computer models and chemical analyses
by Ahmet S. Cakmak, a professor of Civil Engineering at
Princeton and a specialist on the Haghia Sophia.

[14] For the weight of the gold used for the mosaics in Hagia
Sophia, cf. A. Cutler, *"The Industries of Art,"* EHB
557–558.

Chapter Seven

[1] *Technology in World Civilization*, p.7, Arnold Pacey.

[2] This image (or other media file) is in the public domain
because its copyright has expired. *Manners, Custom
and Dress During the Middle Ages and During the
Renaissance Period*, Paul Lacroix. http://www.gutenberg.
org/files/10940/10940-h/images/fig250.png.

[3] http://www.pbs.org/wgbh/nova/vikings/ships.html.

[4] *Science and Technology in World History*, James E.McClellan
and Harold Dorn, p.118.

[5] This image (or other media file) is in the public domain
because its copyright has expired. Taken from the *14th
Century treatise Nong Shu*, written by Wang Zhen in
1313, during the Chinese Yuan Dynasty.

[6] *The Medieval Machine: The Industrial Revolution of the Middle
Ages*, Jean Gimpel, Pimlico 1992.

[7] This image (or other media file) is in the public domain
because its copyright has expired. *Manners, Custom
and Dress During the Middle Ages and During the
Renaissance Period*, Paul Lacroix, http://www.gutenberg.
org/files/10940/10940-h/images/fig250.png.

[8] *Science and Technology in Medieval European Life*, Jeffrey R.
Wigelsworth.

[9] *Science and Technology in Medieval European Life*, Jeffrey R.

Wigelsworth, p.14.

[10] Source: *Technics and Civilization*, Lewis Mumford, 1963, p.439.

[11] http://www.archnet.org/library/dictionary/entry. jsp?entryid=DIA0025&mode=full.

[12] *Practical Reference Library, Volume II*, L. Brent Vaughan Hill. (New York: Dixon, Hanson and Company, 1906), not copyrighted.

[13] *Oxford Handbook of Engineering and Technology in the Classical World*, John Peter Oleson, p.296.

[14] *A Text-Book of the History of Architecture, Seventh Edition*, revised, Hamlin, A. D. F. (Alfred Dwight Foster), 1855-1926, not copyrighted.

[15] *The Cathedral Builders* by Jean Gimpel, p.4.

[16] *The Cathedral Builders* by Jean Gimpel, p.1.

[17] *Outline of Universal History*, Fisher, George Park.

[18] This image (or other media file) is in the public domain because its copyright has expired. *Manners, Custom and Dress During the Middle Ages and During the Renaissance Period*, Paul Lacroix, http://www.gutenberg. org/files/10940/10940-h/images/fig250.png.

[19] Image (or other media file) is in the public domain because its copyright has expired. Jost Amman, professional wood engravers in Germany during the latter half of the 16th Century. *"Eygentliche Beschreibung Aller"* ("Description of All Professions") which was published in Frankfurt in 1568.

[20] *The Cathedral Builders*, Jean Gimpel, p.60.

[21] *The Builder*, John Harvey, p.47, 1973.

[22] *The Cathedral Builders*, Jean Gimpel, p.69, accounts between

1278-1281.

[23] This image (or other media file) is in the public domain because its copyright has expired. G. Dehio and G. von Bezold, *Die Kirchliche Baukunst des abendlandes,* Stuttgart, 1887-1902, plate 383.

[24] *The Construction of Gothic Cathedrals: A Study of Medieval Vault Erection*, John Fitchen.

[25] Drawing of A. M. Beloqui in Corton de las Heras, 1997, p.275.

[26] This image (or other media file) is in the public domain because its copyright has expired. G. Dehio (died 1932) and G. von Bezold (died 1934), *Die Kirchliche Baukunst des abendlandes*, Stuttgart, 1887-1902.

[27] *Building Construction Before Mechanization*, John Fitchen, 1986, p.8.

[28] *Cathedral*, David Macauley, 1985.

[29] James, John, Chartres, *The Masons Who Built a Legend*, Routledge and Kegan Paul, London, 1982.

[30] This image (or other media file) is in the public domain because its copyright has expired. *Arts and Crafts in the Middle Ages*, Julia De Wolf Addison, p.238, http://www. gutenberg.org/files/18212/18212-h/18212-h.htm.

[31] This image (or other media file) is in the public domain because its copyright has expired. Georg Agricola, Zwölf Bücher vom Berg-und Hüttenwesen, übers. v. Carl Schiffner, Berlin 1928.

[32] *The Cathedral Builders*, Jean Gimpel, p.68.

[33] Guo, Qinghua, *"Yingzao Fashi: Twelfth-Century Chinese*

Building Manual," *Architectural History: Journal of the Society of Architectural Historians of Great Britain,* Volume 41, 1998, p.1-13.

[34] *A Text-Book of the History of Architecture Seventh Edition,* revised, Hamlin, A. D. F. (Alfred Dwight Foster), 1855-1926, Not copyrighted.

[35] *Shepp's Photographs of the World,* James W. Shepp, Daniel B. Shepp, Not copyrighted.

[36] *The Cathedral Builders,* Jean Gimpel, p.67.

[37] This image (or other media file) is in the public domain because its copyright has expired. *Plan,* from A. Hamilton Thompson, *Military architecture in England during the middle ages,* London, New York, 1912, p.68.

[38] *Technology in World Civilization,* p.13, Arnold Pacey.

[59] This image (or other media file) is in the public domain because its copyright has expired. 19th Century photo.

[60] This image (or other media file) is in the public domain because its copyright has expired. 19th Century photo.

[61] *World History,* Jeremy Black, p.123.

[62] This image is from Commons: GNU Free Documentation License. http://commons.wikimedia.org/wiki/ Commons:GNU_Free_Documentation_License.

[63] Based on a portfolio of 33 sheets of parchment with 250 drawings from the 1230s, which is in the Bibliothèque Nationale, Paris (MS Fr 19093).

[64] *Building Construction Before Mechanization,* John Fitchen, 1986, p.49.

[65] *Building Construction Before Mechanization,* John Fitchen, 1986, p.49.

[66] *Building Construction Before Mechanization*, John Fitchen, 1986, p.247 #6.

[67] *Building Construction Before Mechanization*, John Fitchen, 1986, p.51.

[68] Source: *Building Construction Before Mechanization*, John Fitchen, 1986, p.17.

[69] *The Cathedral Builders*, Jean Gimpel, p.59.

[70] *The Cathedral Builders*, Jean Gimpel, p.62.

[71] *The Cathedral Builders*, Jean Gimpel, p.42.

[72] *The Cathedral Builders*, Jean Gimpel, p.41,47.

[73] *Building Construction Before Mechanization*, John Fitchen: p.247 #6.

[74] *Building Construction Before Mechanization*, John Fitchen, p.50.

[75] Source: *The Cathedral Builders*, Jean Gimpel, p.52.

[76] Source: *The Construction of Gothic Cathedrals*, John Fitchen, p.276.

Chapter Eight

[1] *Technology in World Civilization*, p.54, Arnold Pacey.

[2] According to medieval historian Philip Daileader in 2007.

[3] Source: *The Day the Universe Changed*, James Burke, 1985.

[4] *The Medieval Machine: The Industrial Revolution of the Middle Ages*, Jean Gimpel, Pimlico 1992.

[5] *Technology in World Civilization*, Arnold Pacey, p.49.

[6] According to Brunelleschi's biographer Antonio Manetti.

[7] An old Italian unit of length, usually about 26 or 27 in. (66 or 68 cm).

[8] *A Text-Book of the History of Architecture Seventh Edition,* revised, Hamlin, A. D. F. (Alfred Dwight Foster), 1855-1926, Not copyrighted.

[9] This image (or other media file) is in the public domain because its copyright has expired.

[10] *A Text-Book of the History of Architecture Seventh Edition,* revised, Hamlin, A. D. F. (Alfred Dwight Foster), 1855-1926, Not copyrighted.

[11] From an order in the book of Migliore di Tommaso, dated October 3, 1419.

[12] This image (or other media file) is in the public domain because its copyright has expired.

[13] *Shepp's Photographs of the World,* James W. Shepp, Daniel B. Shepp, Not Copyrighted.

[14] *Shepp's Photographs of the World,* James W. Shepp, Daniel B. Shepp, Not Copyrighted.

[15] *Shepp's Photographs of the World,* James W. Shepp, Daniel B. Shepp, Not copyrighted.

[16] Dreyer (2006): p.122–124.

[17] This image (or other media file) is in the public domain because its copyright has expired. 17th Century Chinese woodblock print.

[18] *Brunelleschi's Cupola: Past and Present of an Architectural Masterpiece,* by Giovanni Fanelli, Giovanni Fanelli and Michele Fanelli, p.28.

[19] *Brunelleschi's Cupola: Past and Present of an Architectural Masterpiece*, by Giovanni Fanelli, Giovanni Fanelli and Michele Fanelli, p.23.

[20] *Lives of the Most Eminent Painters Sculptors & Architects*, Giorgio Vasari, 1912.

[21] *Brunelleschi's Cupola: Past and Present of an Architectural Masterpiece*, by Giovanni Fanelli, Giovanni Fanelli and Michele Fanelli, p.23.

[22] *Lives of The Most Eminent Painters Sculptors & Architects*, 1912, by Giorgio Vasari.

[23] *Brunelleschi's Cupola: Past and Present of an Architectural Masterpiece*, by Giovanni Fanelli, Giovanni Fanelli and Michele Fanelli, p.23.

[24] *Brunelleschi's Cupola: Past and Present of an Architectural Masterpiece*, by Giovanni Fanelli, Giovanni Fanelli and Michele Fanelli, p.23.

[25] *Brunelleschi's Cupola: Past and Present of an Architectural Masterpiece*, by Giovanni Fanelli, Giovanni Fanelli and Michele Fanelli, p.26.

[26] *Brunelleschi's Cupola: Past and Present of an Architectural Masterpiece*, by Giovanni Fanelli, Giovanni Fanelli and Michele Fanelli, p.25.

[27] *Brunelleschi's Cupola: Past and Present of an Architectural Masterpiece*, Giovanni Fanelli, Giovanni Fanelli and Michele Fanelli, p.23.

[28] *Brunelleschi's Cupola: Past and Present of an Architectural Masterpiece*, Giovanni Fanelli, Giovanni Fanelli and Michele Fanelli, p.27.

[29] *Brunelleschi's Cupola: Past and Present of an Architectural Masterpiece*, Giovanni Fanelli, Giovanni Fanelli and Michele Fanelli, p.27.

Chapter Nine

[1] *Technology in World Civilization*, Arnold Pacey.

[2] This image (or other media file) is in the public domain because its copyright has expired. The image comes from an early 16th-Century book called *Livro das Fortalezas de Duarte Damas.*

[3] A newspaper with pages of a size larger than those of a tabloid, http://www.merriam-webster.com/dictionary/broadsheets.

[4] This image (or other media file) is in the public domain because its copyright has expired. *Prince Henry the Navigator, the Hero of Portugal and of Modern Discovery, 1394-1460,* C. Raymond Beazley.

[5] Spanish currency of one million Maravedis (one cuentos) in 1490 is equivalent to about 308 English Pounds in 1860, or US$ 48,000 in 2005.

[6] This image (or other media file) is in the public domain because its copyright has expired. United States Library of Congress's Prints and Photographs Division under the digital ID cph.3c05453. http://www.loc.gov/rr/print/list/080_columbus.html.

[7] This image (or other media file) is in the public domain because its copyright has expired. United States Library of Congress's Prints and Photographs Division under the digital ID cph.3c05453. http://www.loc.gov/rr/print/list/080_columbus.html.

[8] This is according to Peter Martyr who kept a special account of the voyage.

[9] This image (or other media file) is in the public domain because its copyright has expired. United States Library of Congress's Prints and Photographs Division under the digital ID cph.3c05453. http://www.loc.gov/rr/print/list/080_columbus.html.

[10] This image (or other media file) is in the public domain because its copyright has expired. *Manners, Custom and Dress During the Middle Ages and During the Renaissance Period*, Paul Lacroix. http://www.gutenberg. org/files/10940/10940-h/images/fig250.png.

[11] This image (or other media file) is in the public domain because its copyright has expired. Source based upon the map in Bourne's Spain in America, American Nation Series, Volume III, New York, 1904, Harper.

[12] The Americas received wheat, rice, coffee, bananas, and olives, and horses, cows, pigs, and chickens. They contributed a virulent form of syphilis as well as corn, potatoes, tomatoes, lima beans, squash, peanuts, cassava, cacao, and pineapple.

Chapter Ten

[1] *World History*, Jeremy Black, p.124.

[2] *A short history of progress*, Ronald Wright, p.114.

[3] A good example is 'The Richard II' horary quadrant found in the British Museum, London, UK.

[4] *The Philippine Islands*, 1493-1803, Emma Helen Blair, http:// www.gutenberg.org/files/13255/13255-8.txt.

[5] This image (or other media file) is in the public domain because its copyright has expired. http://www.helmink. com/Antique_Maps_of_America.html.

[6] This image (or other media file) is in the public domain because its copyright has expired. http://www. magellanacademies.com/Strait_of_Magellan.jpg.

[7] This image (or other media file) is in the public domain because its copyright has expired. 1507, G3200 1507, W3 Vault, Library of Congress Rare Book and Special

Collections Division Washington, D.C. 20540-4650, http://
hdl.loc.gov/loc.gmd/g3200.ct000725.

[8] This image (or other media file) is in the public domain
because its copyright has expired. 1581, G3201.S12
1581 .S9, Library of Congress Rare Book and Special
Collections Division Washington, D.C. 20540-4650,
http://hdl.loc.gov/loc.gmd/g3201s.rb000011(Library of
Congress).

[9] Architecture, Jonathon Glancey, Eyewitness Companions DK.

[10] http://www.the-tudors.org.uk/hardwick-hall.htm.

[11] This image (or other media file) is in the public domain
because its copyright has expired. http://www.
oldukphotos.com/graphics/England%20Photos/
Nottinghamshire,%20Mansfield,%20Hardwick%20
Hall%201900's.jpg.

[12] *A Text-Book of the History of Architecture Seventh Edition*,
revised, Hamlin, A. D. F. (Alfred Dwight Foster), 1855-
1926, Not copyrighted.

[13] *Painting of the interior of St. Peter's in Rome*, George Cooke,
1840. (This work of art is in the public domain.) http://chapel.
myweb.uga/edu/images/paintingbig.jpg

[14] *Shepp's Photographs of the World*, James W. Shepp, Daniel
B. Shepp, Not copyrighted.

[15] Pre-revolutionary (1917) Russian postcard of Sait Basil's
Cathedral. It's copyright has expired.

Chapter Eleven

[1] Niall Ferguson, *The Ascent of Money*, 2007.

[2] This image (or other media file) is in the public domain
because its copyright has expired. *Glashütte
Weibersbrunn Im Spessart*, gegründet 1706.

[3] http://www.pbs.org/treasuresoftheworld/taj_mahal/tlevel_2/
t3build_design.html.

[4] http://www.howstuffworks.com/taj-mahal-landmark.htm.

[5] http://www.agraindia.org.uk/taj-mahal/architecture/
building-materials-used.html.

[6] http://www.thetajmahalindia.com/tajmahalinformation.html.

[7] http://www.tce.co.in/Downloads/bro_pdf/tce_world/july06.pdf.

[8] http://www.agraindia.org.uk/taj-mahal/architecture/
building-materials-used.html.

[9] This image (or other media file) is in the public domain
because its copyright has expired. Jackson, William
Henry, 1843-1942, *World's Transportation Commission
photograph collection* (Library of Congress).

[10] This image (or other media file) is in the public domain
because its copyright has expired. *Plan and layout of
the Taj Mahal and Gardens of the Great Mughals*, C.M.
Villiers Stuart, 1913.

[11] *A Text-Book of the History of Architecture Seventh Edition*,
revised, Hamlin, A. D. F. (Alfred Dwight Foster), 1855-
1926, Not copyrighted, Photo by Samuel Bourne, 1860.

[12] *The Story Of Versailles*, Francis Loring Payne, 1919.

[13] This image (or other media file) is in the public domain
because its copyright has expired. Francis Loring Payne,
The Story of Versailles.

[14] Francis Loring Payne, *The Story of Versailles*.

[15] This image (or other media file) is in the public domain
because its copyright has expired. *View of Versailles from
the Avenue de Paris*, ca. 1668, Pierre Patel.

[16] This image (or other media file) is in the public domain because its copyright has expired. Francis Loring Payne, *The Story of Versailles*.

[17] *New Glass Architecture*, Brent Richards, Dennis Gilbert.

[18] This image (or other media file) is in the public domain because its copyright has expired. Francis Loring Payne, *The Story of Versailles*.

[19] Historian A.P.J. Taylor wrote that the Palace of Versailles was one of the fundamental causes.

[20] This image (or other media file) is in the public domain because its copyright has expired. French National Tourist Collection.

[21] *Taj Mahal*, Giles Tillotson, p.73.

Chapter Twelve

[1] http://www.accel-team.com, 2004.

[2] Such as the Combination Acts of 1799/1800 in the UK.

[3] *The Fatal Shore*, Robert Hughes.

[4] The History of Money http://www.pbs.org/wgbh/ nova/moolah/history.html.

[5] This image (or other media file) is in the public domain because its copyright has expired. http://www.mspong. org/cyclopedia/metallurgy_pics.html.

[6] This image (or other media file) is in the public domain because its copyright has expired. Map of 1811, G3200 1811, V2 Vault, Library of Congress Rare Book and Special Collections Division, Washington, D.C. 20540-4650, http://hdl.loc.gov/loc.gmd/g3200.awh00011 (Library of Congress).

[7] http://www.ironbridge.org.uk/our_attractions/

the_iron_bridge_and_tollhouse/history/.

[8] http://www.nationalarchives.gov.uk/pathways/blackhistory/
journeys/voyage_html/iron.htm.

[9] This image (or other media file) is in the public domain
because its copyright has expired. Acc No: AE185.762
Lowry, Wilson (engraver) Robertson, George (artist),
http://www.ironbridge.org.uk/about_us/the_iron_bridge/.

[10] This image (or other media file) is in the public domain
because its copyright has expired. Acc No: 1972.90.
http://www.ironbridge.org.uk/about_us/the_iron_bridge/.

[11] This image (or other media file) is in the public domain
because its copyright has expired. Acc No: 1973.200,
Ellis, William (engraver), Rooker, Michael Angelo
(artist), http://www.ironbridge.org.uk/about_us/
the_iron_bridge/.

[12] *The New Popular Educator, Volume V,* John Lossing Benson,
ed. (London: Cassell & Company Limited, 1891) 5:129, not
copyrighted

[13] This image (or other media file) is in the public domain
because its copyright has expired. Acc No: AE185.771
Dubourg, M. (engraver), http://www.ironbridge.org.uk/
about_us/the_iron_bridge/.

[14] This image (or other media file) is in the public domain
because its copyright has expired. Acc No: 1973.273,
Walker, W. (engraver), Walker, J. (engraver), Burney,
T.F. (artist), http://www.ironbridge.org.uk/about_us/
the_iron_bridge/.

[15] This image (or other media file) is in the public domain.

[16] This image (or other media file) is in the public domain
because its copyright has expired. York Railway
Museum.

[17] http://www.tynelives.org.uk/stephenson/wagon.htm.

[18] This image (or other media file) is in the public domain because its copyright has expired. York Railway Museum.

[19] This image (or other media file) is in the public domain because its copyright has expired. York Railway Museum.

[20] This image (or other media file) is in the public domain because its copyright has expired. *Lives Of The Engineers, The Locomotive*, George and Robert Stephenson, Samuel Smiles, http://www.gutenberg.org/files/27710/27710-h/27710-h.htm.

[21] *Lives Of The Engineers, The Locomotive*. George and Robert Stephenson, Samuel Smiles, p.125, http://www.gutenberg.org/files/27710/27710-h/27710-h.htm.

[22] This image (or other media file) is in the public domain because its copyright has expired. *An Introduction to the Industrial and Social History of England*, Edward P. Cheyney, (Smiles: Life of George Stephenson), http://www.gutenberg.org/files/21660/21660-h/21660-h.htm#page031.

[23] This image (or other media file) is in the public domain because its copyright has expired. Darlington Borough Council.

[24] Stockton and Darlington Railway opening, J.R. Brown, Science Museum.

[25] This image (or other media file) is in the public domain because its copyright has expired. *An Introduction to the Industrial and Social History of England*, Edward P. Cheyney, http://www.gutenberg.org/files/21660/21660-h/21660-h.htm#page031.

[26] This image (or other media file) is in the public domain

because its copyright has expired. *Scientific American Supplement*, No. 460, October 25, 1884. http://www. gutenberg.org/files/11734/11734-h/11734-h.htm.

[27] This image (or other media file) is in the public domain because its copyright has expired. *Encyclopaedia Britannica*, 11th Edition, Volume 4, Part 3, p.537.

[28] This image (or other media file) is in the public domain because its copyright has expired. *Etching of the Menai Bridge* ca: 1825, http://www.anglesey-history.co.uk/ places/bridges/menai.gif.

[29] *The First in the World*, John Wall, according to Francis Mewburn p.11.

[30] *Energy and society: An Introduction*, Harold H. Schobert, p.263.

Chapter Thirteen

[1] Christopher Latham Sholes (1819-1890) was a U.S. mechanical engineer who invented the first practical modern typewriter, patented in 1868.

[2] *New Glass Architecture*, Brent Richards, Dennis Gilbert.

[3] This image (or other media file) is in the public domain because its copyright has expired. *Sir Henry Bessemer, F.R.S, An Autobiography*, 1st edition published in London, 1905, http://www.history.rochester.edu/ehp-book/shb/ hb11.htm.

[4] This image (or other media file) is in the public domain because its copyright has expired. *Agricultural Implements and Machines* in the Collection of the National Museum of History and Technology, Smithsonian, *Studies in History and Technology*, No. 17, Schlebecker, John T.

[5] This image (or other media file) is in the public domain

because its copyright has expired. *Frank Leslie's Illustrated Newspaper*, August 21, 1858.

[6] The British Government gave Field a subsidy of £1,400 a year and loaned the ships to lay the cable.

[7] http://www.gutenberg.org/files/14091/14091-h/14091-h.htm.

[8] This image (or other media file) is in the public domain because its copyright has expired. *Telegraph Connections* (Telegraphen Verbindungen), 1891 Stielers Hand-Atlas, Plate No. 5, Weltkarte in Mercator projection.

[9] *The Pacific tourist*, Author: Williams, Henry T. Collection: Making of America Books.

[10] *A Practical Plan For Building The Pacific Railroad*, T.D. Judah, Civil Engineer, San Francisco, January 1, 1857, Washington, D.C., Henry Polkinhorn, Printer. 1857.

[11] The Big Four consisted of Mark Hopkins, Collis P. Huntington, [Gov.] Leland Stanford, and Charles Crocker. These men were merchants drawn to the west by prospects of finding riches in California.

[12] http://cprr.org/Museum/Eastward, http://www.#Construction%20of%20the%20CPRR.

[13] *The First Transcontinental Railroad*, John Debo Galloway, C. E. (1869-1943), Dorset Press, New York, 1989.

[14] This image (or other media file) is in the public domain because its copyright has expired. This image is available from the Archival Research Catalog of the National Archives and Records Administration under the ARC Identifier 594940.

[15] *The Pacific Tourist Williams' Illustrated Trans-Continental Guide*, Henry T. Williams, Editor, http://cprr.org/Museum/Books/Williams_Pacific_Tourist.html.

[16] This image (or other media file) is in the public domain

because its copyright has expired. General view of the works in progress, Acc No: AE185.3165.69, Berlyn, Peter (author), Fowler, Charles (author).

¹⁷ *New Glass Architecture*, Brent Richards, Dennis Gilbert.

¹⁸ This image (or other media file) is in the public domain because its copyright has expired. General view of the works in progress, Acc No: AE185.3165.69, Berlyn, Peter (author), Fowler, Charles (author).

¹⁹ This image (or other media file) is in the public domain because its copyright has expired. *Tallis' History and Criticism of the Crystal Palace*, 1852.

²⁰ *New Glass Architecture*, Brent Richards, Dennis Gilbert.

²¹ http://www.lib.umd.edu/digital/worldsfairs/essay. jsp?pid=umd:1014.

²² This image (or other media file) is in the public domain because its copyright has expired. Published by C.F. Cheffins, Lithographers, Southhampton Buildings, London, England, 1854 in Snow, John.

²³ http://www.swopnet.com/engr/londonsewers/londontext1.html.

²⁴ *Ancient Transportation*, Michael Woods, Mary B. Woods, p.44.

²⁵ This image (or other media file) is in the public domain because its copyright has expired. Published 1881 in *Young Persons' Cyclopedia of Persons and Places*.

²⁶ This image (or other media file) is in the public domain because its copyright has expired. Scanned from engraving in *"Appleton's Journal of Popular Literature, Science, and Art"*, 1869.

²⁷ Avery, Ralph E. (1913). *"The French Failure"*. *America's Triumph in Panama*, Chicago, IL: L.W. Walter Company.

[28] This image (or other media file) is in the public domain because its copyright has expired. *Paris*, exposition universelle 1889. "74. Paris. Tour Eiffel (juillet 1888)" BnF, Estampes et Photographie, Qe Mat 1 http://expositions.bnf.fr/universelles/.

[29] He was awarded the prestigious Elliot Cresson Medal by the Franklin Institute of Philadelphia in 1890; Gold Medal of the Paris Exposition and the Bronze Medal of the World's Fair in 1893. He was awarded a doctorate for his work in 1890.

[30] http://www-history.mcs.st-andrews.ac.uk/Biographies/Hollerith.html.

[31] http://www.greatwestern.org.uk/m_in_gwr.htm.

[32] *A Practical Plan for Building The Pacific Railroad*, T.D. Judah, Civil Engineer, San Francisco, January 1, 1857, Washington, D.C. Henry Polkinhorn, Printer. 1857.

[33] Source: *Railroad Commissioners' Reports* (Central Pacific Railroad), http://cprr.org/Museum/NARA/index.html.

Chapter Fourteen

[1] Called an auto-typist, it stores and reproduces simple form letters and certain paragraphs using punched paper tape as the storage medium.

[2] *The Guns of August*, Barbara Tuchman, 1962.

[3] http://www.vanderbilt.edu/AnS/Anthro/Anth101/taylorism_and_fordism.htm.

[4] Ground breaking experiments conducted at the Western Electric plant in Chicago from 1927-1932.

[5] Further reading on the Human Relations movement

includes: Kurt Lewin, one of the first researchers to study group dynamics and evolve organizational development; Abraham Maslow presented *"The Hierarchy of Needs"* in *"The U.S. Psychology Review"* in 1943; Douglas McGregor published *"Theory X and Theory Y"* in *"The Human Side of Enterprise"* in 1960; Frederick Herzberg published the *Hygiene-Motivation Theory* in *"The Motivation to Work"* in 1959. (http://www.accel-team.com, 2004).

[6] Source: http://www.economicprincipals.com/issues/02.11.10. html.

[7] Source: http://www.vanderbilt.edu/AnS/Anthro/Anth101/ taylorism_and_fordism.htm.

[8] Source: http://www.asq.org/learn-about-quality/project-Planning-tools/overview/pdca-cycle.html.

[9] This image (or other media file) is in the public domain because its copyright has expired. *"Illustrated History of the Panama Railroad"*, Fessenden Nott Otis, Harper & Brothers, New York, 1861.

[10] This image (or other media file) is in the public domain because its copyright has expired. *"Illustrated History of the Panama Railroad"*, Fessenden Nott Otis, Harper & Brothers, New York, 1861.

[11] Source: Avery, Ralph E. (1913), *"The French Failure"*. *America's Triumph in Panama. Chicago*, IL: L.W. Walter Company.

[12] *The Panama Canal*, José Carlos Rodrigues.

[13] Otis, Fessenden Nott; *llustrated History of the Panama Railroad,* Harper & Brothers, New York, 1861.

[14] This image (or other media file) is in the public domain because its copyright has expired. *"Historical Atlas"* by William R. Shepherd, New York, Henry Holt and Company, 1923. Online source: *University of Texas Perry-*

Casteneda Map Collection, http://www.lib.utexas.edu/
maps/historical/history_shepherd_1923.html.

[15] This image (or other media file) is in the public domain
because its copyright has expired. *Culebra Cut of the
Panama Canal*, 1907.

[16] This image (or other media file) is in the public domain
because its copyright has expired. *From The Panama
Canal, an address to the National Geographic Society*,
Colonel Goethals, February 10, 1911.

[17] This image (or other media file) is in the public domain
because its copyright has expired. Book source:
"Historical Atlas", William R. Shepherd, New York,
Henry Holt and Company, 1923. Online source:
University of Texas Perry-Casteneda Map Collection,
http://www.lib.utexas.edu/maps/historical/history_
shepherd_1923.html.

[18] http://www.canalmuseum.com/documents/
panamacanalhistory023.htm.

[19] http://query.nytimes.com/mem/archive-free/pdf?_r=
1&res=9A03E1DC1638E233A25756C2A9649D94639ED
7CF.

[20] http://www.generalcontractor.com/resources/
articles/hoover-dam.asp.

[21] In today's dollars the bid would be worth more than $577
Million.

[22] This image (or other media file) is in the public domain
because its copyright has expired. The U.S. Bureau of
reclamation.

[23] This image (or other media file) is in the public domain

because its copyright has expired. The U.S. Bureau of reclamation.

[24] *Building the Empire State*, Carol Willis and Donald Friedman.

[25] This image (or other media file) is in the public domain because its copyright has expired. Gottscho, Samuel H., photographer. *Architecture and Interior Design for 20th Century America*, 1935-1955, Library of Congress.

[26] Image courtesy of National Archives and Records Administration.

Chapter Fifteen

[1] Logistics is defined by the modern British Army as: The science of planning and carrying out the movement and maintenance of forces. *Infantry Tactical Doctrine Volume 1*, Pamphlet No.1, The Infantry Company Group, The Fundamentals (1998) p.6-2.

[2] *Building Construction Before Mechanization*, John Fitchen, 1986, p.59.

[3] *Building Construction Before Mechanization*, John Fitchen, 1986, p.51.

[4] *Building Construction Before Mechanization*, John Fitchen, 1986, p.76, p.249 #7.

[5] *Building Construction Before Mechanization*, John Fitchen, p.40.

[6] *Building Construction Before Mechanization*, John Fitchen, 1986, p.247 #6.

[7] *The Cathedral Builders*, Jean Gimpel, p.102.

[8] *Technology in World Civilization*, Arnold Pacey, p.42.

[9] The population of the world circa 1 was 200-300 million

people and in the Roman Empire under Augustus about 45 million (15% of the world's population) with 4 million Roman citizens.

[10] *Pyramid Quest*, Robert M Schoch, 2005, p.90.

[11] *Pyramid*, David Macauley, 1975.

[12] *A Social Study of the Emperor's Freedmen and Slaves*. P.R.C. Weaver, Familia Caesaris. Cambridge, University Press, 1972.

[13] *History of Communications*, http://www.nathan. com/projects/current/comtimeline.html.

[14] *Vitruvius The Ten Books On Architecture*, Vitruvius Pollio.

[15] *Building Construction Before Mechanization*, John Fitchen, 1986, p.40.

[16] *Building Construction Before Mechanization*, John Fitchen, 1986, p.86.

[17] *Origins of Modern PM*, Patrick Weever, December 2007, http://www.mosaicprojects.com.au/PDF_Papers/ P050_Origins_of_Modern_PM.pdf.

[18] *The Roman Pantheon—The Triumph of Concrete*, David Moore, P.E..

[19] *Fabric Rolls and Documents Of York Minster* (Paperback), illustrated by John Browne, James Raine (Author).

References

[1] The International Project Management Association (IPMA) was founded in 1967 and the Project Management Institute was founded in 1969.

[2] *The Secrets of the Great Pyramid*, Bob Brier, Jean-Pierre Houdin, 2008, HarperCollins, p.69.

[3] The Farther Reaches of Human Nature.

[4] *Technics and Civilization*, Lewis Mumford, 1963.

[5] The first North American explorers, Stone Age, http://www.pbs.org/wgbh/nova/transcripts/3116_stoneage.html.

[6] Dates, definitions and examples of the earliest known usage of these words can be found in the Merriam-Webster Dictionary.

[7] According to David Nye, Jacob Bigelow, who was a Harvard professor of medicine, formulated the word 'technology' in 1828. In his *Elements of Technology*, Bigelow "encouraged the fusion of science and art, which he felt was characteristic of industrial society" (Nye, 1994, p.46).

[8] "Upon the request of the poet Coleridge in 1833, Whewell invented the English word "scientist". Before this time the only terms in use were "natural philosopher" and "man of science."" See Snyder, Laura J., "William Whewell", *The Stanford Encyclopedia of Philosophy* (Spring 2001 Edition), Edward N. Zalta (ed.), http://plato.stanford.edu/archives/spr2001/entries/whewell/.

[9] *History of engineering*, http://www.creatingtechnology.org/history.htm#1.

[10] Paul Allen, *History of PM*, http://members.aol.com/AllenWeb/history.html.

[11] Based on a portfolio of 33 sheets of parchment with 250 drawings from the 1230s, which is in the Bibliothèque Nationale, Paris (MS Fr 19093).

[12] *Sketchbook of Villard de Honnecourt.* This image (or other media file) is in the public domain because its copyright has expired.

[13] *Technology in World Civilization*, Arnold Pacey, p.42.

[14] *Science and Technology in World History*, James E. McClellan and Harold Dorn, p.118.

[15] *The Day the Universe Changed*, James Burke, 1985, p.113.

[16] Derived from experiment and observation rather than theory.

[17] *Project Lessons from the Great Escape*, Multi-Media Publications Inc.

[18] Project Management Body of Knowledge.

[19] *Architecture*, Jonathon Glancey, Eyewitness companions DK.

[20] http://www.historyworld.net/wrldhis/PlainTextHistories.asp?groupid=1431&HistoryID=ab23.

[21] *The Development of Double Entry Bookkeeping and its Relevance in Today's Business Environment*, Regina Libina, Pace University, 2005.

[22] Based on the six simple machines defined by Renaissance scientists as the lever, wheel and axel, pulley, inclined plane, wedge, and screw.

[23] *A History of Accounting Thought*, Michael Chatfield, Chapter 6, New York: Dryden Press, 1977.

[24] Source: *The Secret of the Great Pyramid*, Bob Brier and Jean-Pierre Houdin, p.23-24.

[25] This image (or other media file) is in the public domain because its copyright has expired. http://pymd.com/Saqqara-Pyramids-Djoser.htm.

[26] This image (or other media file) is in the public domain because its copyright has expired. http://www.guardians.net/egypt/pyramids/dahshur/Sneferu/TheBentPyramid.htm#about.

[27] *Building Construction Before Mechanization*, John Fitchen, 1986, p.4.

[28] Adult life expectancy was about 35 for men and 30 for women. For a Pharaoh it would have been between 60 to 70. Source: Filer, Joyce (1996). *Disease*, Austin, Texas: University of Texas Press. ISBN 0-292-72498-5.

[29] *Science and Technology in World History*, James E. McClellan and Harold Dorn.

[30] Source: *Pyramid*, David Macauley, 1975.

[31] http://www.touregypt.net/featurestories/pyramidworkforce.htm.

[32] *A New Pyramid Age*, Philip Coppens.

[33] Who Built the Pyramids? Not slaves. Archeaologist Mark Lehner, digging deeper, discovers a city of privileged workers, by Jonathan Shaw, http://harvardmagazine. com/2003/07/who-built-the-pyramids.html.

[34] A team of construction managers from international architectural, engineering, and construction management firm DMJM performed a forensic analysis to determine the construction methods and management techniques that were employed by the ancient Egyptians in constructing the Great Pyramid.

[35] Source: http://www.smithsonianmag.com/science-nature/ pyramid.html?c=y&page=2; http://www.archaeology. org/0705/etc/pyramid.html; http://www.usatoday.com/ tech/science/2007-05-16-pyramid-theory_N.htm.

[36] This image is a computer generated image in the public domain.

[37] This image is not copyrighted. *Shepp's Photographs of the World*, James W. Shepp, Daniel B. Shepp.

[38] *The Secret of the Great Pyramid*, p.14.

[39] *The Great Wall of China*, William Edgar Geil.

[40] This image (or other media file) is in the public domain because its copyright has expired. http://bygones.ebaypix. net/china1805map.jpg.

[41] This image (or other media file) is in the public domain because its copyright has expired. Photograph of The Great Wall of China from 1907, Herbert George Ponting. http://www.geocities.com/blackinkal4/ RoyalGeographicalSociety_Asia_2.html.

[42] *Technology in World Civilization*, Arnold Pacey p.42.

[43] *Science and Technology in World History*, James E.McClellan and Harold Dorn, p.25.

[44] This image (or other media file) is in the public domain because its copyright has expired. Photograph of Stonehenge 1910. http://www.oldukphotos.com/wiltshire-salisbury.htm.

[45] http://www.mystae.com/restricted/streams/thera/phoenicians.html.

[46] *National Geographic: Visual History of the World*, p.37.

[47] This image (or other media file) is in the public domain because its copyright has expired. http://www.crystalinks.com/meso.html.

[48] *The Secrets of the Great Pyramid*, Bob Brier, Jean-Pierre Houdin, p.69.

[49] *The Secrets of the Great Pyramid*, Bob Brier, Jean-Pierre Houdin, p.69.

[50] *Pyramid*, David Macauley, 1975.

[51] *Building Construction Before Mechanization*, John Fitchen, 1986, p.4.

[52] http://www.all-art.org/Visual%20History/3.htm.

[53] *Science and Technology in World History*, James E. McClellan and Harold Dorn.

[54] *Egypt the Black Land*, Paul Jordan, 2000, p.151.

[55] *Ancient Labor's Untold Story: Evidence of Workers' Organization from 3000 BCE to 550 CE in the Mediterranean World*, Dr. Charles Micallef, 2008.

[56] *Building Construction Before Mechanization*, John Fitchen, p.6.

[57] Source: *The Development of Double Entry Bookkeeping and its Relevance in Today's Business Environment*, Regina Libina, Pace University, 2005.

[58] Book-keeping system in which a person charges themselves with receipts and credits themselves with payments. Used extensively prior the advent of double-entry book-keeping.

[59] *Architecture*, Jonathon Glancey, Eyewitness companions DK.

[60] *Battle*, R.R. Grant, 2005.

[61] *A History of Warfare* by John Keegan, 1994, p.169.

[62] This image (or other media file) is in the public domain because its copyright has expired. *The Ten Books on Architecture*, Vitruvius Pollio.

[63] *A History of Accounting Thought*, Michael Chatfield, Chapter 9. New York: Dryden Press, 1977.

[64] Chatfield, Michael. *A History of Accounting Thought*, Chapter 6. New York: Dryden Press, 1977.

[65] A black siliceous stone used to ascertain the purity of gold and silver. The streaks of metal left behind on the touchstone are treated with nitric acid, which dissolves impurities, and thus, when the streaks are compared, the contrast between pure and impure metal is heightened. Source: http://www.britannica.com/EBchecked/ topic/600649/touchstone.

[66] *A History of Accounting Thought*, Chatfield, Michael, Chapter 10, New York: Dryden Press, 1977.

[67] This image (or other media file) is in the public domain because its copyright has expired. Historical Atlas by William R. Shepherd, New York, Henry Holt and Company, 1923. *University of Texas Perry-Casteneda Map Collection*, http://www.lib.utexas.edu/maps/ historical/history_shepherd_1923.html.

[68] Jeffrey Hurwit, a professor of art history at the University of Oregon and expert on the architecture of the Acropolis, http://www.pbs.org/wgbh/nova/parthenon/hurwit.html.

[69] Pericles, *The World Book Encyclopedia*, 1968, vol. 15, p.255. The funding of the project seems to have been quite complex. Aside from these sources of revenue there were many others including public moneys, tax revenues, harbor fees, boards of commissioners and judges contributing as well.

[70] *The Acropolis in the Age of Pericles*, Jeffrey M. Hurwit, University of Oregon.

[71] http://faculty.frostburg.edu/phil/forum/Athens.htm.

[72] Chisel marks from about 200 different stonemasons have been identified. http://www.pbs.org/wgbh/nova/ parthenon.

[73] Plutarch's Life of Pericles.

[74] *A Text-Book of the History of Architecture Seventh Edition*, revised, Hamlin, A. D. F. (Alfred Dwight Foster), 1855-1926, Not copyrighted.

[75] http://www.pbs.org/wgbh/nova/parthenon/hurwit.html.

[76] *A Text-Book of the History of Architecture*, Seventh Edition, revised, Hamlin, A. D. F. (Alfred Dwight Foster), 1855-1926, Not copyrighted.

[77] This image (or other media file) is in the public domain because its copyright has expired. http://www.mcah. columbia.edu/arthum/publicportfolio.cgi?view=1960.

[78] This image (or other media file) is in the public domain because its copyright has expired. http://www.mcah. columbia.edu/arthum/publicportfolio.cgi?view=1960.

[79] *Secrets of the Parthenon*, PBS Airdate: January 29, 2008. http://www.pbs.org/wgbh/nova/transcripts/3502_ partheno.html.

[80] *Secrets of the Parthenon*, PBS Airdate: January 29, 2008. http://www.pbs.org/wgbh/nova/transcripts/3502_ partheno.html.

[81] ~~*Plutarch's Life of Pericles*. http://classics.mit.edu/Plutarch/~~ pericles.html.

[82] De architectura (Latin: On architecture).

[83] This image (or other media file) is in the public domain because its copyright has expired. *Historical Atlas* by William R. Shepherd, New York, Henry Holt and Company, 1923. *University of Texas Perry-Casteneda Map Collection*, http://www.lib.utexas.edu/maps/ historical/history_shepherd_1923.html.

[84] *A History of Accounting Thought*, Chapter 12, Michael Chatfield, New York: Dryden Press, 1977.

[85] Title: *Manners, Custom and Dress During the Middle Ages and During the Renaissance Period*, Paul Lacroix.

[86] *A History of Accounting Thought*, Chapter 13, Michael Chatfield, New York: Dryden Press, 1977.

[87] *Ancient and Medieval Banking and Business in the Roman World*, by Jean Andreau, Translated by Janey Lloyd, Cambridge: Cambridge University Press, 1999.

[88] De architectura (Latin: "On architecture").

[89] *The Roman Army*, by Pat Southern, p.103.

[90] *A Text-Book of the History of Architecture Seventh Edition*, revised, Hamlin, A. D. F. (Alfred Dwight Foster), 1855-1926, Not copyrighted.

[91] *A Text-Book of the History of Architecture Seventh Edition*, revised, Hamlin, A. D. F. (Alfred Dwight Foster), 1855-1926, Not copyrighted.

[92] *A Text-Book of the History of Architecture Seventh Edition,* revised, Hamlin, A. D. F. (Alfred Dwight Foster), 1855-1926, Not copyrighted.

[93] The Flavian Amphitheatre, Commonly Called The Colosseum at Rome: Its History and Substructures Compared with Other Amphitheatres, John Henry Parker, C.B. Hon. M.A. Oxon., F.S.A. London.

[94] *Building Construction Before Mechanization,* John Fitchen, 1986, p.51.

[95] *Origins of Modern PM,* Patrick Weever, December 2007, http://www.mosaicprojects.com.au/PDF_Papers/P050_Origins_of_Modern_PM.pdf.

[96] *The Roman Pantheon—The Triumph of Concrete,* written by David Moore, P.E.

[97] *The Common People of Ancient Rome Studies of Roman Life and Literature,* Frank Frost Abbott.

[98] Manumission is the act of freeing a slave, done at the will of the owner. *Merriam-Webster Online Dictionary,* 2009.

[99] *Ancient Labor's Untold Story: Evidence of Workers' Organization from 3000 BCE to 550 CE in the Mediterranean World,* Dr. Charles Micallef, 2008.

[100] The earliest permanent amphitheatre would appear to be that built c.80 BC by Marcus Porcius at Pompeii, which in plan differs very little from that of the flavian amphitheatre.

[101] *The Upside of Down: Catastrophe, Creativity, and the Renewal of Civilization,* Thomas Homer-Dixon, and Ms. Karen Frecker, an energy analyst based in Toronto, Ontario, Canada.

[102] The Colosseum measures 48 meters (157 feet/165 Roman feet) high, 189 meters (615feet) long, and 165 meters (510 feet) wide, and covers an area of 6 acres.

[103] http://www.roman-colosseum.info/colosseum/index.htm.

[104] This image (or other media file) is in the public domain. Source: Profile of Colosseum with seating areas named in english. Created by Ningyou. http://commons.wikimedia.org/.

[105] *The Colosseum*, by Keith Hopkins and Mary Beard, 2006.

[106] *A Text-Book of the History of Architecture Seventh Edition*, revised, Hamlin, A. D. F. (Alfred Dwight Foster), 1855-1926, Not copyrighted.

[107] This image (or other media file) is in the public domain because its copyright has expired. Postcard circa 1960.

[108] This image (or other media file) is in the public domain because its copyright has expired.

[109] This image is not copyrighted.

[110] *Roman Architecture from Augustus To Hadrian The Colosseum: An Analysis Of The Inherent Political And Architectural Significance.* C.J Lyes.

[111] *A Text-Book of the History of Architecture Seventh Edition*, revised, Hamlin, A. D. F. (Alfred Dwight Foster), 1855-1926, Not copyrighted.

[112] *The Common People of Ancient Rome Studies of Roman Life and Literature*, Frank Frost Abbott.

[113] *A Text-Book of the History of Architecture Seventh Edition*, revised, Hamlin, A. D. F. (Alfred Dwight Foster), 1855-1926, Not copyrighted.

[114] *Oxford Handbook of Engineering and Technology in the Classical World,* John Peter Oleson.

[115] If the dome was treated as a series of arch segments, then the weight of the step rings near the base had a beneficial structural effect, acting somewhat like a buttress.

[116] *A Text-Book of the History of Architecture Seventh Edition,* revised, Hamlin, A. D. F. (Alfred Dwight Foster), 1855-1926, Not copyrighted.

[117] This image (or other media file) is in the public domain because its copyright has expired. Source: 18th Century, painted by Giovanni Paolo Panini.

[118] This image is not copyrighted.

[119] This image (or other media file) is in the public domain because its copyright has expired. http://worldheritagesite.org/sites/pontdugard.html.

[120] According to Liber IV.

[121] This image (or other media file) is in the public domain because its copyright has expired. Painting by John Soane, 1814.

[122] This image (or other media file) is in the public domain because its copyright has expired. *Encyclopaedia Britannica,* 11th Edition, Volume 4, Part 3, p.585.

[123] This image (or other media file) is in the public domain because its copyright has expired. Painting by William Bell Scott, 1857. This image (or other media file) is in the public domain because its copyright has expired.

[124] *Encyclopaedia Britannica,* 11th Edition, Volume 4, Part 3, p.586, Figure 3.

[125] http://www.aboutscotland.co.uk/hadrian/wall.html.

[126] *The Roman Pantheon—The Triumph of Concrete,* David Moore, P.E..

[127] *Building Construction Before Mechanization,* John Fitchen, 1986, p.51.

[128] *The Colosseum,* Keith Hopkins and Mary Bend, p.144.

[129] The population of the world circa 1 was 200-300 million people and in the Roman Empire under Augustus about 45 million (15% of the world's population) with 4 million Roman citizens.

[130] *The Roman Army*, Pat Southern, p.229.

[131] *Roman Builders – A Study in Architectural Process*, p.27, Rabun Taylor, 2004, Harvard University.

[132] *Vitruvius The Ten Books On Architecture*, Vitruvius Pollio.

[133] *Building Construction Before Mechanization*, John Fitchen, 1986, p.40.

[134] *The Roman Pantheon—The Triumph of Concrete*, David Moore, P.E..

[135] *Roman Builders – A Study in Architectural Process*, p.28, Rabun Taylor, 2004, Harvard University.

[136] *The Upside of Down: Catastrophe, Creativity, and the Renewal of Civilization*, Thomas Homer-Dixon.

[137] This image (or other media file) is in the public domain because its copyright has expired. *Historical Atlas* by William R. Shepherd, New York, Henry Holt and Company, 1923. *University of Texas Perry-Casteneda Map Collection*, http://www.lib.utexas.edu/maps/historical/history_shepherd_1923.html.

[138] *The History of Money*, http://www.pbs.org/wgbh/nova/moolah/history.html.

[139] *Nature*, Volume 443, p.28, September 2006, Virginia Hughes. Incorporates the investigation of the *Hagia Sophia Project* by Ahmet Çakmak, a professor emeritus in earthquake engineering at Princeton University.

[140] *A Text-Book of the History of Architecture Seventh Edition*, revised, Hamlin, A. D. F. (Alfred Dwight Foster), 1855-1926. This image (or other media file) is in the public domain because its copyright has expired.

[141] *A Text-Book of the History of Architecture Seventh Edition*, revised, Hamlin, A. D. F. (Alfred Dwight Foster), 1855-1926. This image (or other media file) is in the public domain because its copyright has expired.

[142] *Master-builders of Byzantium*, Robert Ousterhout, University of Pennsylvania Museum of Archaeology and Anthropology.

[143] *Shepp's Photographs of the World*, James W. Shepp Daniel, B. Shepp, Not copyrighted.

[144] *Shepp's Photographs of the World*, James W. Shepp Daniel, B. Shepp, Not copyrighted.

[145] This image (or other media file) is in the public domain because its copyright has expired. http://www.philographikon.com/printsmexico.html.

[146] *Building Construction Before Mechanization*. John Fitchen, 1986, p.49.

[147] This image (or other media file) is in the public domain because its copyright has expired. *University of Texas Libraries*, http://www.lib.utexas.edu/maps/historical/history_europe.html.

[148] *Revival and Decline of Greek Mathematics*. Carl Benjamin Boyer, 1991.

[149] Research using computer models and chemical analyses by Ahmet S. Cakmak, a professor of Civil Engineering at Princeton and a specialist on the Haghia Sophia.

[150] *For the weight of the gold used for the mosaics in Hagia Sophia*, cf. A. Cutler, "The Industries of Art," EHB 557-58.

[151] *Technology in World Civilization*, p.7. Arnold Pacey.

[152] This image (or other media file) is in the public domain because its copyright has expired. *Manners, Custom and Dress During the Middle Ages and During the Renaissance Period*, Paul Lacroix, http://www.gutenberg. org/files/10940/10940-h/images/fig250.png.

[153] http://www.pbs.org/wgbh/nova/vikings/ships.html.

[154] *Science and Technology in World History*, James E. McClellan and Harold Dorn, p.118.

[155] This image (or other media file) is in the public domain because its copyright has expired. Taken from the *14th Century Treatise Nong Shu*. Written by Wang Zhen in 1313, during the Chinese Yuan Dynasty.

[156] *The Medieval Machine: The Industrial Revolution of the Middle Ages*, Jean Gimpel, Pimlico 1992.

[157] This image (or other media file) is in the public domain because its copyright has expired. *Manners, Custom and Dress During the Middle Ages and During the Renaissance Period*, Paul Lacroix, http://www.gutenberg. org/files/10940/10940-h/images/fig250.png.

[158] *Science and Technology in Medieval European Life*, Jeffrey R. Wigelsworth.

[159] *Science and Technology in Medieval European Life*, Jeffrey R. Wigelsworth, p.14.

[160] *Technics and Civilization*, Lewis Mumford, 1963, p.439.

[161] http://www.archnet.org/library/dictionary/entry.jsp?entry_ id=DIA0025&mode=full.

[162] *A Text-Book of the History of Architecture Seventh Edition*, revised. Hamlin, A. D. F. (Alfred Dwight Foster), 1855-1926, Not copyrighted.

[163] *Oxford Handbook of Engineering and Technology in the Classical World*, John Peter Oleson, p.296.

[164] *A Text-Book of the History of Architecture, Seventh Edition*, revised. Hamlin, A. D. F. (Alfred Dwight Foster), 1855-1926, Not copyrighted.

[165] *The Cathedral Builders*, Jean Gimpel, p.4.

[166] *The Cathedral Builders*, Jean Gimpel, p.1.

[167] *Outline of Universal History*, Fisher, George Park.

[168] This image (or other media file) is in the public domain because its copyright has expired. *Manners, Custom and Dress During the Middle Ages and During the Renaissance Period*, Paul Lacroix, http://www.gutenberg. org/files/10940/10940-h/images/fig250.png.

[169] This image (or other media file) is in the public domain because its copyright has expired. Jost Amman, professional wood engravers in Germany during the latter half of the 16th Century. Source: *"Eygentliche Beschreibung Aller"* ("Description of All Professions") which was published in Frankfurt in 1568.

[170] *The Cathedral Builders*, Jean Gimpel, p.60.

[171] *The Builder*, John Harvey, p.47, 1973.

[172] *The Cathedral Builders*, Jean Gimpel, p.69, accounts between 1278-1281.

[173] This image (or other media file) is in the public domain because its copyright has expired. G. Dehio and G. von Bezold, Die Kirchliche Baukunst des abendlandes, Stuttgart, 1887-1902, plate 383.

[174] *The Construction of Gothic Cathedrals: A Study of Medieval Vault Erection*, John Fitchen.

[175] Drawing of A. M. Beloqui in Corton de las Heras 1997, p.275.

[176] This image (or other media file) is in the public domain because its copyright has expired. G. Dehio (died 1932) and G. von Bezold (died 1934), Die Kirchliche Baukunst des abendlandes, Stuttgart, 1887-1902.

[177] *Building Construction Before Mechanization* by John Fitchen, 1986, p.8.

[178] *Cathedral*, David Macauley, 1985.

[179] James, John, Chartres, *The Masons Who Built a Legend*, Routledge and Kegan Paul, London, 1982.

[180] This image (or other media file) is in the public domain because its copyright has expired. *Arts and Crafts in the Middle Ages*, Julia De Wolf Addison, p.238, http://www.gutenberg.org/files/18212/18212-h/18212-h.htm.

[181] This image (or other media file) is in the public domain because its copyright has expired. Georg Agricola, Zwölf Bücher vom Berg-und Hüttenwesen, übers. v. Carl Schiffner, Berlin 1928.

[182] *The Cathedral Builders*, Jean Gimpel, p.68.

[183] *Guo, Qinghua. "Yingzao Fashi: Twelfth-Century Chinese Building Manual,* "Architectural History: Journal of the Society of Architectural Historians of Great Britain", Volume 41, 1998: 1-13.

[184] *A Text-Book of the History of Architecture Seventh Edition*, revised, Hamlin, A. D. F. (Alfred Dwight Foster), 1855-1926, Not copyrighted.

[185] *Shepp's Photographs of the World*, James W. Shepp, Daniel B. Shepp, Not copyrighted.

[186] *The Cathedral Builders*, Jean Gimpel, p.67.

[187] This image (or other media file) is in the public domain because its copyright has expired. *Plan, from A. Hamilton Thompson, Military Architecture in England During the Middle Ages,* (London, New York, 1912), p.68.

[188] *Technology in World Civilization,* p.13, Arnold Pacey.

[189] This image (or other media file) is in the public domain because its copyright has expired. 19th Century photo.

[190] This image (or other media file) is in the public domain because its copyright has expired. 19th Century photo.

[191] *World History,* Jeremy Black, p.123.

[192] This image is from *Commons: GNU Free Documentation License,* http://commons.wikimedia.org/wiki/ Commons:GNU_Free_Documentation_License.

[193] Based on a portfolio of 33 sheets of parchment with 250 drawings from the 1230s, which is in the Bibliothèque Nationale, Paris (MS Fr 19093).

[194] *Building Construction Before Mechanization,* John Fitchen, 1986, p.49.

[195] *Building Construction Before Mechanization,* John Fitchen, 1986, p.49.

[196] *Building Construction Before Mechanization,* John Fitchen, 1986, p.247 #6.

[197] *Building Construction Before Mechanization,* John Fitchen, 1986, p.51.

[198] *Building Construction Before Mechanization,* John Fitchen, 1986, p.17.

[199] *The Cathedral Builders,* Jean Gimpel, p.59.

[200] *The Cathedral Builders,* Jean Gimpel, p.62.

[201] *The Cathedral Builders*, Jean Gimpel, p.42.

[202] *The Cathedral Builders*, Jean Gimpel, p41 & 47.

[203] *Building Construction Before Mechanization, John Fitchen,* p.247 #6.

[204] *Building Construction Before Mechanization,* John Fitchen, p.50.

[205] *The Cathedral Builders*, Jean Gimpel, p.52.

[206] *The Construction of Gothic Cathedrals*, John Fitchen, p.276.

[207] *Technology in World Civilization*, p.54, Arnold Pacey.

[208] According to medieval historian Philip Daileader in 2007.

[209] *The Day the Universe Changed*, James Burke, 1985.

[210] *The Medieval Machine: The Industrial Revolution of the Middle Ages*, Jean Gimpel, Pimlico 1992.

[211] *Technology in World Civilization*, Arnold Pacey, p.49.

[212] According to Brunelleschi's biographer Antonio Manetti.

[213] An old Italian unit of length, usually about 26 or 27 in. (66 or 68 cm).

[214] *A Text-Book of the History of Architecture Seventh Edition,* revised, Hamlin, A. D. F. (Alfred Dwight Foster), 1855-1926, Not copyrighted.

[215] This image (or other media file) is in the public domain because its copyright has expired.

[216] *A Text-Book of the History of Architecture Seventh Edition,* revised, Hamlin, A. D. F. (Alfred Dwight Foster), 1855-1926, Not copyrighted.

[217] From an order in the book of *Migliore di Tommaso*, dated October 3, 1419.

[218] This image (or other media file) is in the public domain because its copyright has expired.

[219] S*hepp's Photographs of the World*, James W. Shepp, Daniel B. Shepp, Not copyrighted.

[220] *Shepp's Photographs of the World*, James W. Shepp, Daniel B. Shepp, Not copyrighted.

[221] *Shepp's Photographs of the World*, James W. Shepp, Daniel B. Shepp, Not copyrighted.

[222] Dreyer (2006): 122–124.

[223] This image (or other media file) is in the public domain because its copyright has expired. 17th Century Chinese woodblock print.

[224] *Brunelleschi's Cupola: Past and Present of an Architectural Masterpiece*, Giovanni Fanelli, Giovanni Fanelli and Michele Fanelli, p.28.

[225] Source: *Brunelleschi's Cupola: Past and Present of an Architectural Masterpiece*, Giovanni Fanelli, Giovanni Fanelli and Michele Fanelli, p.23.

[226] *Lives of the Most Eminent Painters Sculptors & Architects*, Giorgio Vasari, 1912.

[227] *Brunelleschi's Cupola: Past and Present of an Architectural Masterpiece*, Giovanni Fanelli, Giovanni Fanelli and Michele Fanelli, p.23.

[228] *Lives of The Most Eminent Painters Sculptors & Architects, 1912*, Giorgio Vasari.

[229] *Brunelleschi's Cupola: Past and Present of an Architectural Masterpiece*, Giovanni Fanelli, Giovanni Fanelli and Michele Fanelli, p.23.

[230] *Brunelleschi's Cupola: Past and Present of an Architectural Masterpiece*, Giovanni Fanelli, Giovanni Fanelli and Michele Fanelli, p.23.

[231] *Brunelleschi's Cupola: Past and Present of an Architectural Masterpiece*, Giovanni Fanelli, Giovanni Fanelli and Michele Fanelli, p.26.

[232] *Brunelleschi's Cupola: Past and Present of an Architectural Masterpiece*, Giovanni Fanelli, Giovanni Fanelli and Michele Fanelli, p.25.

[233] *Brunelleschi's Cupola: Past and Present of an Architectural Masterpiece*, Giovanni Fanelli, Giovanni Fanelli and Michele Fanelli, p.23.

[234] *Brunelleschi's Cupola: Past and Present of an Architectural Masterpiece*, Giovanni Fanelli, Giovanni Fanelli and Michele Fanelli, p.27.

[235] *Brunelleschi's Cupola: Past and Present of an Architectural Masterpiece*, Giovanni Fanelli, Giovanni Fanelli and Michele Fanelli, p.27.

[236] *Technology in World Civilization*, Arnold Pacey.

[237] This image (or other media file) is in the public domain because its copyright has expired. The image comes from an early 16th Century book titled: *Livro das Fortalezas de Duarte Damas*.

[238] A newspaper with pages of a size larger than those of a tabloid. http://www.merriam-webster.com/dictionary/broadsheets.

[239] This image (or other media file) is in the public domain because its copyright has expired. *Prince Henry the Navigator, the Hero of Portugal and of Modern Discovery, 1394-1460*, C. Raymond Beazley.

[240] Spanish currency of one million Maravedis (one cuentos) in 1490 is equivalent to about 308 English Pounds in 1860, or US$ 48,000 in 2005.

[241] This image (or other media file) is in the public domain because its copyright has expired. United States Library of Congress's Prints and Photographs Division under the digital ID cph.3c05453, http://www.loc.gov/rr/print/list/080_columbus.html.

[242] This image (or other media file) is in the public domain because its copyright has expired. United States Library of Congress's Prints and Photographs Division under the digital ID cph.3c05453, http://www.loc.gov/rr/print/list/080_columbus.html.

[243] This is according to Peter Martyr who kept a special account of the voyage.

[244] This image (or other media file) is in the public domain because its copyright has expired. United States Library of Congress's Prints and Photographs Division under the digital ID cph.3c05453, http://www.loc.gov/rr/print/list/080_columbus.html.

[245] This image (or other media file) is in the public domain because its copyright has expired. *Manners, Custom and Dress During the Middle Ages and During the Renaissance Period*, Paul Lacroix, http://www.gutenberg.org/files/10940/10940-h/images/fig250.png.

[246] This image (or other media file) is in the public domain because its copyright has expired. Source based upon the map in Bourne's Spain in America, American Nation Series, Volume III, New York, 1904, Harper.

[247] The Americas received wheat, rice, coffee, bananas, and olives, and horses, cows, pigs, and chickens. They contributed a virulent form of syphilis as well as corn, potatoes, tomatoes, lima beans, squash, peanuts, cassava, cacao, and pineapple.

[248] *World History*, Jeremy Black, p.124.

[249] *A Short History of Progress*, Ronald Wright, p.114.

[250] A good example is 'The Richard II' horary quadrant found in the British Museum, London, UK.

[251] *The Philippine Islands*, 1493-1803, Emma Helen Blair, http://www.gutenberg.org/files/13255/13255-8.txt.

[252] This image (or other media file) is in the public domain because its copyright has expired. http://www.helmink.com/Antique_Maps_of_America.html.

[253] This image (or other media file) is in the public domain because its copyright has expired. http://www.magellanacademies.com/Strait_of_Magellan.jpg.

[254] This image (or other media file) is in the public domain because its copyright has expired. 1507, G3200 1507. W3 Vault, Library of Congress Rare Book and Special Collections Division Washington, D.C. 20540-4650, http://hdl.loc.gov/loc.gmd/g3200.ct000725.

[255] This image (or other media file) is in the public domain because its copyright has expired. 1581, G3201.S12 1581 .S9, Library of Congress Rare Book and Special Collections Division Washington, D.C. 20540-4650, http://hdl.loc.gov/loc.gmd/g3201s.rb000011 (Library of Congress).

[256] *Architecture*, Jonathon Glancey, Eyewitness companions DK.

[257] http://www.the-tudors.org.uk/hardwick-hall.htm.

[258] This image (or other media file) is in the public domain because its copyright has expired. http://www.oldukphotos.com/graphics/England%20Photos/Nottinghamshire,%20Mansfield,%20Hardwick%20Hall%201900's.jpg.

259 *A Text-Book of the History of Architecture Seventh Edition,* revised, Hamlin, A. D. F. (Alfred Dwight Foster), 1855-1926, Not copyrighted.

260 *A Text-Book of the History of Architecture Seventh Edition,* revised, Hamlin, A. D. F. (Alfred Dwight Foster), 1855-1926, NOT copyrighted.

261 *Shepp's Photographs of the World,* James W. Shepp Daniel B. Shepp, Not copyrighted.

262 This image (or other media file) is in the public domain because its copyright has expired. 17th Century engraving.

263 *Niall Ferguson the Ascent of Money,* 2007.

264 This image (or other media file) is in the public domain because its copyright has expired. *Glashütte Weibersbrunn im Spessart,* gegründet 1706.

265 http://www.pbs.org/treasuresoftheworld/taj_mahal/tlevel_2/t3build_design.html.

266 http://www.howstuffworks.com/taj-mahal-landmark.htm.

267 http://www.agraindia.org.uk/taj-mahal/architecture/building-materials-used.html.

268 http://www.thetajmahalindia.com/tajmahalinformation.html.

269 http://www.tce.co.in/Downloads/bro_pdf/tce_world/july06.pdf.

270 http://www.agraindia.org.uk/taj-mahal/architecture/building-materials-used.html.

271 This image (or other media file) is in the public domain because its copyright has expired. Jackson, William Henry, 1843-1942. *World's Transportation Commission photograph collection* (Library of Congress).

[272] This image (or other media file) is in the public domain because its copyright has expired. *Plan and Layout of the Taj Mahal and Gardens of the Great Mughals,* C.M. Villiers Stuart, 1913.

[273] *A Text-Book of the History of Architecture,* Seventh Edition, revised, Hamlin, A. D. F. (Alfred Dwight Foster), 1855-1926, Not copyrighted. Photo by Samuel Bourne, 1860.

[274] *The Story Of Versailles,* Francis Loring Payne, 1919.

[275] This image (or other media file) is in the public domain because its copyright has expired. Francis Loring Payne, *The Story of Versailles.*

[276] Francis Loring Payne, *The Story of Versailles.*

[277] This image (or other media file) is in the public domain because its copyright has expired. View of Versailles from the Avenue de Paris, ca. 1668, Pierre Patel.

[278] This image (or other media file) is in the public domain because its copyright has expired. Francis Loring Payne, *The Story of Versailles.*

[279] *New Glass Architecture,* Brent Richards, Dennis Gilbert.

[280] This image (or other media file) is in the public domain because its copyright has expired. Francis Loring Payne, *The Story of Versailles.*

[281] Historian A.P.J. Taylor wrote that the Palace of Versailles was one of the fundamental causes.

[282] This image (or other media file) is in the public domain because its copyright has expired. *French National Tourist Collection.*

[283] *Taj Mahal,* Giles Tillotson, p.73.

[284] http://www.accel-team.com, 2004.

285 Such as the Combination Acts of 1799/1800 in the UK.

286 *The Fatal Shore*, Robert Hughes.

287 *The History of Money*, http://www.pbs.org/wgbh/nova/moolah/history.html.

288 This image (or other media file) is in the public domain because its copyright has expired. http://www.mspong.org/cyclopedia/metallurgy_pics.html.

289 This image (or other media file) is in the public domain because its copyright has expired. Map of 1811, G3200 1811, V2 Vault, Library of Congress Rare Book and Special Collections Division Washington, D.C. 20540-4650, http://hdl.loc.gov/loc.gmd/g3200.awh00011 (Library of Congress).

290 http://www.ironbridge.org.uk/our_attractions/the_iron_bridge_and_tollhouse/history/.

291 http://www.nationalarchives.gov.uk/pathways/blackhistory/journeys/voyage_html/iron.htm.

292 This image (or other media file) is in the public domain because its copyright has expired. Acc No: AE185.762 Lowry, Wilson (engraver) Robertson, George (artist),. http://www.ironbridge.org.uk/about_us/the_iron_bridge/.

293 This image (or other media file) is in the public domain because its copyright has expired. Acc No: 1972.90. http://www.ironbridge.org.uk/about_us/the_iron_bridge/.

294 This image (or other media file) is in the public domain because its copyright has expired. Acc No: 1973.200, Ellis, William (engraver), Rooker, Michael Angelo (artist), http://www.ironbridge.org.uk/about_us/the_iron_bridge/.

[295] This image (or other media file) is in the public domain because its copyright has expired. Acc No: AE185.771 Dubourg, M. (engraver), http://www.ironbridge.org.uk/about_us/the_iron_bridge/.

[296] This image (or other media file) is in the public domain because its copyright has expired. Acc No: AE185.771 Dubourg, M. (engraver), http://www.ironbridge.org.uk/about_us/the_iron_bridge/.

[297] This image (or other media file) is in the public domain because its copyright has expired. Acc No: 1973.273, Walker, W. (engraver), Walker, J. (engraver), Burney, T.F. (artist), http://www.ironbridge.org.uk/about_us/the_iron_bridge/.

[298] This image (or other media file) is in the public domain because its copyright has expired. http://www.ironbridge.org.uk/about_us/the_iron_bridge/.

[299] This image (or other media file) is in the public domain because its copyright has expired. York Railway Museum.

[300] http://www.tynelives.org.uk/stephenson/wagon.htm.

[301] This image (or other media file) is in the public domain because its copyright has expired. York Railway Museum.

[302] This image (or other media file) is in the public domain because its copyright has expired. York Railway Museum.

[303] This image (or other media file) is in the public domain because its copyright has expired. *Lives Of The Engineers. The Locomotive,* George and Robert Stephenson, By Samuel Smiles. http://www.gutenberg.org/files/27710/27710-h/27710-h.htm.

[304] *Lives Of The Engineers. The Locomotive*, George and Robert Stephenson, Samuel Smiles, p.125. http://www.gutenberg.org/files/27710/27710-h/27710-h.htm.

[305] This image (or other media file) is in the public domain because its copyright has expired. *An Introduction to the Industrial and Social History of England*, Edward P. Cheyney. (Smiles: Life of George Stephenson), http://www.gutenberg.org/files/21660/21660-h/21660-h.htm#page031.

[306] This image (or other media file) is in the public domain because its copyright has expired. Darlington Borough Council.

[307] Stockton and Darlington Railway opening, J.R. Brown, Science Museum.

[308] This image (or other media file) is in the public domain because its copyright has expired. *An Introduction to the Industrial and Social History of England*, Edward P. Cheyney, http://www.gutenberg.org/files/21660/21660-h/21660-h.htm#page031.

[309] This image (or other media file) is in the public domain because its copyright has expired. *Scientific American Supplement*, No. 460, October 25, 1884, http://www.gutenberg.org/files/11734/11734-h/11734-h.htm.

[310] This image (or other media file) is in the public domain because its copyright has expired. *Encyclopaedia Britannica*, 11th Edition, Volume 4, Part 3, p.537.

[311] This image (or other media file) is in the public domain because its copyright has expired. *Etching of the Menai Bridge* ca: 1825, http://www.anglesey-history.co.uk/places/bridges/menai.gif.

[312] *The First in the World*, John Wall, according to Francis Mewburn, p.11.

[313] *Energy and society: an Introduction*, Harold H. Schobert, p.263.

[314] Christopher Latham Sholes (1819-1890) was a U.S. mechanical engineer who invented the first practical modern typewriter, patented in 1868.

[315] *New Glass Architecture*, Brent Richards, Dennis Gilbert.

[316] This image (or other media file) is in the public domain because its copyright has expired. Sir Henry Bessemer, F.R.S, *An Autobiography*, 1st edition, published in London 1905, http://www.history.rochester.edu/ehp-book/shb/hb11.htm.

[317] This image (or other media file) is in the public domain because its copyright has expired. Agricultural Implements and Machines in the Collection of the National Museum of History and Technology Smithsonian Studies in History and Technology, No. 17, Schlebecker, John T.

[318] This image (or other media file) is in the public domain because its copyright has expired. Frank Leslie's *Illustrated Newspaper*, August 21, 1858.

[319] The British Government gave Field a subsidy of £1,400 a year and loaned the ships to lay the cable.

[320] http://http://www.gutenberg.org/files/14091/14091-h/14091-h.htm.

[321] This image (or other media file) is in the public domain because its copyright has expired. *Telegraph Connections* (Telegraphen Verbindungen), 1891 Stielers Hand-Atlas, Plate No. 5, Weltkarte in Mercator projection.

[322] *The Pacific tourist*, Williams, Henry T. Collection: Making of America Books.

[323] *A Practical Plan For Building The Pacific Railroad,* T.D. Judah, Civil Engineer, San Francisco, January 1, 1857, Washington, D.C. Henry Polkinhorn, Printer 1857.

[324] The Big Four consisted of Mark Hopkins, Collis P. Huntington, [Gov.] Leland Stanford, and Charles Crocker. These men were merchants drawn to the west by prospects of finding riches in California.

[325] http://cprr.org/Museum/Eastward. html#Construction%20 of%20the%20CPRR.

[326] *The First Transcontinental Railroad*, John Debo Galloway, C. E. (1869-1943) Dorset Press, New York, 1989.

[327] This image (or other media file) is in the public domain because its copyright has expired. This image is available from the Archival Research Catalog of the National Archives and Records Administration under the ARC Identifier 594940.

[328] *The Pacific Tourist Williams' Illustrated Trans-Continental Guide*, Henry T. Williams, Editor, http://cprr.org/ Museum/Books/Williams_Pacific_Tourist.html.

[329] This image (or other media file) is in the public domain because its copyright has expired. *General view of the works in progress*, Acc No: AE185.3165.69, Berlyn, Peter (author), Fowler, Charles (author).

[330] *New Glass Architecture*, Brent Richards, Dennis Gilbert.

[331] This image (or other media file) is in the public domain because its copyright has expired. *General view of the works in progress*, Acc No: AE185.3165.69, Berlyn, Peter (author), Fowler, Charles (author).

[332] This image (or other media file) is in the public domain because its copyright has expired. *Tallis' History and criticism of the Crystal Palace*, 1852.

[333] *New Glass Architecture*, Brent Richards, Dennis Gilbert.

[334] http://www.lib.umd.edu/digital/worldsfairs/essay. jsp?pid=umd:1014.

[335] This image (or other media file) is in the public domain because its copyright has expired. Published by C.F. Cheffins, Lithographers, South Hampton Buildings, London, England, 1854 in Snow, John.

[336] http://www.swopnet.com/engr/londonsewers/londontext1.html.

[337] *Ancient Transportation*, Michael Woods, Mary B. Woods, p.44.

[338] This image (or other media file) is in the public domain because its copyright has expired. Published 1881 in Young Persons' Cyclopedia of Persons and Places.

[339] This image (or other media file) is in the public domain because its copyright has expired. Scanned from engraving in *"Appleton's Journal of Popular Literature, Science, and Art"*, 1869.

[340] Avery, Ralph E. (1913). *"The French Failure"*. *America's Triumph in Panama*. Chicago, IL: L.W. Walter Company.

[341] This image (or other media file) is in the public domain because its copyright has expired. *Paris, Exposition Universelle*, 1889. "74. Paris. Tour Eiffel (juillet 1888)" BnF, Estampes et Photographie, Qe Mat 1, http://expositions.bnf.fr/universelles/.

[342] He was awarded the prestigious Elliot Cresson Medal by the Franklin Institute of Philadelphia in 1890; Gold Medal of the Paris Exposition and the Bronze Medal of the World's Fair in 1893. He was awarded a doctorate for his work in 1890.

[343] http://www-history.mcs.st-andrews.ac.uk/Biographies/Hollerith.html.

[344] http://www.greatwestern.org.uk/m_in_gwr.htm.

[345] *A Practical Plan for Building The Pacific Railroad*, T.D. Judah, Civil Engineer, San Francisco, January 1, 1857, Washington, D.C. Henry Polkinhorn, Printer. 1857.

346 *Railroad Commissioners' Reports* (Central Pacific Railroad), http://cprr.org/Museum/NARA/index.html.

347 Called an autotypist, it stores and reproduces simple form letters and certain paragraphs using punched paper tape as the storage medium.

348 *The Guns of August*, Barbara Tuchman, 1962.

349 http://www.vanderbilt.edu/AnS/Anthro/Anth101/taylorism_ and_fordism.htm.

350 Groundbreaking experiments conducted at the Western Electric plant in Chicago from 1927 to 1932.

351 Further reading on the Human Relations movement includes: Kurt Lewin, one of the first researchers to study group dynamics and evolve organizational development; Abraham Maslow presented "*The Hierarchy of Needs*" in "*The U.S. Psychology Review*" in 1943; Douglas McGregor published "*Theory X and Theory Y*" in "*The Human Side of Enterprise*" in 1960; Frederick Herzberg published the *Hygiene-Motivation Theory* in "*The Motivation to Work*" in 1959. (www.accel-team.com, 2004).

352 http://www.economicprincipals.com/issues/02.11.10.html.

353 http://www.vanderbilt.edu/AnS/Anthro/Anth101/taylorism_ and_fordism.htm.

354 www.asq.org/learn-about-quality/project-planning-tools/ overview/pdca-cycle.html.

355 This image (or other media file) is in the public domain because its copyright has expired. *Illustrated History of the Panama Railroad* by Fessenden Nott Otis, Harper & Brothers, New York, 1861.

356 This image (or other media file) is in the public domain because its copyright has expired. *Illustrated History of the Panama Railroad*, Fessenden Nott Otis, Harper & Brothers, New York, 1861.

[357] Avery, Ralph E. (1913). *The French Failure, America's Triumph in Panama*, Chicago, IL: L.W. Walter Company.

[358] *The Panama Canal*, José Carlos Rodrigues.

[359] Otis, Fessenden Nott; *llustrated History of the Panama Railroad*, Harper & Brothers, New York, 1861.

[360] This image (or other media file) is in the public domain because its copyright has expired. *Historical Atlas* by William R. Shepherd, New York, Henry Holt and Company, 1923, *University of Texas Perry-Casteneda Map Collection*, http://www.lib.utexas.edu/maps/historical/history_shepherd_1923.html.

[361] This image (or other media file) is in the public domain because its copyright has expired. *Culebra Cut of the Panama Canal*, 1907.

[362] This image (or other media file) is in the public domain because its copyright has expired. *From The Panama Canal, an Address to the National Geographic Society*, Colonel Goethals, February 10, 1911.

[363] This image (or other media file) is in the public domain because its copyright has expired. *Historical Atlas*, William R. Shepherd, New York, Henry Holt and Company, 1923, *University of Texas Perry-Casteneda Map Collection*, http://www.lib.utexas.edu/maps/historical/history_shepherd_1923.html.

[364] http://www.canalmuseum.com/documentspanamacanal history023.htm.

[365] http://query.nytimes.com/mem/archive-free/pdf?_r= 1&res=9A03E1DC1638E233A25756C2A9649D94639ED 7CF.

[366] http://www.generalcontractor.com/resources/articles/hoover-dam.asp.

[367] In today's dollars the bid would be worth more than $577 million.

[368] This image (or other media file) is in the public domain because its copyright has expired. The U.S. Bureau of reclamation.

[369] This image (or other media file) is in the public domain because its copyright has expired. The U.S. Bureau of reclamation.

[370] *Building the Empire State*, Carol Willis and Donald Friedman.

[371] This image (or other media file) is in the public domain because its copyright has expired. Gottscho, Samuel H., photographer. *Architecture and Interior Design for 20th Century America*, 1935 - 1955, Library of Congress.

[372] Image courtesy of National Archives and Records Administration.

[373] Logistics is defined by the modern British Army as: The science of planning and carrying out the movement and maintenance of forces. *Infantry Tactical Doctrine Volume 1, Pamphlet No.1, The Infantry Company Group, The Fundamentals* (1998) pp.6-2.

[374] *Building Construction Before Mechanization*, John Fitchen, 1986, p.59.

[375] *Building Construction Before Mechanization*, John Fitchen, 1986, p.51.

[376] *Building Construction Before Mechanization*, John Fitchen, 1986, p.76, p.249 #7.

[377] *Building Construction Before Mechanization*, John Fitchen, p.40.

[378] *Building Construction Before Mechanization*, John Fitchen, 1986, p.247 #6.

[379] *The Cathedral Builders*, Jean Gimpel, p.102.

[380] *Technology in World Civilization*, Arnold Pacey, p.42.

[381] The population of the world circa 1 was 200-300 million people and in the Roman Empire under Augustus about 45 million (15% of the world's population) with 4 million Roman citizens.

[382] *Pyramid Quest*, Robert M Schoch, 2005, p.90.

[383] *Pyramid*, David Macauley 1975.

[384] *A Social Study of the Emperor's Freedmen and Slaves.* P.R.C. Weaver, Familia Caesaris. Cambridge, University Press, 1972.

[385] History of Communications, http://www.nathan.com/projects/current/comtimeline.html.

[386] *Vitruvius The Ten Books On Architecture*, Vitruvius Pollio.

[387] *Building Construction Before Mechanization*, John Fitchen, 1986, p.40.

[388] *Building Construction Before Mechanization*, John Fitchen, 1986, p.86.

[389] *Origins of Modern PM, Patrick Weever*, December 2007, http://www.mosaicprojects.com.au/PDF_Papers/P050_Origins_of_Modern_PM.pdf.

[390] *The Roman Pantheon—The Triumph of Concrete*, David Moore, P.E..

[391] *Fabric Rolls and Documents Of York Minster* (Paperback), illustrated By John Browne, James Raine (Author).

[392] *Colosseum Calculation Assumptions continued.. The Upside of Down: Catastrophe, Creativity, and the Renewal of Civilization*, Thomas Homer-Dixon.

Index

About the Author

As the author behind the "Lessons from History" series, Mark Kozak-Holland brings years of experience as a consultant who helps Fortune-500 companies formulate projects that leverage emerging technologies. Since 1985 he has been straddling the business and IT worlds, making these projects happen. He is a certified business consultant, the author of several books, and a noted speaker. As a historian, Kozak-Holland seeks out the wisdom of the past to help others avoid repeating mistakes and to capture time-proven techniques. His lectures have been very popular at gatherings of project managers and CIOs.

Mark is very passionate about history and sees its potential use as an educational tool in business today. As a result, he has been developing the "Lessons from History" series for organisations, applying today's Information Technology (IT) to common business problems. It is written for primarily business and IT professionals looking for inspiration for their projects. It uses relevant historical case studies to examine how historical projects and emerging technologies of the past solved complex problems.

For thousands of years people have been running projects that leveraged emerging technologies of the time, to create unique and wonderful structures like the pyramids, buildings, or bridges. Similarly, people have gone on great expeditions and journeys and have raced their rivals in striving to be first, e.g., circumnavigating the world or conquering the poles. These were all forms of projects that required initiation, planning and design, production, implementation, and breakout.

The series looks at historical projects and then draws comparisons to challenges encountered in today's projects. It outlines the stages involved in delivering a complex project, providing a step-by-step guide to the project deliverables. It vividly describes the crucial lessons from historical projects and complements these with some of today's best practices.

This makes the whole learning experience more memorable. The series should inspire the reader, as these historical projects were achieved with a less sophisticated emerging technology.

Email: **mark.kozak-holl@sympatico.ca**

Web Site: **www.lessons-from-history.com**

LESSONS FROM
HISTORY

About the Series

This series is for primarily business and IT professionals looking for inspiration for their projects. Specifically, business managers responsible for solving business problems, or Project Managers (PMs) responsible for delivering business solutions through IT projects.

This series uses relevant historical case studies to examine how historical projects and emerging technologies of the past solved complex problems. It then draws comparisons to challenges encountered in today's IT projects.

This series benefits the reader in several ways:

- It outlines the stages involved in delivering a complex IT project providing a step-by-step guide to the project deliverables.

- It vividly describes the crucial lessons from historical projects and complements these with some of today's best practices.

- It makes the whole learning experience more memorable.

The series should inspire the reader as these historical projects were achieved with a lesser (inferior) technology.

Website: **http://www.lessons-from-history.com/**

Did you like this book?

If you enjoyed this book, you will find more interesting books at

www.MMPubs.com

Please take the time to let us know how you liked this book. Even short reviews of 2-3 sentences can be helpful and may be used in our marketing materials. If you take the time to post a review for this book on Amazon.com, let us know when the review is posted and you will receive a free audiobook or ebook from our catalog. Simply email the link to the review once it is live on Amazon.com, with your name, and your mailing address—send the email to orders@mmpubs. com with the subject line "Book Review Posted on Amazon."

If you have questions about this book, our customer loyalty program, or our review rewards program, please contact us at info@mmpubs.com.

Multi-Media Publications Inc.
Oshawa, Ontario, Canada

Avoiding Project Disaster: Titanic Lessons for IT Executives

Imagine you are in one of *Titanic's* lifeboats. As you look back at the wreckage site, you wonder what could have happened. What were the causes? How could things have gone so badly wrong?

Titanic's maiden voyage was a disaster waiting to happen as a result of the compromises made in the project that constructed the ship. This book explores how modern executives can take lessons from a nuts-and-bolts construction project like *Titanic* and use those lessons to ensure the right approach to developing online business solutions. Looking at this historical project as a model will prove to be incisive as it cuts away the layers of IT jargon and complexity.

Avoiding Project Disaster is about delivering IT projects in a world where being on time and on budget is not enough. You also need to be up and running around the clock for your customers and partners. This book will help you successfully maneuver through the ice floes of IT management in an industry with a notoriously high project failure rate.

ISBN: 1-895186-73-0   (paperback)

http://www.mmpubs.com/disaster

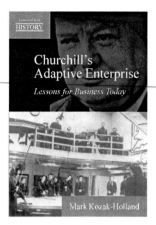

Churchill's Adaptive Enterprise: Lessons for Business Today

This book analyzes a period of time from World War II when Winston Churchill, one of history's most famous leaders, faced near defeat for the British in the face of sustained German attacks. The book describes the strategies he used to overcome incredible odds and turn the tide on the impending invasion.

The historical analysis is done through a modern business and information technology lens, describing Churchill's actions and strategy using modern business tools and techniques. Aimed at business executives, IT managers, and project managers, the book extracts learnings from Churchill's experiences that can be applied to business problems today. Particular themes in the book are knowledge management, information portals, adaptive enterprises, and organizational agility.

ISBN: 1-895186-19-6 (paperback)

http://www.mmpubs.com/churchill

Project Lessons from The Great Escape (Stalag Luft III)

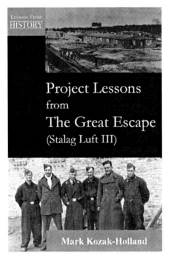

While you might think your project plan is perfect, would you bet your life on it? In World War II, a group of 220 captured airmen did just that—they staked the lives of everyone in the camp on the success of a project to secretly build a series of tunnels out of a prison camp their captors thought was escape proof. The prisoners formally structured their work as a project, using the project organization techniques of the day. This book analyzes their efforts using modern project management methods and the nine knowledge areas of the *Guide to the Project Management Body of Knowledge (PMBoK® Guide)*. Learn from the successes and mistakes of a project where people really put their lives on the line.

ISBN: 9781895186802 (paperback)

http://www.mmpubs.com/escape

Agile Leadership and the Management of Change: Project Lessons from Winston Churchill and the Battle of Britain

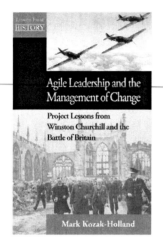

Around the turn of the millennium, there was a British poll that asked who was the most influential person in all of Britain's history. The winner: Winston Churchill. What distinguished him were his leadership qualities: his ability to create and share a powerful vision, his ability to motivate the population in the face of tremendous fear, and his ability to get others to rally behind him and quickly turn his visions into reality. By any measure, Winston Churchill was a powerful leader.

What many don't know, however, was how Churchill used his leadership skills to restructure the British military, government, and even the British manufacturing sector to get ready for an imminent enemy invasion in early 1940.

Learn how Churchill acted as the head project manager of a massive change project that affected the daily lives of millions of people. Learn about his change management and agile management techniques and how they can be applied to today's projects.

ISBN: 9781554890354   (paperback)

http://www.mmpubs.com/

Polaris: Lessons in Risk Management

Risk management is one of the most important practices that a manager can employ to help drive a successful outcome from a project. Good risk management allows organizations to proactively respond to risks.

Unfortunately, many managers believe risk management to be too time consuming or too complicated. Some find it to be shrouded in mystery.

This book is designed to demystify risk management, explaining introductory and advanced risk management approaches in simple language. To illustrate the risk management concepts and techniques, this book uses real-life examples from a very influential project that helped change the course of world history -- the project that designed and built the Polaris missile and accompanying submarine launch system that became a key deterrent to a Soviet nuclear attack during the Cold War. The Polaris design and construction project employed many risk management approaches, inventing one that is still widely used today.

Containing a foreword by James R. Snyder, one of the founders of the Project Management Institute (PMI), this book is structured to align with the risk management approach described in PMI's *Guide to the Project Management Body of Knowledge (PMBOK^R Guide)*.

ISBN: 9781554890972   (paperback)

http://www.mmpubs.com/

Lightning Source UK Ltd.
Milton Keynes UK
UKOW03n1937250813

215903UK00004B/25/P